3-D fibrous assemblies

The Textile Institute and Woodhead Publishing

The Textile Institute is a unique organisation in textiles, clothing and footwear. Incorporated in England by a Royal Charter granted in 1925, the Institute has individual and corporate members in over 90 countries. The aim of the Institute is to facilitate learning, recognise achievement, reward excellence and disseminate information within the global textiles, clothing and footwear industries.

Historically, The Textile Institute has published books of interest to its members and the textile industry. To maintain this policy, the Institute has entered into partnership with Woodhead Publishing Limited to ensure that Institute members and the textile industry continue to have access to high calibre titles on textile science and technology.

Most Woodhead titles on textiles are now published in collaboration with The Textile Institute. Through this arrangement, the Institute provides an Editorial Board which advises Woodhead on appropriate titles for future publication and suggests possible editors and authors for these books. Each book published under this arrangement carries the Institute's logo.

Woodhead books published in collaboration with The Textile Institute are offered to Textile Institute members at a substantial discount. These books, together with those published by The Textile Institute that are still in print, are offered on the Woodhead website at: www.woodheadpublishing.com. Textile Institute books still in print are also available directly from the Institute's website at: www.textileinstitutebooks.com.

A list of Woodhead books on textile science and technology, most of which have been published in collaboration with The Textile Institute, can be found at the end of the contents pages.

Woodhead Publishing in Textiles: Number 74

3-D fibrous assemblies

Properties, applications and modelling of three-dimensional textile structures

Jinlian HU

The Textile Institute

CRC Press
Boca Raton Boston New York Washington, DC

WOODHEAD PUBLISHING LIMITED
Cambridge England

Published by Woodhead Publishing Limited in association with The Textile Institute
Woodhead Publishing Limited, Abington Hall, Granta Park,
Great Abington, Cambridge CB21 6AH, England
www.woodheadpublishing.com

Published in North America by CRC Press LLC, 6000 Broken Sound Parkway, NW,
Suite 300, Boca Raton, FL 33487, USA

First published 2008, Woodhead Publishing Limited and CRC Press LLC

British Library Cataloguing in Publication Data
A catalogue record for this book is available from the British Library.

Library of Congress Cataloging in Publication Data
A catalog record for this book is available from the Library of Congress.

Woodhead Publishing ISBN 978-1-84569-377-0 (book)
Woodhead Publishing ISBN 978-1-84569-498-2 (e-book)
CRC Press ISBN 978-1-4200-7986-9
CRC Press order number: WP7986

The publishers' policy is to use permanent paper from mills that operate a sustainable forestry policy, and which has been manufactured from pulp which is processed using acid-free and elementary chlorine-free practices.
Furthermore, the publishers ensure that the text paper and cover board used have met acceptable environmental accreditation standards.

Typeset by SNP Best-set Typesetter Ltd., Hong Kong
Printed by TJ International Limited, Padstow, Cornwall, England

Contents

Woodhead Publishing in Textiles

1 **Watson's textile design and colour Seventh edition**
Edited by Z. Grosicki

2 **Watson's advanced textile design**
Edited by Z. Grosicki

3 **Weaving Second edition**
P. R. Lord and M. H. Mohamed

4 **Handbook of textile fibres Vol 1: Natural fibres**
J. Gordon Cook

5 **Handbook of textile fibres Vol 2: Man-made fibres**
J. Gordon Cook

6 **Recycling textile and plastic waste**
Edited by A. R. Horrocks

7 **New fibers Second edition**
T. Hongu and G. O. Phillips

8 **Atlas of fibre fracture and damage to textiles Second edition**
J. W. S. Hearle, B. Lomas and W. D. Cooke

9 **Ecotextile '98**
Edited by A. R. Horrocks

10 **Physical testing of textiles**
B. P. Saville

11 **Geometric symmetry in patterns and tilings**
C. E. Horne

12 **Handbook of technical textiles**
Edited by A. R. Horrocks and S. C. Anand

13 **Textiles in automotive engineering**
W. Fung and J. M. Hardcastle

14 **Handbook of textile design**
J. Wilson

15 **High-performance fibres**
Edited by J. W. S. Hearle

16 **Knitting technology Third edition**
D. J. Spencer

17 **Medical textiles**
Edited by S. C. Anand

18 **Regenerated cellulose fibres**
Edited by C. Woodings

19 **Silk, mohair, cashmere and other luxury fibres**
Edited by R. R. Franck

20 **Smart fibres, fabrics and clothing**
Edited by X. M. Tao

21 **Yarn texturing technology**
J. W. S. Hearle, L. Hollick and D. K. Wilson

22 **Encyclopedia of textile finishing**
H-K. Rouette

23 **Coated and laminated textiles**
W. Fung

24 **Fancy yarns**
R. H. Gong and R. M. Wright

25 **Wool: Science and technology**
Edited by W. S. Simpson and G. Crawshaw

26 **Dictionary of textile finishing**
H-K. Rouette

27 **Environmental impact of textiles**
K. Slater

28 **Handbook of yarn production**
P. R. Lord

29 **Textile processing with enzymes**
Edited by A. Cavaco-Paulo and G. Gübitz

30 **The China and Hong Kong denim industry**
Y. Li, L. Yao and K. W. Yeung

31 **The World Trade Organization and international denim trading**
Y. Li, Y. Shen, L. Yao and E. Newton

32 **Chemical finishing of textiles**
W. D. Schindler and P. J. Hauser

33 **Clothing appearance and fit**
J. Fan, W. Yu and L. Hunter

34 **Handbook of fibre rope technology**
H. A. McKenna, J. W. S. Hearle and N. O'Hear

35 **Structure and mechanics of woven fabrics**
J. Hu

36 **Synthetic fibres: nylon, polyester, acrylic, polyolefin**
Edited by J. E. McIntyre

37 **Woollen and worsted woven fabric design**
E. G. Gilligan

38 **Analytical electrochemistry in textiles**
P. Westbroek, G. Priniotakis and P. Kiekens

39 **Bast and other plant fibres**
R. R. Franck

40 **Chemical testing of textiles**
Edited by Q. Fan

41 **Design and manufacture of textile composites**
Edited by A. C. Long

42 **Effect of mechanical and physical properties on fabric hand**
Edited by Hassan M. Behery

43 **New millennium fibers**
T. Hongu, M. Takigami and G. O. Phillips

44 **Textiles for protection**
Edited by R. A. Scott

45 **Textiles in sport**
Edited by R. Shishoo

46 **Wearable electronics and photonics**
Edited by X. M. Tao

47 **Biodegradable and sustainable fibres**
Edited by R. S. Blackburn

65 **Shape memory polymers and textiles**
J. Hu

66 **Environmental aspects of textile dyeing**
Edited by R. Christie

67 **Nanofibers and nanotechnology in textiles**
Edited by P. Brown and K. Stevens

68 **Physical properties of textile fibres Fourth edition**
W. E. Morton and J. W. S. Hearle

69 **Advances in apparel production**
Edited by C. Fairhurst

70 **Advances in fire retardant materials**
Edited by A. R. Horrocks and D. Price (forthcoming)

71 **Polyesters and polyamides**
Edited by B. L. Deopora, R. Alagirusamy, M. Joshi and B. S. Gupta

72 **Advances in wool**
Edited by N. A. G. Johnson and I. Russell (forthcoming)

73 **Military textiles**
Edited by E. Wilusz

74 **3-D fibrous assemblies: Properties, applications and modelling of three-dimensional textile structures**
J. Hu

75 **Medical textiles 2007**
Edited by J. Kennedy, A. Anand, M. Miraftab and S. Rajendran (forthcoming)

76 **Fabric testing**
Edited by J. Hu

77 **Biomimetics and textiles**
Edited by A. Abbott and M. Ellison (forthcoming)

78 **Friction in textiles**
Edited by B. S. Gupta

Preface

3-D woven, knitted, braided or stitched fibrous assemblies are textile architectures having fibres oriented so that both the in-plane and transverse tows are interlocked to form an integrated structure that might have a unit cell with comparable dimensions in all three orthogonal directions. This integrated architecture provides improved stiffness and strength in the transverse direction and impedes the separation of in-plane layers in comparison to traditional 2-D fabrics. Recent automated manufacturing techniques have substantially reduced costs and significantly improved the potential for large-scale production of such structures.

I always felt that a comprehensive book on 3-D fibrous assemblies would provide a great support for the industry and for research/educational institutions in understanding the concepts of these new fabric architectures to develop new products for specific applications such as in composites, medical, sports, geotechnical and aerospace fields.

Hence, this book is the culmination of research into 3-D textiles and their applications to composites and other related areas and is based on thorough and detailed compilation of various topics related to 3-D fibrous assemblies. The detailed information provided in this book has been the collective compilation of works of various researchers in the field, mainly from our own research studies, especially on modelling of multiaxial warp-knitted and multilayer woven fabrics.

The book consists of nine chapters covering the topics from introduction of 3-D fibrous assemblies such as 3-D woven, knitted, non-woven and braided fabrics to the advanced modelling of these fabrics for various applications.

Chapter 1 introduces the various 3-D fibrous assemblies such as 3-D woven fabrics, multiaxial warp-knitted structures, 3-D braided, stitched and non-woven fabrics and their general structures. The principle of production of these fabrics with their advantages over 2-D fabrics is also discussed.

Chapter 2 is an overview of the various applications of these fabrics in composites, medical, aerospace and other fields. Specific applications of 3-D woven, knitted, braided and non-woven fabrics in composite manufacture are discussed in detail.

In **Chapter 3** a detailed discussion of the structures and manufacturing methods of multiaxial warp-knitted fabrics is presented. The advantages of these fabrics over other fabrics are discussed. In addition, detailed information on various mechanical properties such as tearing, shear, compression and flexural behaviour of these fabrics is also presented in this chapter.

Chapter 4 provides in-depth information on structures of various multilayer woven fabrics and their manufacturing methods. Various routes to manufacture these fabrics are discussed in detail with suitable illustrations. At the end, detailed information is available on their mechanical properties such as tensile, shear and compression with a brief section on applications of these fabrics.

Chapters 5 to 9 are devoted to the modelling of various preforms of multiaxial warp knits and multilayer woven fabrics. In **Chapter 5** an attempt has been made to discuss in detail some of the modelling techniques used to understand the tensile behaviour of multiaxial warp-knitted fabrics. A macroscopic approach dealing with fabric structure under uniaxial tensile deformation is developed for a greater understanding of the geometry of these fabrics. A model for uniaxial tensile deformation is obtained, which is justified by Instron 4466 tensile testing. In addition, a formula for calculating the tensile modulus of the fabric in any direction is presented.

Chapter 6 describes the modelling methods used to analyze the bending behaviour of MWK fabrics. An elaborate description and interpretation on the bending properties of MWK fabrics are presented, which are based on many bending hysteresis curves obtained on KES-FB-2. Further, a predictive bending model to assess the MWK fabrics based on KES-F experiments in different bending directions is described.

In **Chapter 7** a model for the prediction of the formability of a multiaxial warp-knitted (MWK) fabric to a 3-D surface is described. Fibre and yarn movement during fabric forming can cause adverse effects such as wrinkling and thinning, which will lead to a decrease of the mechanical properties of the finished composite. In addition, the high level of waste generated by subsequent trimming operations is unacceptable. Hence there was a need to establish a model to predict the deformation and possible wrinkles of MWK fabrics during the forming process in order to enable waste-free design and defect predictability. For this purpose, a detailed characterization of the forming behaviour of MWK fabrics containing two bias inserting yarns (TBMWK fabric) is discussed in this chapter.

Chapter 8 presents a comprehensive study on the permeability modelling of multilayer woven fabrics. A framework for flow permeability measurement in resin transfer moulding is discussed. The Darcy law is used to model the flow through the reinforcement fibres where permeability is a measure of the resistance of the fibres to the flow. Two types of woven fabrics, one fabricated by the monofilament method so as to eliminate the effects of other factors such as fibre bundle on permeability, and the other by using multifilament yarns, are used for modelling. Two models are developed to explain the permeability of these two different varieties of multilayer woven fabrics.

Finally, in **Chapter 9** a detailed theoretical analysis for in-plane impregnation in multilayer woven fabrics is reported in order to understand the mechanism of void formation. Unlike the previous approaches, where the void is formed in the plane of one layer of woven fabric, the void formation in the cross-section of multilayer woven fabrics is presented. Based on two simplified unit cells, which were identified from two typical multiple modes of multilayer woven fabrics, a mathematical model is developed to analyze the formation and size of voids. The flow front and void formation processes are also numerically simulated using the control-volume method.

I sincerely feel that the readers will find the book both interesting and informative as it contains the state of the art information on 3-D fibrous assemblies and their application in various fields.

Professor Jinlian Hu
Hong Kong Polytechnic University

Acknowledgements

The content of this book is based mainly on the original work produced by my research students at Hong Kong Polytechnic University. A number of people have contributed their original work, especially towards modelling of multiaxial warp-knitted and multilayer woven fabrics.

I am extremely grateful to Dr K. Murugesh Babu for his consistent help and hard work during the editing and preparation of this book. His outstanding reviewing and editing skills combined with sincere efforts have made this book a meaningful piece of work.

My sincere acknowledgements are due to Dr Jiang Yaming for his research work on multiaxial warp-knitted fabrics, and to Dr Shao Xueming and Dr Liu Yi for their research works on multilayer woven fabrics, especially on the modelling behaviour of these fabrics. Chapters 5 to 9 represent the original works of the above people and their outstanding contribution to these chapters.

I would like to acknowledge the generous support of the Hong Kong Government for funding support from the Innovation and Technology Commission over the past several years. From these industry-guided projects, we can realize the significance of 3-D fibrous assemblies to the advanced technical textiles field as well as to the research and academic world.

Finally, I wish to express my gratitude to Kathryn Picking of Woodhead Publishing Limited for her consistent help and understanding.

1
Introduction to three-dimensional fibrous assemblies

Abstract: Three-dimensional (3-D) textiles are those materials that have a system or systems in all three axes of plane. These materials offer particular properties, such as interlaminar shearing force, mechanical and thermal stability along all three axes of space, that are not achievable with other reinforcements. The development of three-dimensional textiles has taken place rapidly over the past two decades. It can be credited largely to the growth of another technology: composite materials, which combine fibres and a matrix. An understanding of the production methods and structures of these 3-D fibrous assemblies would go a long way in design, process control, process optimization, quality control, clothing manufacture and development of new techniques for specific end uses. This chapter introduces various 3-D woven, knitted, non-woven, braided and stitched fabrics with their brief description and advantages.

Key words: three-dimensional (3-D) textiles, 3-D woven fabrics, 3-D knitted fabrics, 3-D non-woven fabrics, 3-D braided fabrics.

1.1 Introduction: concepts of three-dimensional fibrous assemblies

Textile structures such as in woven, knitted, non-woven and braided fabrics are being widely used in advanced structures in the aerospace, automobile, geotechnical and marine industries. In addition, they are finding wide application as medical implants such as scaffolds, artificial arteries, nerve conduits, heart valves, bones, sutures, etc. This is because they possess outstanding physical, thermal and favourable mechanical properties, particularly light weight, high stiffness and strength, good fatigue resistance, excellent corrosion resistance and dimensional stability. In addition, they act as attractive reinforcing materials in various composite applications with low fabrication cost and easy handling (Tan *et al.*, 1997). With high-end applications such as in aerospace, the orientation of the fibrous reinforcement is becoming more and more important from a load-bearing point of view, as is the need for placing the reinforcement oriented in the third dimension (Alagirusamy *et al.*, 2006).

Textile fabrics, termed preforms in composites and other applications, consist of various reinforcing fabrics such as wovens, knits, braids and non-wovens. Two-dimensional fabrics have allowed us to drape bed, board and body in a profusion of texture, pattern and colour over the centuries. The development of advanced fibres has led engineers to consider textiles for high-performance applications such as in construction and aeronautics. These fabrics have been relatively well developed in terms of production, analysis and application and some of them have long been used in structural composite fields (Chou and Ko, 1989; Mohamed, 1990). However, the strength of these traditional fabrics is anisotropic, manifesting itself primarily in the direction of the fibre orientations. Most of these 2-D textile structures retain the inherent weakness of laminated composites that are susceptible to delamination.

To extend the use and value of textiles into industrial and engineering applications, which typically require strength in more than two directions, textile designers have bound together layers of textiles and exploited the chemical properties of fibres and binders to create novel non-woven textiles whose fibres are not restricted to two-dimensional arrangements. More recently, they have taken the next step: finding ways to manufacture true three-dimensional (3-D) textiles. Hence, 3-D fabrics have been introduced to respond to the needs of a number of industrial requirements such as composites capable of withstanding multidirectional stresses.

The development of 3-D textiles has taken place rapidly over the past two decades. It can be credited largely to the growth of another technology: composite materials, which combine fibres and a matrix. Textile engineers have been challenged to develop strong fibre architectures and new manufacturing processes for building textile structures in three dimensions, as these 3-D fabrics hold great promise for use in industry, construction, transportation and even military and space applications. They are often made into a near net shape so that the overall manufacturing cost can be very low for certain applications (Mohamed, 1990).

An understanding of the production methods and structures of these 3-D fibrous assemblies would go a long way in the design, process control, process optimization, quality control, clothing fabrication and the development of new techniques for specific end uses. The interrelationship between their structure and various properties may be of great help in designing new types of 3-D structures for the construction, medical, sports and aerospace industries.

1.2 Two-dimensional structures (two-dimensional fabrics)

1.2.1 Two-dimensional wovens

Weaving is the most widely used textile manufacturing technique and accounts for the majority of the two-dimensional (2-D) fabric produced

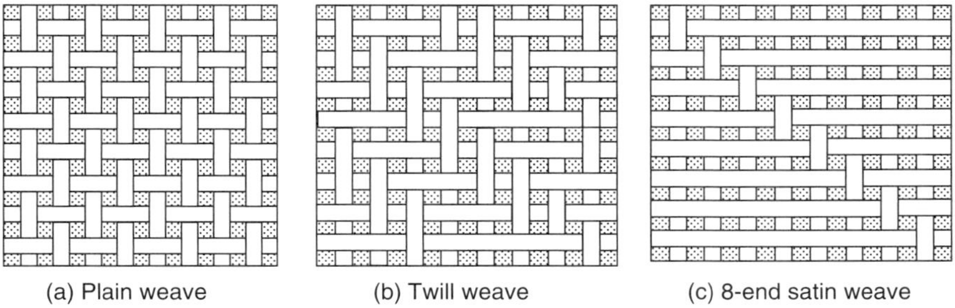

1.1 Basic weaves.

(Stobbe and Mohamed, 2003). Woven structures have the greatest history of application in textile manufacturing. Conventional woven fabrics consist of two sets of yarns mutually interlaced into a textile fabric structure. The threads that run along the length of the fabric are called warp or ends, while the threads that run along the width of the fabric from selvedge to selvedge are referred to as weft or picks. Warp and weft yarns are mutually positioned at an angle of 90°. The number of warp and weft yarns per unit length is called the warp and weft density. The warp and weft yarns in a woven fabric can be interlaced in various ways, called a weave structure. The structure in which warp yarns alternately lift and go over across one weft yarn and vice versa is the simplest woven structure, called plain weave (Fig. 1.1(a)). Other common structures are twill and satin weave. Twill is a weave that produces diagonal lines on the face of a fabric (Fig. 1.1(b)). The direction of the diagonal lines viewed along the warp direction can be from upwards to the right or to the left, making Z or S twill respectively. Compared to plain weave of the same cloth parameters, twills have longer floats, fewer intersections and a more open construction. A weave in which the binding places are arranged to produce a smooth fabric surface free from twill lines is called satin (Fig. 1.1(c)). The distribution of interlacing points must be as random as possible to avoid twill lines. The smallest repeat of satin weave is 5, while the most popular weaves are satins of 5 and 8 repeats. The 5-ends satin is most frequently used for technical applications for providing firm fabric, although having a moderate cover factor.

Triaxial woven fabrics

A triaxial woven structure consists of three systems of threads: one system for weft and two systems for warp. This fabric has three layers of material at any point, and is thus stronger than a rectangular woven fabric made using the same elements. Warp threads in a basic triaxial fabric are interlaced at 60° and the structure is fairly open with a diamond-shaped centre (Fig. 1.2(a)). A modification of basic triaxial fabric is basket weave, which forms a closer structure with different characteristics (Fig. 1.2(b)).

(a) Basic triaxial weave (b) Basket weave

1.2 Triaxial fabrics.

Triaxial fabrics possess exceptional mechanical properties in several directions. Since the interlacing points are fixed into the fabric structure, these fabrics exhibit high shear resistance (Lee *et al.*, 2002).

1.2.2 Two-dimensional knits

Knitted fabrics are textile structures assembled from basic construction units called loops. There exist two basic technologies for manufacturing knitted structures: weft and warp-knitted technology.

Weft-knitted fabric

The repeating unit of the knitted fabric is called the loop. The feature of weft-knitted fabric is that the loops of one row of fabric are formed from the same yarn. A horizontal row of loops in a knitted fabric is called a course, and a vertical row of loops is called a wale. In weft-knitted fabrics the loops are formed successively along the fabric width. The feature of weft-knitted fabric is that the neighbouring loops of one course are created of the same yarn. The simplest weft knit structure produced by the needles of one needle-bed machine is called plain knit or jersey knit (Fig. 1.3(a)). Plain knit has a different appearance on each side of the fabric. A structure produced by the needles of both needle beds is called a rib structure or double jersey (Fig. 1.3(b)) and has the same appearance on both sides of the fabric.

Warp-knitted fabric

In warp-knitted technology every loop in the fabric structure is formed from a separate yarn called the warp, introduced mainly in the longitudinal fabric direction. The most characteristic feature of warp-knitted fabric (Fig. 1.3(c)) is that neighbouring loops of one course are not created from

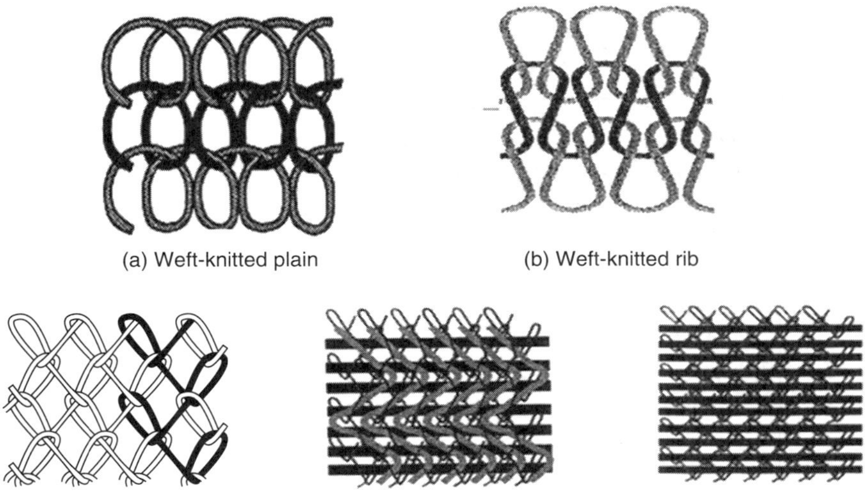

(a) Weft-knitted plain (b) Weft-knitted rib

(c) Basic warp-knit structure (d) Weft-inserted warp-knit (e) Multibar weft-inserted warp-knit

1.3 Schematic representations of weft- and warp-knitted structures.

the same yarn. While weft-knitted technology is most commonly used in clothing manufacture, warp-knitted technology is substantially engaged in manufacturing structures for technical applications. Of special interest for technical applications are structures with inserted weft yarns, called weft-inserted warp-knitted fabric (Fig. 1.3(d)), and a multibar weft-knitted fabric (Fig. 1.3(e)).

1.2.3 Two-dimensional non-wovens

Non-woven fabrics are broadly defined as a sheet or web structure bonded together by entangling fibre or filaments either mechanically, thermally or chemically. They form a sheet, web or batt of directionally or randomly oriented fibres, bonded by friction and/or cohesion and/or adhesion, excluding paper and products, that is woven, knitted, tufted, stitch-bonded (incorporating bonding yarns or filaments) or felted by wet milling, whether or not additionally needled. The fibres may be of natural or artificial origin. They may form staple or continuous filaments. They are engineered to provide specific properties such as absorbency, liquid repellency, resilience, stretch, softness, strength, flame retardancy, washability, cushioning, filtering, bacterial barrier and sterility. A basic non-woven structure is shown in Fig. 1.4.

1.4 Basic non-woven fabric.

1.2.4 Two-dimensional braids

A braid is a textile structure formed by interlacing two or more sets of yarns resulting from the carriers rotating in clockwise and counter-clockwise directions (Brunnschweiler, 1953). Braiding has been conventionally used for applications such as shoelaces, ropes, etc. However, in recent years, fibre-reinforced composites and medical implants have become interesting applications for braiding. This has been achieved by employing 3-D preforms for such applications (Zhang *et al.*, 1997).

Braided textile structures are manufactured by intertwining or orthogonally interlacing two (or more) sets of yarns to form an integral structure in a tubular form. One set of yarns is called the axial yarns while the other is called the braided yarns. Hence, the structures of braided fabrics consist of parallel axial yarns, interconnected with braided yarns that are placed along complex spatial orientations. There are three typical braid structures: diamond, regular and hercules. A regular diamond structure is shown in Fig. 1.5(a). It is obtained when the yarns cross alternately over and under the yarns running in the opposite direction. The repeat notation is 1/1. Using this notation, the regular braid structure has notation 2/2 and hercules 3/3.

Braids are mostly produced in a regular structure, generally in a tubular form of biaxial yarn direction. By inserting longitudinally oriented yarns (middle-end-fibre) into the structure, triaxial braid is obtained (Fig. 1.5(b)). Moreover, in the centre of the tubular braid, additional fibres called axial fibres can be inserted. When the number of braiding fibre bundles is the same, the tubular braid increases the fibre volume fraction more than the flat braid (Fig. 1.5(c)). The main feature of the braid is the angle of intertwining, which can vary between 10° and 80° and depends on the yarn

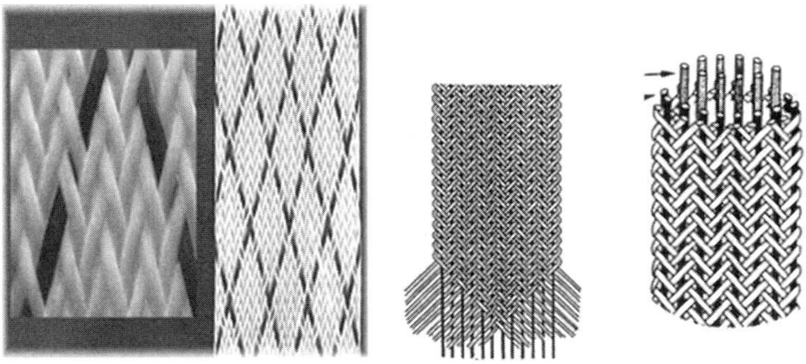

(a) Regular diamond (b) Triaxial braid (c) Triaxial braid with axial fibres

1.5 Typical braided structures.

fineness, the type of structure (biaxial or triaxial), the cover factor (tightness of the structure) and the volume ratio of the longitudinal yarns.

1.3 Limitations of two-dimensional textile structures

Polymer laminates reinforced with a 2-D layer fibre structure have been used with outstanding success for over 50 years in maritime craft, for about 30 years in aircraft and for nearly 20 years in high-performance automobiles and civil infrastructure such as buildings and bridges. Despite the use of 2-D laminates over a long period, their use in many structural applications has been limited by manufacturing problems and some inferior mechanical properties. The manufacture of laminates can be expensive because of the high labour requirement in the manual lay-up of piles (Mouritz *et al.*, 1999).

The application of 2-D laminates in some critical structures in aircraft and automobiles has also been restricted by their inferior impact damage resistance and low through-thickness mechanical properties when compared to traditional aerospace and automotive materials such as aluminium alloys and steel. The low through-thickness properties, such as stiffness and fatigue resistance, have impeded the use of 2-D laminates in thick structures subjected to high through-thickness and interlaminar shear stresses. An additional problem is that many 2-D laminates have low resistance to delamination cracking under impact loading because of their poor interlaminar fracture toughness. As a consequence of this, their post-impact in-plane mechanical properties can be severely degraded, particularly their compression strength and fatigue performance. While these properties can be improved to a certain extent by the use of toughened resins of fibre interleaves, these solutions are usually expensive and do not overcome many of the problems associated with the manufacture of laminates (Mouritz *et al.*, 1999).

1.4 Three-dimensional structures (three-dimensional fabrics)

1.4.1 Definition of three-dimensional fibrous assemblies

Three-dimensional woven, braided or stitched fibrous assemblies are textile architectures having fibres oriented so that both the in-plane and transverse yarns are interlocked to form an integrated structure that has a unit cell with comparable dimensions in all three orthogonal directions, i.e., the 3-D structure basically consists of in-plane yarns for stiffness and strength and z-binder yarns for through-thickness reinforcement (Yang *et al.*, 2004). In other words, 3-D textiles are those materials that have a system or systems in all three orthogonal planes.

These materials offer particular properties, such as interlaminar shear force, mechanical and thermal stability along all three spatial axes, that are not achievable with other reinforcements. This integrated architecture provides improved stiffness and strength in the transverse direction and impedes the separation of in-plane layers in comparison to traditional 2-D fabrics. Because of their high transverse strength, high shear stiffness, low delamination tendency and near-net-shape manufacture, textile composites from weaving, knitting and braiding have received tremendous attention recently (Xuekum Sun and Changjie Sun, 2004). Recent automated manufacturing techniques have substantially reduced costs and significantly improved the potential for large-scale production.

Optimal orientations, fibre combinations and distributions of yarns have yet to be fully developed and perfected for 3-D fabrics subjected to impact loading conditions. For example, current body armour relies on ceramic plates to defeat penetrators. The rigidity and brittleness of these materials limit their use to military fighting applications. In addition, over time, environmental degradation and accidental mechanical impact damage the ceramic and render it ineffective. Hence, there are ample opportunities for substitute materials, and innovative concepts that combine hybrid 3-D fabrics with other materials such as ceramic and possibly new nano-scale materials are needed. The optimal combinations of these materials need to be determined along with new methodologies to ascertain how to utilize the inherent mechanisms (friction, micro-cracking, fibre breakage, fibre bridging, etc.) of these systems for energy dissipation and strengthening.

1.4.2 Comparison of three-dimensional with two-dimensional fabrics

- The absence of interlacing between the warp and filling yarns allows 3-D fabrics to bend and internally shear rather easily, without buckling within the in-plane reinforcement, which is not the case in 2-D fabrics.
- The presence of z-direction reinforcement in 3-D fabrics is an obvious advantage, as the dramatic improvement in composite transverse strength and impact damage tolerance is well documented. For example, tests of laminates made from these preforms have shown a 10–30% increase in short-beam shear strength over 2-D textile laminates.
- Three-dimensional fabrics exhibit improved compression after impact strength, reduced delamination area, and increased number of sub-perforation energy blows required to penetrate the panel.
- Composites made from 3-D preforms exhibit high fibre content (per cent by weight). Although somewhat lower percentages can be expected, the fibre content is still higher than in composites made from 2-D fabrics (Malik and Parmar, 2006).

1.4.3 Three-dimensional woven fabrics

Most woven fabrics are 2-D biaxial woven structures wherein two sets of yarns, warp and weft (fill), intersect and interlace at right-angles with one another. The biaxial fabrics can provide fairly balanced properties and good stability in the warp and weft directions. However, they exhibit a relatively low modulus or resistance to extension when deformed on the bias (45° to warp and weft) as compared with deformation in the warp or weft directions. There are many weave patterns for 2-D biaxial woven fabrics; the three most common weave structures are plain weave, twill weave and satin weave.

The use of 3-D woven fabrics as the reinforcing medium for composites, termed woven preforms, is becoming a popular choice. Because of the 3-D integrated fibre assemblage, such structures are less prone to delamination and can offer high impact resistance. They have emerged as a new class of lightweight material that has potential applications in the aerospace, maritime, infrastructure and medical fields. Three-dimensional woven fabrics are formed to near net shape with substantial thickness and additional yarns in the through-thickness direction. This distinguishes them from 2-D fabrics which possess a high width to thickness ratio and are typically layered to form a thick structure. They have been found to have better delamination resistance and damage resistance than 2-D woven laminates. The fibre structure of a 3-D woven composite consists of in-plane warp and weft yarn

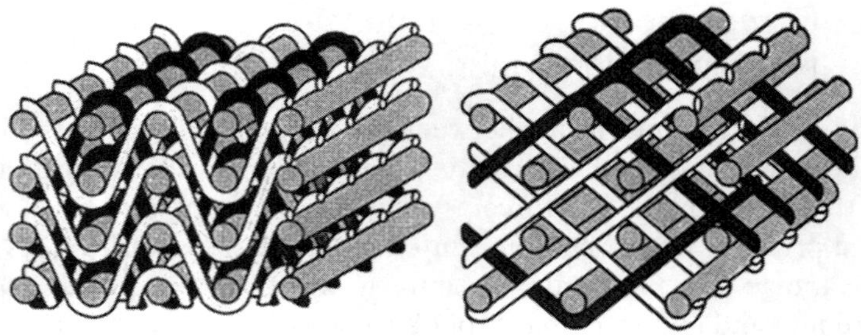

1.6 Typical 3-D woven structure.

layers interlaced in the through-thickness direction by *z*-binder yarns. A typical 3-D woven structure is shown in Fig. 1.6 (Mohamed, 1990).

In a 3-D woven fabric, the warp and weft yarns are bound together by a series of warp binder yarns. Various arrangements of yarn placement are used to produce a wide range of multilayer 3-D reinforcements for composite applications (Yi and Ding, 2004). These are broadly categorized under two distinct headings:

- integrated structures, in which binder yarns link one layer to any other layer within the textile structure;
- interlinked structures, in which the binder yarns link the outer two layers, top to bottom.

In 3-D weaving, a number of the warp (or 0° direction) yarns provide through-the-thickness reinforcement to consolidate the preform. The through-thickness yarns are arranged in different areas and levels of the reinforcement according to the net shape and mechanical properties required. These through-thickness tows have been shown to provide increases in interlaminar shear strength of composite components (Quinn *et al.*, 2001). In general, 3-D weaving refers to the following:

- making of fabrics with substantial thickness by layering
- enabling shedding and weft insertion both horizontally and vertically
- creating 3-D woven shapes, e.g., a dome shape.

Classification of three-dimensional woven fabrics

Three-dimensional woven fabrics are produced principally by the multiple-warp weaving method, which has long been used for the manufacture of double and triple cloths for bags, waddings and carpets. The 3-D woven fabrics produced by using either multiwarp weaving technology or conventional weaving technology can be broadly classified as follows:

- 3-D solid
  - multilayer
  - orthogonal
  - angle interlock
- 3-D hollow
  - flat surface
  - uneven surface
- 3-D shell
  - by weave combination
  - by differential take-up
  - by moulding
- 3-D nodal.

Three-dimensional woven solid fabrics produced by the multilayer principle are characterized by the presence of several layers of yarns woven together by using different interlacing techniques. The layers can also be stitched together and the stitching arrangement may have either a self-stitching or a central stitching arrangement. Wadding yarns may also be incorporated into the structure for specific applications. It is important to control the yarn ratio in the fabric to obtain specific properties (Xiaogang Chen, 2006).

Three-dimensional solid orthogonal fabrics feature straight warp, weft and vertical yarns in their design. The number of layers of straight weft yarn is always one more than that of warp yarn, and the amount of vertical yarn depends on the binding weave. Two options are possible: ordinary and enhanced. It is possible to have a variable interlinking depth.

The 3-D solid angle interlock principle involves the binding of straight warp yarns by interlocking warp yarns. Warp yarns can be bound to different depth. As in orthogonal fabrics, wadding yarns may be used in the structure. The structures of various 3-D solid woven fabrics are presented in Fig. 1.7 (Xiaogang Chen, 2006).

Three-dimensional hollow structures are of two types: one with a flat surface and the other with an uneven surface. The hollow structures with a flat surface are based on the multilayer principle but with different fabric section lengths. Such structures should be self-opening under the right conditions. It is possible to have multilevel cells in their structure. The hollow fabrics with uneven surface structures are based on the multilayer principle and are created by joining and separating adjacent layers. Such structures need opening. The cells created in the structure are generally of hexagonal shape. Different 3-D hollow structures are shown in Fig. 1.8 (Xiaogang Chen, 2006).

Three-dimensional shell structures can be created either by using different weave patterns or by employing a discrete take-up in the loom to

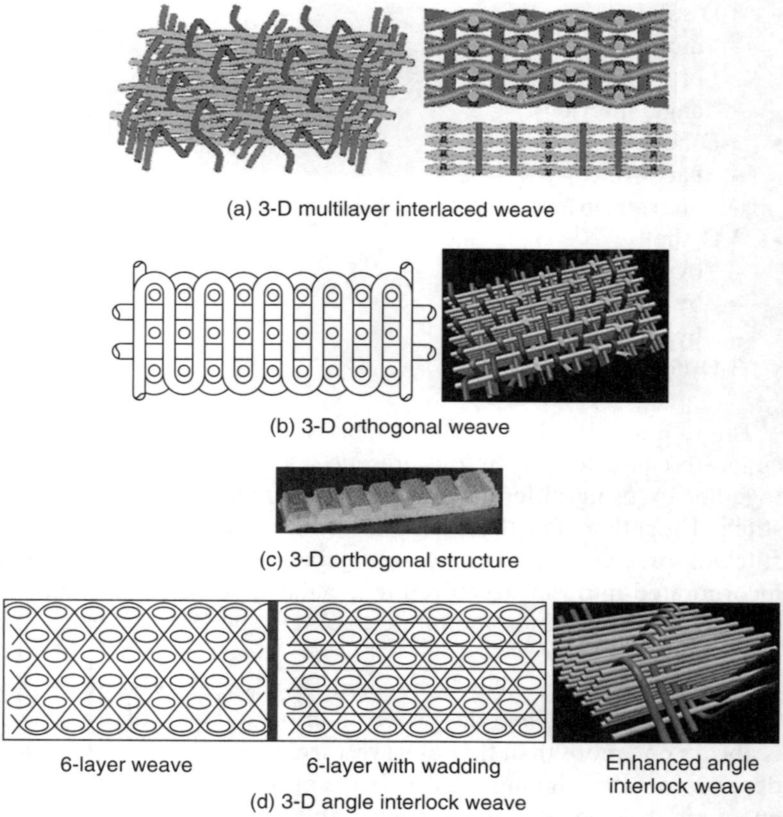

1.7 3-D solid woven structures (adapted from Xiaogang Chen, 2006).

modify their structure to have a hollow-type surface. Sometimes it is possible to mould these fabrics to obtain a special moulded structure for technical or industrial applications. Three different types of hollow structures are shown in Fig. 1.9.

Three-dimensional nodal fabrics (Fig. 1.10) refer to joining tubes to obtain certain special types of structures for industrial applications. The walls of the tubes may be 3-D solid structures themselves. All tubes must be in the same plane (x–y). The design procedure involves creating a nodal design in 2-D space, flattening, area segmentation and assignment of weaves for different sections (Xiaogang Chen, 2006).

A 3-D weave contains multiple planes of nominally straight warp and weft yarns that are connected together by warp weavers to form an integral structure. The most common classes are shown in Fig. 1.11. Within each class, there are several parameters that can be varied.

(a) 3-D hollow flat surface

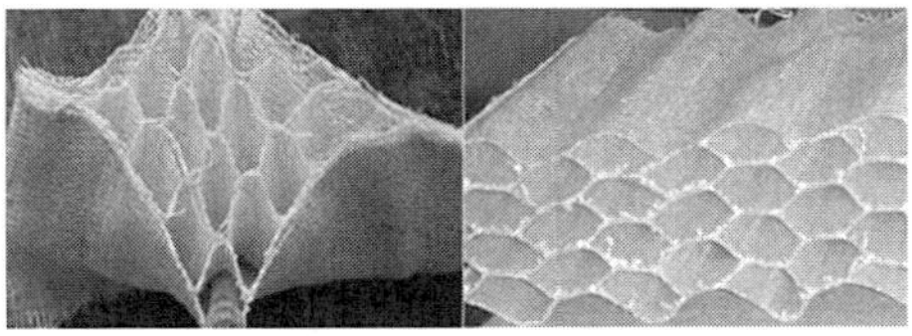

(b) 3-D hollow with hexagonal cells

(c) 3-D hollow uneven surface

1.8 3-D woven hollow structures (adapted from Xiaogang Chen, 2006).

The angle interlock type of structure is a variation of 2-D weaving wherein more than two yarns are in the thickness direction. The angle interlock structure allows the preform to be up to 10 cm in thickness. In this fabric, warp yarns are used to bind several layers of weft yarns together. Weft yarns could be used for binding as well. In place of warp or weft yarns, an additional third yarn may also be used as binder. Stuffer yarns, which are straight, can be used to increase fibre volume fraction and in-plane strength.

Angle interlock weaves can be categorized by the number of layers that the warp weavers penetrate. Figure 1.11(a) shows a through-thickness interlock fabric, in which the warp weavers pass though the entire thickness. Figure 1.11(c) and (d) show layer-to-layer interlock patterns, where a given weaver connects only two planes of weft yarns but the weavers collectively bind the entire thickness. Various intermediate combinations can be fabricated, with the weavers penetrating a specified number of layers, with the warp weavers passing through the thickness orthogonal to both in-plane directions, as shown in Fig. 1.11(b). Interlock weaves are sometimes manufactured without straight warp yarns (stuffers) to produce a composite reinforced predominantly in one direction. They may also be fabricated with weft rather than warp yarns used for interlock. A major limitation of 3-D weaves is the difficulty of introducing bias-direction yarns to achieve in-plane isotropy. One solution is to stitch additional 2-D fabric plies oriented at 45° onto the woven preform.

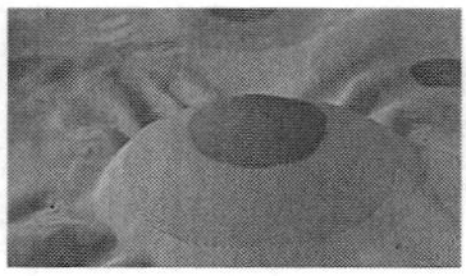

(a) 3-D shell by using different weaves

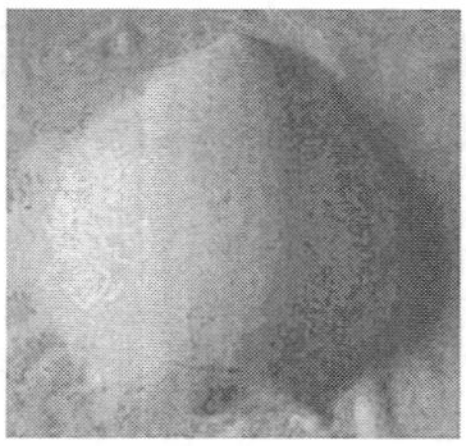

(b) 3-D shell by discrete take-up

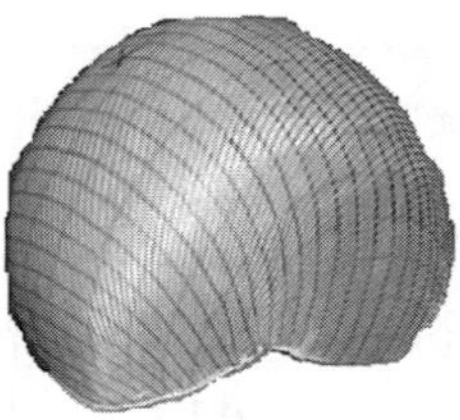

(c) 3-D shell by moulding

1.9 3-D woven shell structures.

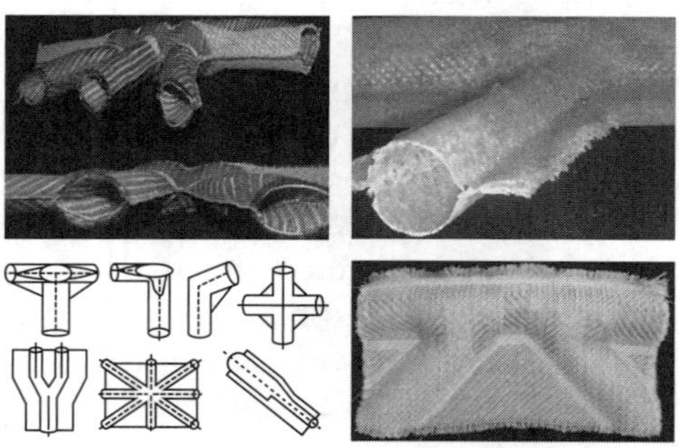

1.10 3-D woven nodal structures (adapted from Xiaogang Chen, 2006).

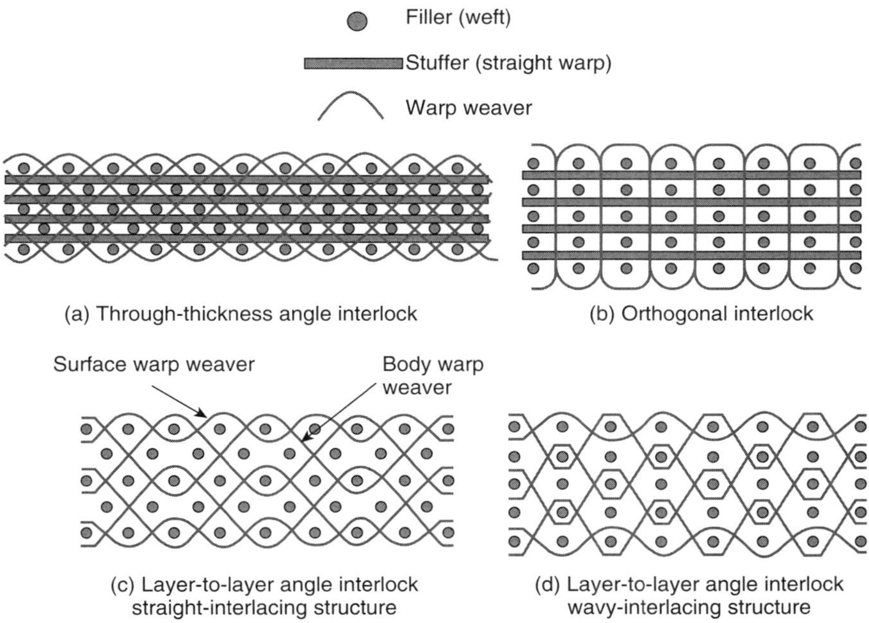

1.11 3-D weave patterns.

Textile technology has been involved in the construction of 3-D shells, but these are rather flexible structures for clothing and similar purposes. For production of strong 3-D fabrics, cutting, sewing and joints should be avoided as much as possible. True 3-D weaving can be accomplished on special machines as already developed. The simplest case of such a structure, in which the warp is crossed by two sets of wefts, is shown in Fig. 1.12(a). A woven structure in which the multilayer warp moves between the top and bottom of the material, thus providing x and y directions at an angle of 45°, and with weft yarns crossing between the warp providing the z direction, is shown in Fig. 1.12(b).

In general, a variety of 3-D structures can be accomplished. For example, the warp yarns can go only part-way across the whole material (Fig. 1.13(a)), or one set of warp yarns can be introduced in the axial direction and one set angled, thus providing yarns in four different directions (Fig. 1.13(b)). The latter structure gives higher control of the directional properties of the material. Shaping of 3-D weaves can be accomplished by varying the width of the warp layers, thus creating the required cross-section, for example in the form of a T-beam (Fig. 1.13(c)) (Demoski and Bogoeva-Gaceva, 2005).

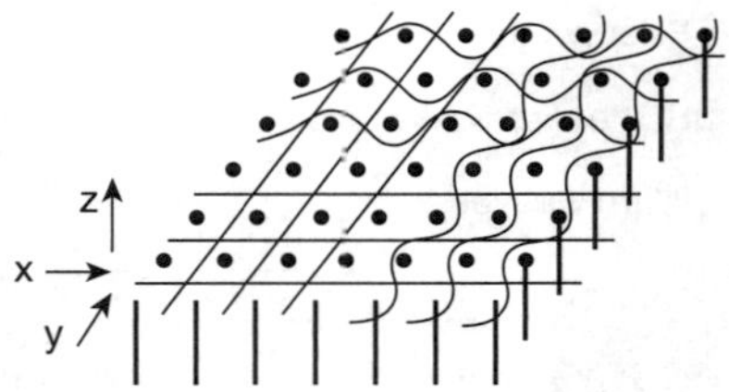

(a) Multilayer 3-D weave with warps in *x* and *y* directions

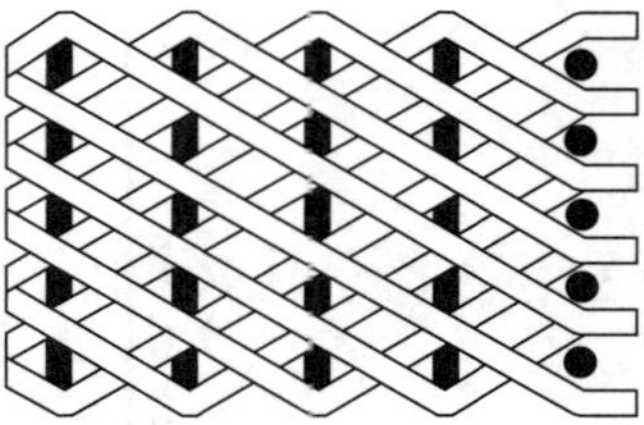

(b) Weft in *z* direction

1.12 Two types of multilayer 3-D woven fabric (a) warp in the *z* direction and wefts inserted in the *x* and *y* directions, (b) warps at 45° in the *x* and *y* directions and weft in the *z* direction.

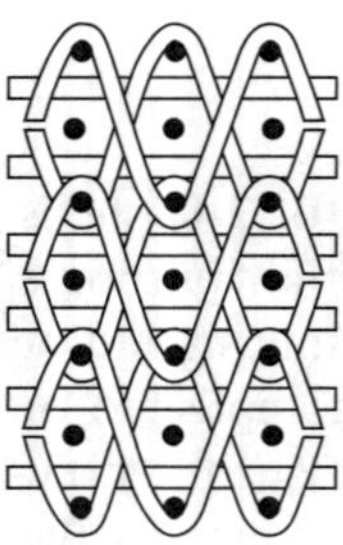

(a) Partial interlocking of layers

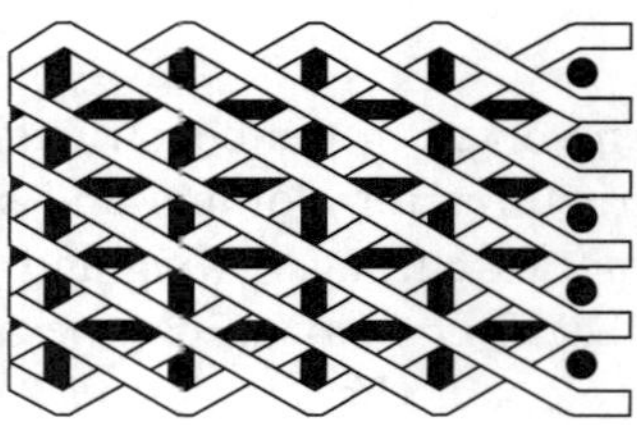

(b) Two sets of warp yarns

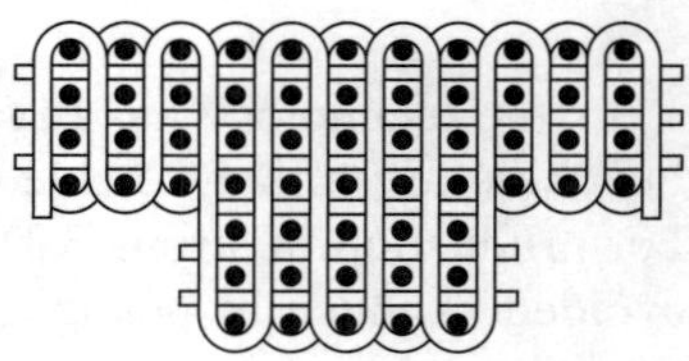

(c) Shaping of 3-D weaving by varying warp layers

1.13 Variants of 3-D multilayer weaving.

Advantages of three-dimensional woven structures

- 3-D weaving can produce complex near-net shaped preforms.
- 3-D woven composites with a complex geometry can be less expensive to produce.
- 3-D weaving allows the tailoring of properties for specific applications.
- 3-D woven composites show better delamination resistance and damage tolerance.

- 3-D woven composites show higher tensile strain-to-failure values.
- 3-D woven composites exhibit higher interlaminar fracture toughness properties.

1.4.4 Three-dimensional knitted fabrics

In the search for methods to reduce composite manufacturing costs, textile preforms, including knitted structures, are receiving increasing interest in the composites industry. While conformability and productivity are obvious attributes for knitted preforms, the availability of a broad range of micro- and macrostructural geometries has only recently been recognized. The non-linearity of knitting loops, severe bending of yarns during the knitting process and limited fibre packing density, resulting in the formation of resin pockets within a knitting loop, prevent knits from being considered for structural applications.

Knitting is the interlocking of one or more yarns through a series of loops (also called stitches). Knitted fabrics are considered 3-D due to their non-planar configuration of the loops in the structure. They are also known as multiaxial–multilayer structures and are fabrics bonded by a loop system, consisting of one or several yarn layers stretched in parallel. Multilayers of linear yarns are assembled in warp (0°), weft (90°) and bias ($\pm\theta$) directions to provide structural integrity and through-thickness reinforcement (Du and Ko, 1996).

Three-dimensional knitted fabrics are produced by weft or warp knitting. An example of a weft knit is the near-net-shape structure knitted under computer control by the pressure foot process. In a collapsed form this preform has been used for carbon–carbon aircraft brakes. While weft-knitted structures have applications in limited areas, multiaxial warp knit (MWK) 3-D structures are more promising and have undergone a great deal of development in recent years. MWK fabrics generally possess up to four different load-bearing yarn systems arranged so that each can take on stress and strain virtually in all directions. Since these load-bearing yarns lie straight in the fabric, with no crimp, the physical parameters of the individual yarn system are fully utilized (Kaufmann, 1991). Multiaxial–multilayer warp knits (MWK) are also termed non-crimp structures since the presence of knitted loops is to perform the function of holding layers of uncrimped inlay yarns. These yarn layers may have different orientation and different yarn densities of single ends. MWK fabrics are used to reinforce different matrices, since the combination of multidirectional fibre layers and matrices has proved capable of absorbing and distributing extraordinarily high strain forces (Padaki *et al.*, 2006). Some of the 3-D knit structures are shown in Fig. 1.14.

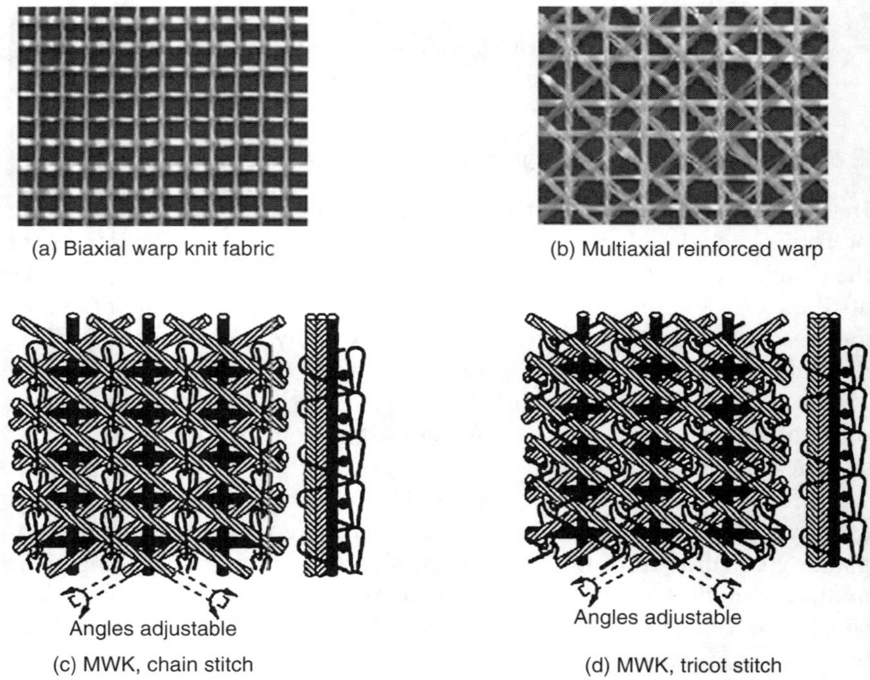

1.14 Knit structures.

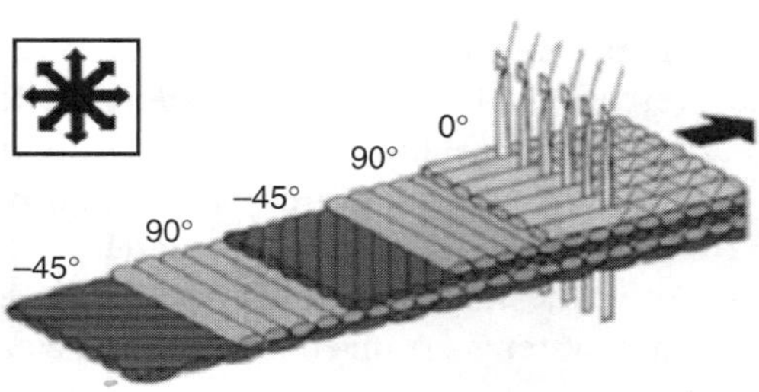

1.15 Multiaxial warp-knit system.

The MWK fabric system consists of warp (0°), weft (90°) and bias ($\pm\theta$) yarns held together by a chain or tricot stitch through the thickness of the fabric, as illustrated in Fig. 1.15. Theoretically, the MWK can be made to as many layers of multiaxial yarns as needed, but current commercially available machines allow four layers (the Mayer system) of 0°, 90°, $\pm\theta$ for insertion yarns, or at most eight layers (the LIBA system) of 0°, 90°, three ($\pm\theta$) insertion yarns, to be stitched together. All layers of insertion yarns are placed in perfect order each on top of the other in the knitting process. Each layer shows the uniformity of the uncrimped parallel yarns. The insertion yarns usually possess a much higher linear density than the stitch yarns and are therefore the major load-bearing component of the fabric.

Advantages of three-dimensional knitted structures

- 3-D knitted preforms have better formability because they are more drapable.
- 3-D knitting can produce more complex near-net-shape preforms.
- Some types of 3-D knitting can be done on existing automatic machines with little modification.
- 3-D knitted sandwich composites have a lower specific density.
- Some types of 3-D knitted composites have higher impact damage tolerance and energy absorption (crash) properties.

1.4.5 Three-dimensional braided fabrics

Braiding is a textile process that is known for its simplicity and versatility. It is an old technique but experiencing resurgence in interest because of its diverse applications. The conventional 2-D braiding (or Maypole braiding) is a simple textile process. It is composed of two sets of yarn carriers rotating on a circular track. One set rotates in the clockwise direction while the other set rotates in the counter-clockwise direction and interlaces with the first one to form a tubular preform. The yarn carriers move through two sinusoidal slots in the track plate by means of horn gears. Several layers each with a specified braiding angle can be serially superimposed in order to form a multilayer braid. But the problem of these multilayer braids consists in the interlaminar weakness or in other words in its sensitivity to delamination. Three-dimensional braiding overcomes this problem by introducing reinforcing yarns of materials that transversely connect the different layers during the process.

Three-dimensional braiding technology is an extension of the well-established 2-D braiding technology wherein the fabric is constructed by intertwining or orthogonal interlacing of two or more yarn systems to form an integral structure. This extension of 2-D to 3-D braiding has opened up new opportunities in the near-net-shape manufacture of high-damage-resistant structures. At present most composite manufacture uses 2-D textiles in the form of simple 2-D braids, fabrics and unidirectional plies as basic reinforcement. Additional steps such as cutting, stacking of plies and stitching are then needed to complete the final textile reinforcement. These secondary operations are expensive and could be eliminated if suitable 3-D textile preforms could be produced using cost-sensitive mass production. Three-dimensional rotary braiding is an automated production method that could provide an attractive solution for the manufacture of certain

1.16 Braid structure.

net-shaped 3-D-textile preforms (Bigaud *et al.*, 2005). A typical 3-D braided structure is shown in Fig. 1.16.

Three-dimensional braiding is one of the textile processes in which a wide variety of complex structural shapes can be produced in an integrated manner, resulting in highly damage-resistant structural preforms, as these preforms can withstand axial, flexural and torsional loads (Rawal *et al.*, 2005).

Principle of three-dimensional braiding

The 3-D braids are produced by a number of processes including the track and column (3-D circular loom) method (Brown and Ashton, 1989), the two-step braiding method (Popper and McConnell, 1987), and a variety of displacement braiding techniques. The basic braiding motion includes the alternate x and y displacement of yarn carriers followed by a compacting motion. The proper positioning of the carriers and the joining of various rectangular groups through selected carrier movements accomplish shape formation.

A generalized schematic of a 3-D braiding process is shown in Fig. 1.17. Axial yarns, if present in a particular braid, are fed directly into the structure from packages located below the track plate. Braiding yarns are fed from bobbins mounted on carriers that move on the track plate. The pattern produced by the motion of the braiders relative to each other and the axial yarns establishes the type of braid being termed, as well as the microstructure.

Three-dimensional rotary braiding is based on the well-known 2-D rotary braiding concept but now uses the horn gears arranged in a flat array as shown in Fig. 1.18(a). Each horn gear is equipped with a special clutch–brake mechanism, which allows a controlled stop or rotation of each single horn gear and the attached bobbins. Grooves in the machine working plate guide the bobbins driven by the horn gears. Switches, however, located between each pair of horn gears, can be activated to transfer the bobbin to an adjacent horn gear or to be kept. According to the status of the clutch–brake mechanism and switches, any arbitrary movement of the bobbins is

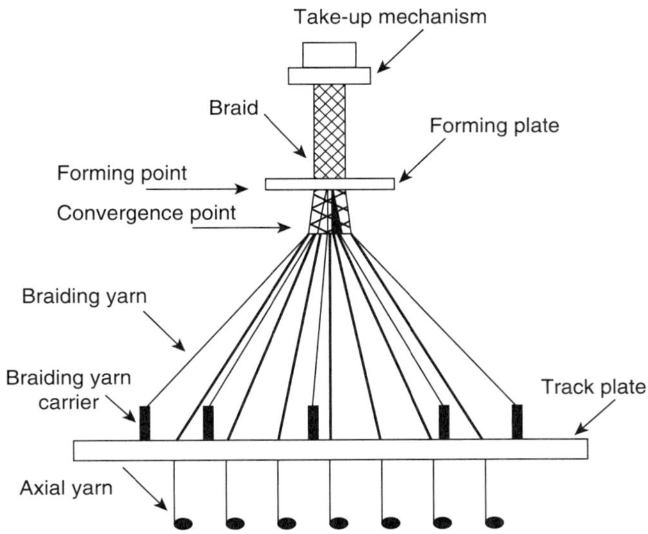

1.17 Schematic drawing of a generalized 3-D braider.

possible. The principle of formation of 3-D braided fabrics is shown in Fig. 1.18(b).

For the insertion of standing ends into the braid, those yarns are led through input tubes positioned between the horn gears or through the horn gear axles (Fig. 1.18(a)). In this way it is possible with this technique to produce braids with almost any fibre orientations and cross-section geometry in near net shape with minimum waste. Due to the possibility of changing the number of 'active' bobbins, the cross-sectional area and with that the geometry of the braid can be varied online (Schneider *et al.*, 2000). Another 3-D braider consisting of star-shaped rotors arranged in a matrix of multiple rows and columns was presented by Tsuzuki *et al.* (1991). In this machine (Fig. 1.19), four yarn carriers can surround a rotor and can move in four diagonal directions which are determined by the rotation of the rotors. The addition of axial yarns and the addition and subtraction of braider yarns allows for changes in fabric geometry and the ability to braid complex shapes.

Advantages of three-dimensional braided structures

- 3-D braiding has the ability to produce complex near-net-shape preforms.
- 3-D braiding processes can be automatically controlled, which increases production and preform quality.
- 3-D braided composites with a complex shape can be inexpensive and simple to manufacture.

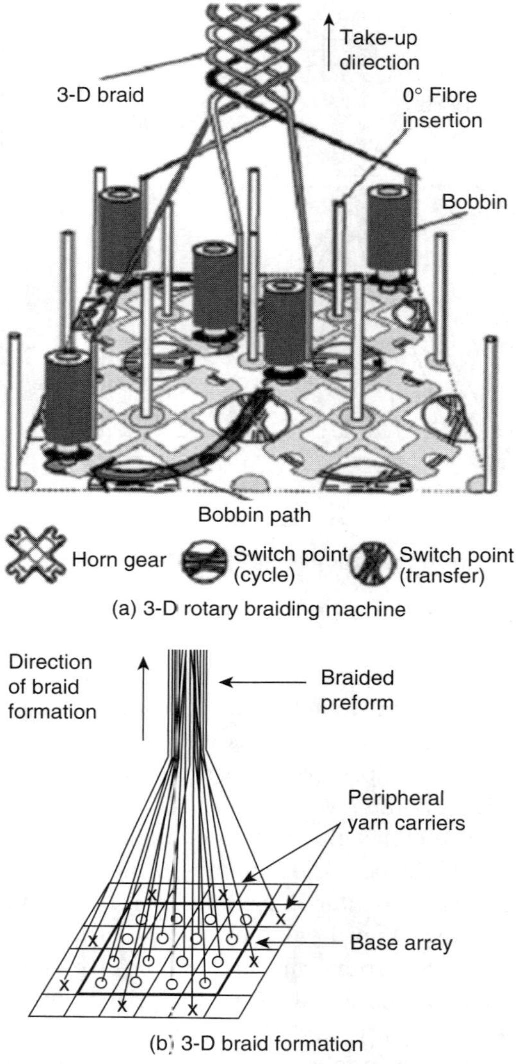

1.18 Principle of 3-D rotary braiding.

- 3-D braided composites have better delamination resistance and impact damage tolerance.
- 3-D braided composites are less sensitive to notches.

1.4.6 Three-dimensional stitched fabrics

The development of stitch-bonded, multiaxial fabrics has allowed for faster fabrication of parts with better physical and mechanical properties. Parts

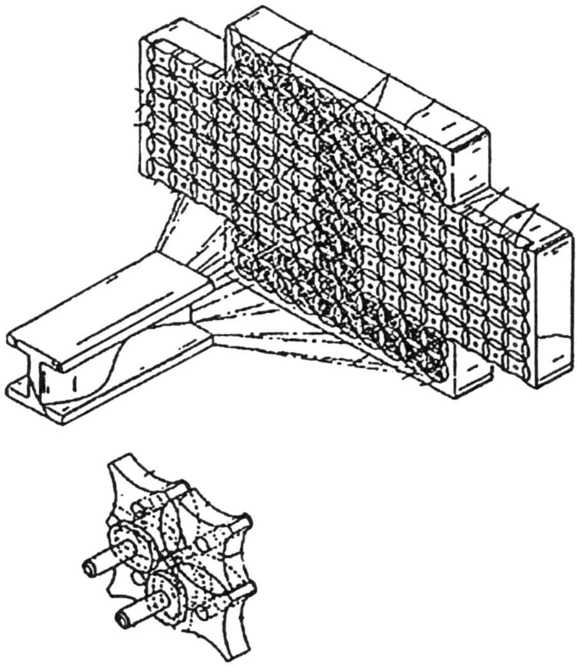

1.19 Solid braiding apparatus of Tsuzuki *et al.* (1991).

made from these structures have led to cost-effective solutions for a variety of applications including marine, transportation, infrastructure, sports and recreation, and aerospace. The cost-effective solution begins with engineering the laminate requirements at the point of fabric manufacture. Stitch-bonding of fabric is essentially an automated process and is highly efficient compared to a shop-fabricated laminate using unidirectional or woven fabrics.

Conventional fabrics are made by weaving fibres in two perpendicular directions (warp and weft). Weaving bends the fibres, reducing the maximum strength and stiffness that can be attained. When cut, fabrics also tend to fray, making them difficult to handle. Stitched fabrics offer several advantages over conventional woven fabrics. In the simplest case, woven fabrics can be replaced by stitched fabrics, maintaining the same fibre count and orientation.

Stitch-bonded multi-plies

Multi-plies are fabrics consisting of one or more parallel and drawn layers of threads that can have different orientations. Stitch-bonded, multiaxial fabrics or, simply, stitched fabrics consist of several layers of unidirectional fibre bundles held together by a non-structural stitching thread, usually polyester. The fibres in each layer can be input at almost any angle between 0° and 90°. The entire fabric may be made of a single material, or different materials can be used in each layer for a hybrid fabric. A typical stitch-bonded fabric structure is shown in Fig. 1.20.

Multiaxial multi-plies have versatile properties such as drawn thread orientation, different angles between the layers, manifold layer composition and arbitrary mass. A stitch-bonded multiaxial multi-ply consists of several layers of reinforcing threads and a mesh structure – the warp knit. Up to eight layers can be combined with the orientation of the layers arranged as necessary (for example 0°, 90°, +45°, –45°: Fig. 1.21).

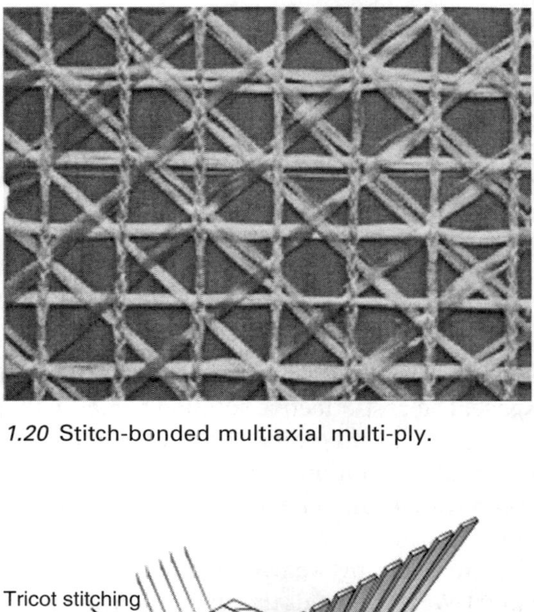

1.20 Stitch-bonded multiaxial multi-ply.

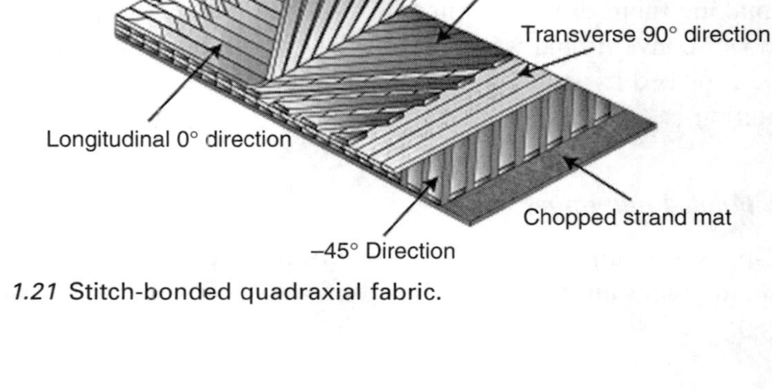

1.21 Stitch-bonded quadraxial fabric.

A typical quadraxial ply stack includes 0°, 90°, +45° and –45° plies. They are often made balanced (equal weight on all axes) but can also be tailored to suit a particular load case, such as for a typical boat-bottom panel where bending occurs mostly in the transverse direction. In this case quads are designed with more 90° fibre than the other axis. The 0° orientation is called the warp system and corresponds to the work direction. The other layers are called weft systems (Potluri *et al.*, 2003).

Stitch-bonded fabrics offer greater range and flexibility compared to woven fabrics, especially in the field of multiaxial (three plies or more). Multiaxial reinforcements can be engineered to meet specific requirements and perform multiple tasks such as providing good surface finish, impact and abrasion resistance, and structural integrity, all in one fabric (Hausding *et al.*, 2006).

Advantages of three-dimensional stitched fabrics

- inexpensive and simple to manufacture
- handling preforms (plies prevented from moving)
- better impact damage tolerance
- improved delamination resistance to ballistic impact and blast loading
- better interlaminar fatigue resistance
- improved joint strength under monotonic and cyclic loading.

1.4.7 Three-dimensional non-woven fabrics

Non-wovens are widely used in technical applications such as fitted filters, preforms for composites and geotechnical equipment. Three-dimensional shaped non-woven products are currently constructed from flat webs. In addition to the high cost of the conversion processes, irregularity is inevitably introduced into the final product because of joints. There is a long history of 3-D non-woven reinforcements, primarily in carbon–carbon composites. Orthogonal 3-D materials are fabricated by fixing a series of yarns in one direction (or rods which will later be withdrawn and replaced by yarns) and then inserting planar yarns in the two orthogonal directions around the fixed yarns.

Most of the processes described in the literature are based on production of 3-D non-wovens using the regular manufacturing processes, i.e. needle punching, spun bonding, melt blowing, air laying, etc. Needle felts in 3-D form are a type of substrate that acquires its dimension or shape after fabric/web formation is complete (i.e. it is an additional process) via a moulding or thermoforming process. Air-laid technologies also exist that utilize air streams to blow or lay fibres on screens or moulds, thus providing a 3-D form during fabric/web formation. Melt blowing has similar capabilities. Molten polymer is extruded between two high-velocity laminar sheets of air, and fibre is collected on a drum (Wang *et al.*, 2007).

Gong *et al.* (2003) have described a method for producing 3-D non-wovens directly from fibres, thus eliminating the conversion processes required for many applications. The 3-D fibrous web is formed by air-laying and is then consolidated by heat through-air bonding. The process was based on the air-laying principle for web formation of thermal through-air bonding for web consolidation (Fig. 1.22(a)). The fibre-opening unit was modified from a roller card with a 1 m working width. Opened fibres were stripped off the main cylinder of the opening unit by and dispersed in airflow, then carried by the airflow through a transport duct to perforated 3-D moulds, which move across the machine during web formation. After moving out of the web-forming area, the 3-D web, carried on the mould surface, was moved into the bonding section for consolidation. After evaluating numerous bonding techniques, it was found that the thermal through-air method was the most appropriate because it could be readily adapted to suit different shapes and is very economical. In the bonding section, the 3-D web was stationary and hot air was drawn through it by a suction fan to bond the fibres. The time for which the 3-D web stays in the bonding chamber (the dwell time) is an important parameter that influences the bonding effect and also the cycle time. A minimum time was needed in order for the hot airflow to stabilize and for the temperature around the 3-D web to reach the desired level. Because the web-forming process was con-

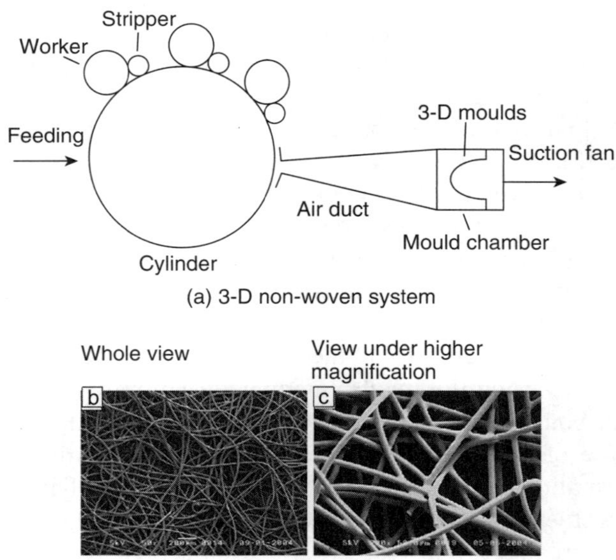

1.22 Examples of 3-D non-woven structures.

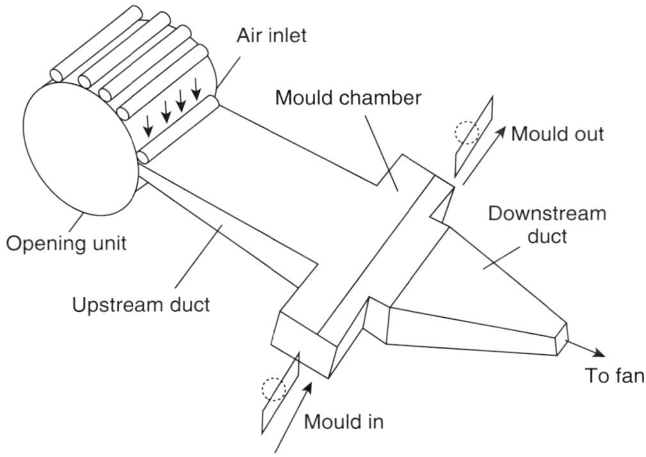

1.23 3-D non-woven web forming by air-laying principle.

tinuous while the bonding is intermittent, fibre feed, mould movement and the bonding process had to be closely coordinated so that the area weight of web and the effects of bonding meet the preset requirements. Examples of 3-D non-woven structures are presented in Fig. 1.22(b) and (c).

Production of 3-D non-wovens by the air-laying principle has been described by Ravirala and Gong (2003). In this process the 3-D non-woven products were produced directly from staple fibres using the air-laying principle. The fibres were deposited onto 3-D porous moulds before the fibrous web was consolidated to produce the final product. The web-forming process is schematically shown in Fig. 1.23.

The machine width was 1245 mm and the length of the upstream air duct was 1600 mm from the opening unit to the mould chamber. The fibres opened by the opening unit were transported by airflow through the upstream duct to the mould chamber, which had a vertical depth of 300 mm. The fibres were then deposited on the 3-D porous moulds as they moved across the machine in the mould chamber. As in flat webs, the distribution of fibres in the 3-D web is a key factor in determining the performance of the final product. However, in the process of forming flat webs, the angle between the airflow and the deposition surface is broadly the same over the web formation zone, while this angle varies greatly at different points of the 3-D mould surface. Because of flow angle variations, even when the airflow distribution is perfectly uniform, fibre deposition will be uneven around the 3-D mould. In order to produce an even product, the airflow must be regulated according to the shape of the mould.

Production of 3-D fabrics by non-woven technology introduces through-thickness reinforcement without causing significant in-plane fibre damage

(Kamiya *et al.*, 2000). In an extensive review on recent advances in the fabrication and design of 3-D textile preforms, the above authors have presented different techniques of production of 3-D non-woven fabrics. A technique presented by Yasui *et al.* (1994) is shown in Fig. 1.24. In this technique, an array of pipes is arranged with predetermined spacing on a base plate. A line of yarn is then looped back and forth widthwise through the array of pipes. A second layer can then be formed by looping the yarn in a biased direction. In this manner, many layers of various orientations can be produced. The through-thickness yarns can then be introduced by stitching (or knitting) needles which are inserted into each pipe and pushed through the thickness of the fibre bed. As shown in Fig. 1.25, the yarns are looped over a selvedge yarn at the bottom of the fibre array, which effectively binds the preform together. By changing the base plate, various 3-D shapes can be formed.

Three-dimensional non-woven structures can also be obtained by means of needle-punching two pre-needled non-woven layers with a defined space between, realized by mean of a spacer (Vasile *et al.*, 2006). Napco® is a technology that enables the manufacture of a 3-D spacer non-woven fabric with large hollow spaces, using 3-D Web Linker®, a special machine devel-

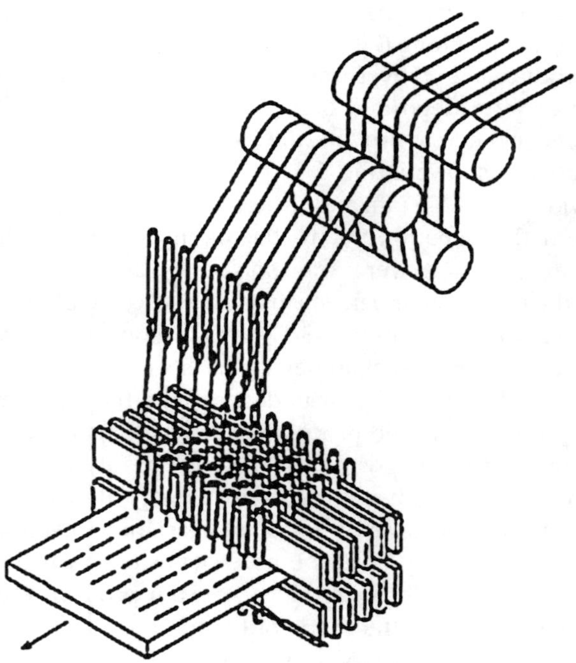

1.24 Non-woven apparatus (Yasui *et al.*, 1994).

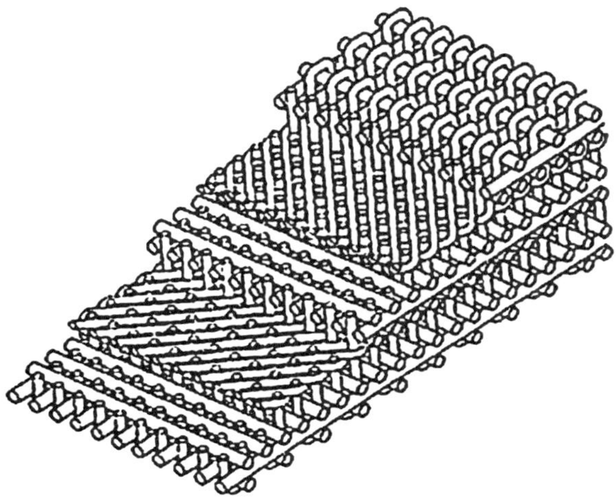

1.25 Non-woven fibre architecture.

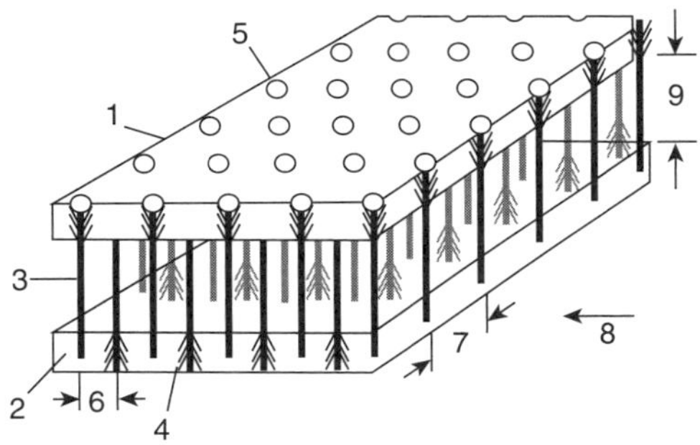

1.26 Structure of Napco® 3-D non-woven fabric: 1 – top layer; 2 – bottom layer; 3 – connecting layer (bridge fibres from 1); 4 – bridge fibres from 2; 5 – needle stitch; 6 – distance between bridge fibres depending on stitch depth; 7 – distance between bridge fibres depending on needle density; 8 – take-out direction; 9 – product thickness depending on the spacer's width.

oped by Laroche. In this process, special barb or fork needles arranged in rows penetrate simultaneously from both sides of the two pre-needled non-wovens, creating fibre bridges formed by fibre bundles, and so the structure of the layers should contain enough unbounded fibres of sufficiently great fibre length. In Fig. 1.26, a schematic representation of a Napco® structure is shown, and a cross-section of the working zone of a 3-D Web Linker® machine that produces 3-D structures starting from two pre-needled mono-layers A and B can be seen in Fig. 1.27. This technology allows the production of unfilled 3-D non-wovens, as well as filled products (e.g. granulates, powders, tubes, paste, foam, textile wastes, etc.) for composites.

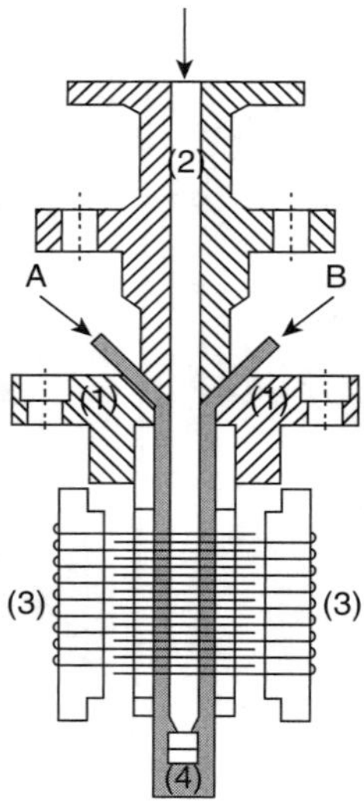

1.27 Machine cross-section: A and B – pre-needled non-woven monolayers; 1 – stripper plate; 2 – spacer tables; 3 – needle area; 4 – fibre bridges.

1.5 Conclusions

Three-dimensional fibrous assemblies constitute a whole family of textile structures manufactured using weaving, knitting, braiding and non-woven methods. They comprise structural preforms, which are fully integrated continuous fibre assemblies having multiaxial in-plane and out-of-plane fibre orientation. More specifically, a 3-D fabric is one that is fabricated by a textile process resulting in three or more yarn diameters in the thickness direction, with fibres oriented in three orthogonal planes. Among the large family of textile structures, 3-D fibrous assemblies have attracted the most serious interest in the aerospace industry and served as a catalyst in stimulating the revival of interest in textile composites. The engineering applications of 3-D composites originate from the use of carbon–carbon composites in aerospace. With the experience gained from 3-D carbon–carbon composites and the recent progress in fibre technology, the class of 3-D fabric structures is increasingly being recognized as a serious candidate for structural composites.

The ability to take complexity and labour out of manual composites fabrication processes through the innovative automation of engineered preforms is a key to more widespread use of composites. The trend is towards more control over fibre orientation and architecture while increasing productivity. This trend continues with the 3-D fibre assemblies discussed in this chapter. The 3-D weaving, knitting, braiding and non-woven processes and resulting preforms offer many advantages in both performance and economics. These 3-D fabrics will continue to gain acceptance as more companies recognize the value these materials offer. Technocrats using 3-D preforms could efficiently and accurately design totally new materials, novel manufacturing processes and new fabric structures to accelerate the fabric development process and foster innovation.

1.6 References

Alagirusamy R, Fangueiro R, Ogale V and Padaki N (2006), Hybrid yarns and textile preforming for thermoplastic composites, *Textile Progress*, **38**, 4, 1–71.

Bigaud D, Dreano L and Hamelin P (2005), Models of interactions between process, microstructure and mechanical properties of composite materials – a study of the interlock layer-to-layer braiding technique, *Composite Structures*, **67**, 99–114.

Brown R T and Ashton C H (1989), Automation of 3D braiding machines, paper presented at *4th Textile Structural Composites Symposium*, Philadelphia, PA, 24–26 July.

Brunnschweiler D (1953), Braids and braiding, *Journal of the Textile Institute*, **44**, 9, 666–685.

Chou T-W and Ko F K (eds) (1989), *Textile Structural Composites*, Elsevier, New York.

Demoski G and Bogoeva-Gaceva G (2005), Textile structures for technical textiles, Part II: Types and features of textile assemblies, *Bulletin of the Chemists and Technologists of Macedonia*, **24**, 1, 77–86.

Du G-W and Ko F (1996), Analysis of multiaxial warp-knit preforms for composite reinforcement, *Composites Science and Technology*, **56**, 3, 253–260.

Gong R H, Dong Z and Porat I (2003), Novel technology for 3D nonwovens, *Textile Research Journal*, **73**, 2, 120–123.

Hausding J, Engler T, Franzke G, Köckritz U and Cherif C (2006), Plain stitch-bonded multi-plies for textile reinforced concrete, *AUTEX Research Journal*, **6**, 2, June.

Kamiya R, Cheeseman B A, Popper P and Chou T-W (2000), Some recent advances in the fabrication and design of three-dimensional textile preforms: a review, *Composites Science and Technology*, **60**, 33–47.

Kaufmann J R (1991), in *Proceedings of Fibre-Tex 1991, The Fifth Conference on Advanced Engineering Fibres and Textile Structures for Composites*, NASA Conference Publication 3176, Raleigh, NC, 15–17 October, 77–86.

Lee L, Rudov-Clark S, Mouritz A P, Bannister M K and Herszberg I (2002), Effect of weaving damage on the tensile properties of three-dimensional woven composites, *Composite Structures*, **57**, 405–413.

Malik T and Parmar S (2006), 3D fabrics – an overview, www.fibre2fashion.com

Mohamed M H (1990), Three-dimensional textiles, *American Scientist*, **78**, 6, 530–541.

Mouritz A P, Bannister M K, Falzon P J and Leong K H (1999), Review of applications for advanced three-dimensional fibre textile composites, *Composites*, Part A, **30**, 1445–1461.

Padaki N V, Alagirusamy R and Sugun B S (2006), Knitted preforms for composite applications, *Journal of Industrial Textiles*, **35**, 4, 295–321.

Popper P and McConnell R (1987), A new 3D braid for integrated parts manufacture and improved delamination resistance – the 2-step process, *32nd International SAMPE Symposium and Exhibition*, 6–9 April, 92–102.

Potluri P, Kusak E and Reddy T Y (2003), Novel stitch-bonded sandwich composite structures, *Composite Structures*, **59**, 251–259.

Quinn J P, Mellhegger R, Mellhagger A and Rogers P (2001), A modified system for design and analysis of 3D woven preforms, Proceedings of International Conference for manufacturing of advanced composites, Belfast, Northern Ireland, 306–312.

Ravirala N and Gong G H (2003), Effects of mold porosity on fibre distribution in a 3D nonwoven process, *Textile Research Journal*, **73**, 7, 588–592.

Rawal A, Potluri P and Steele C (2005), Geometrical modeling of the yarn paths in three-dimensional braided structures, *Journal of Industrial Textiles*, **35**, 2, October, 115–135.

Schneider M, Pickett A K and Wulfhorst B (2000), A new rotary braiding machine and CAE procedures to produce efficient 3D-braided textiles for composites, *45th International SAMPE Symposium*, 21–25 May, Long Beach, CA.

Stobbe D and Mohamed M (2003), 3D woven composites: Cost and performance viability in commercial applications, *48th International SAMPE Symposium*, 11–15 May, Long Beach, CA.

Tan P, Tong L and Steven G P (1997), Modelling for predicting the mechanical properties of textile composites – a review, *Composites*, Part A, **28A**, 903–922.

Tsuzuki M, Kimbara M, Fukuta K and Machii A (1991), Three-dimensional fabric woven by interlacing threads with rotor driven carriers, US Patent 5,067,525, 26 November.

Vasile S, Van Langenhove L and De Meulemeester S (2006), Effect of production process parameters on different properties of a nonwoven spacer produced on a 3D Web Linker®, *Fibres and Textiles in Eastern Europe*, **14**, 4, October/December, 58–74.

Wang X Y, Gong R H, Dong Z and Porat I (2007), Abrasion resistance of thermally bonded 3D nonwoven fabrics, *Wear*, **262**, 424–431.

Xiaogang Chen (2006), 3D woven fabrics I: Design and manufacture, *Engineering of 3D Fabrics and their Applications Workshop*, Manchester, 6 February.

Xuekum Sun and Changjie Sun (2004), Mechanical properties of three-dimensional braided composites, *Composite Structures*, **65**, 485–492.

Yang C, Kim Y K, Qidwai U A and Wilson A R (2004), Related strength properties of 3D fabrics, *Textile Research Journal*, **74**, 7, July, 634–639.

Yasui Y, Anahara M, Hori F and Takeuchi J (1994), Method of producing fabric reinforcing matrix for composites, US Patent 5,327,621, 12 July.

Yi H L and Ding X (2004), Conventional approach on manufacturing 3D woven preforms used for composites, *Journal of Industrial Textiles*, **34**, 1, July, 39–50.

Zhang Q, Beale D, Adanur S, Broughton R M and Walker R P (1997), Structural analysis of a two-dimensional braided fabric, *Journal of the Textile Institute*, **88**, 1, 41–52.

2
Applications of three-dimensional textiles

Abstract: In an attempt to overcome many of the problems with the manufacture and mechanical properties of laminates, considerable attention has been given over the past 30 years to the development of advanced structures reinforced with 3-D fibre architectures. Among the large family of textile structures, 3-D fabrics have attracted the most serious interest in the aerospace industry and served as a catalyst in stimulating the revival of interest in textile composites. Considering the importance of 3-D fabrics, in this chapter some of the applications of 3-D woven, knitted, non-woven and braided fabrics to advanced composites, medical textiles, sports, geotextiles, space and protective garments are presented.

Key words: textile composites, multiaxial warp-knitted (MWK) fabrics, 3-D woven fabrics, applications of 3-D fabrics, geotextiles, automotives, sports, aerospace industry.

2.1 Introduction

Textile reinforcements have received widespread use as preforms in composites due to their flexibility to accommodate various reinforcing requirements. With the developments that have taken place in the aerospace industry, geotechnical fields, the composites industry and implantable medical devices, the use of high-performance textiles as reinforcements is increasing rapidly (Alagirusamy *et al.*, 2006). Developments in the field of preforming have led to the production of preforms with fibres oriented in different directions by weaving, knitting and braiding individually or in combination. Since reinforcements play a major role in dominating the mechanical properties of composites, the continuity and integrity of the architecture of fibre preforms becomes a main concern in advanced composites. Polymer laminates reinforced with a two-dimensional (2-D) layered fibre structure have been used with outstanding success for over 50 years in maritime craft, for about 30 years in aircraft and for nearly 20 years in high-performance automobiles and civil infrastructure laminates such as buildings and bridges.

Despite the use of 2-D laminates over a long period, their use in many structural applications has been limited by manufacturing problems and by some inferior mechanical properties. The manufacture of laminates can be expensive because of the high labour requirement in the manual lay-up of piles. As a result, many complex components need to be built from a number of machined laminate parts that must then be joined by co-curing, adhesive bonding or mechanical fastening. This is a major problem in the aircraft industry, where structures such as wings need to be made from a large number of smaller composite parts such as skin panels, stiffeners and stringers rather than being fabricated as a single integrated structure. Hence the use of 2-D structures in aircraft and automobiles has been restricted because of their inferior impact damage resistance and low through-thickness mechanical properties. In addition, these structures have low resistance to delamination cracking under impact loading because of their poor interlaminar fracture toughness.

In an attempt to overcome many of the problems with the manufacturing and mechanical properties of laminates, considerable attention has been given over the past 30 years to the development of advanced structures reinforced with 3-D fibre architectures. The development of advanced 3-D textile composites for specialized aircraft components began in the late 1960s, and since then these materials have attracted increasing attention because of their potential uses in aircraft, marine vessels, civil infrastructure and medical fields (Mouritz *et al.*, 1999). Considering the importance of 3-D fabrics, this chapter will present some of the applications of 3-D woven, knitted, non-woven and braided fabrics to advanced composites, medical textiles, sports, geotextiles, space and protective garments.

2.2 Application of three-dimensional fabrics to composites

Three-dimensional fabrics for structural composites are fully integrated continuous fibre assemblies having multiaxial in-plane and out-of-plane fibre orientation. More specifically, a 3-D fabric is one that is fabricated by a textile process resulting in three or more yarn diameters in the thickness direction with fibres oriented in three orthogonal planes. Among the large family of textile structures, 3-D fabrics have attracted the most serious interest in the aerospace industry and have served as a catalyst in stimulating the revival of interest in textile composites. The engineering applications of 3-D fabrics for composites date back to the late 1960s, and have their origin in aerospace carbon–carbon composites. Since most of these early applications were for high-temperature and ablative environments, carbon–carbon composites were the principal materials (Ko, 1989). In recent years, composites fabricated using reinforcements made with textile preforming processes, combined with cost-effective composite fabrication such as resin infusion, are being researched and developed at an ever-increasing pace. The increasing interest and use of textile composites, particularly 3-D textile composites, is attributed to improved performance due to controlled fibre distributions and lower cost through the use of automated textile processing equipment (Stobbe and Mohamed, 2003).

The expansion of the interest in 3-D fabrics for resin, metal and ceramic matrix composites is a direct result of the current trend in the expansion of the use of composites from secondary to primary load-bearing applications in automobiles, building infrastructures, surgical implants, aircraft and space structures (Chou and Ko, 1989). This requires a substantial improvement in the damage tolerance and reliability of composites. In addition, it is also desirable to reduce the cost and broaden the usage of composites from aerospace to automotive applications. This calls for the development of capability for quantity production and the direct formation of structural shapes. In order to improve the damage tolerance of composites, a high level of through-thickness and interlaminar strength is required. The reliability of a composite depends on the uniform distribution of the materials and consistency of interfacial properties. The structural integrity and handleability of the reinforcing materials for the composite is critical for large-scale, automated production. A method for direct formation of the structural shapes would therefore greatly simplify the laborious hand lay-up composite formation process. With the experience gained in 3-D carbon–carbon composites and the recent progress in fibre technology, the class of 3-D fabric structures such as woven, knitted, non-woven and braided forms is increasingly being recognized as important materials for advanced structural composites.

The attractions of advanced composites include their high ratio of strength to weight, their resistance to elongation under strain and their low coefficient of thermal expansion. These properties derive from the qualities of the reinforcement fibre and matrix material and of the interface between them and from the reinforcement provided by the fibre architecture (Mohamed, 1990). With the proper selection of materials and manufacturing technology, composites can be designed to outperform metals in any application.

The degree of a composite structural component dictates which fibre preform manufacturing technique should be employed. Knowledge of the load-carrying structural shape and fibre orientation determines which textile manufacturing technique is best suited for preform fabrication. It is obvious that an integrated, systematic approach, ranging from microstructural design, preform processing to performance characterization, is indispensable in the utilization of textile composites (Kamiya *et al.*, 2000).

2.2.1 Classification of textile preforms

There is a large family of textile preforming methods suitable for composite manufacturing (Ko, 1989). The key criteria for the selection of textile pre-

Table 2.1 Fibre architecture for composites

Level	Reinforcement system	Textile construction	Fibre length	Fibre orientation	Fibre entanglement
I	Discrete	Chopped fibre	Discontinuous	Uncontrolled	None
II	Linear	Filament yarn	Continuous	Linear	None
III	Laminar	Simple fabric	Continuous	Planar	Planar
IV	Integrated	Advanced fabric	Continuous	3-D	3-D

forms for structural composites are (a) the capability for in-plane multiaxial reinforcement, (b) through-thickness reinforcement and (c) the capability for formed shape and/or net shape manufacturing. Depending on the processing and end-use requirements, some or all of these features are required.

On the basis of structural integrity and fibre linearity and continuity, fibre architecture can be classified into four categories: discrete, continuous, planar interlaced (2-D) and fully integrated (3-D) structures. In Table 2.1 the nature of the various levels of fibre architecture is summarized (Scardino, 1989).

A discrete fibre system such as a whisker or fibre mat has no material continuity; the orientation of the fibres is difficult to control precisely, although some aligned discrete fibre systems have recently been introduced. The structural integrity of the fibrous preform is derived mainly from inter-fibre friction. The strength translation efficiency, or the fraction of fibre strength translated to the non-aligned fibrous assembly of the reinforcement system, is quite low.

The second category of fibre architecture is the continuous filament, or unidirectional (0°) system. This architecture has the highest level of fibre continuity and linearity, and consequently has the highest level of property translation efficiency and is very suitable for filament-wound and angle ply tape lay-up structures. The drawback of this fibre architecture is its intra- and interlaminar weakness owing to the lack of in-plane and out-of-plane yarn interlacings.

A third category of fibre reinforcement is the planar interlaced and interloped system. Although the intralaminar failure problem associated with the continuous filament system is addressed with this fibre architecture, the interlaminar strength is limited by the matrix strength owing to the lack of through-thickness fibre reinforcement.

The fully integrated system forms the fourth category of fibre architecture wherein the fibres are oriented in various in-plane and out-of-plane directions. With the continuous filament yarn, a 3-D network of yarn bundles is formed in an integral manner. The most attractive feature of the integrated structure is the additional reinforcement in the through-thickness direction which makes the composite virtually delamination-free. Another interesting aspect of many of the fully integrated structures such as 3-D woven, knits, braids and non-wovens is their ability to assume complex structural shapes.

From the point of view of preform fabrication and the macrostructural geometry of the textile preforms, textile structural reinforcements can be classified according to the axis of fibre or yarn introduction and geometric dimension (Fukuta *et al.*, 1984). As shown in Table 2.2, the axis, or the direction of yarn introduction, is divided into 0 (or 'non-axial'), monoaxial, biaxial, triaxial and multiaxial systems for reinforcements; in the last case, yarns are introduced from four or more directions.

2.2.2 Processing of textile composites

The first processing step is the formation of yarns from fibres. Figure 2.1 illustrates scales in one textile process. The part shown is an integrally

Table 2.2 Types of reinforcement

Dimension \ Axis		0 Non-axial	1 Monoaxial	2 Biaxial	3 Triaxial	4~ Multiaxial
1-D			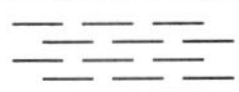 Roving yarn			
2-D		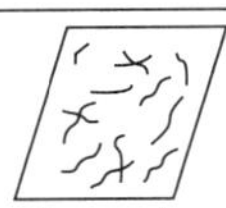 Chopped strand mat	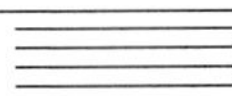 Pre-impregnation sheet	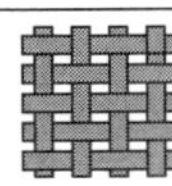 Plain weave	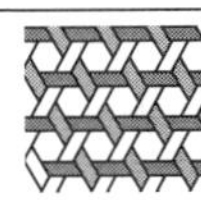 Triaxial weave	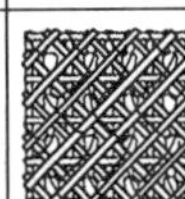 Multiaxial weave, knit
3-D	Linear element		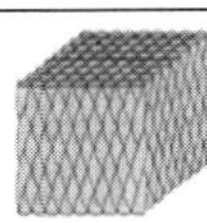 3-D braiding	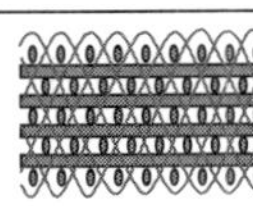 Multi-ply weave	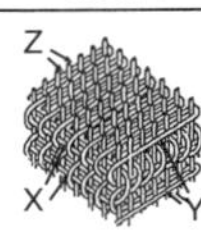 Triaxial 3-D weave	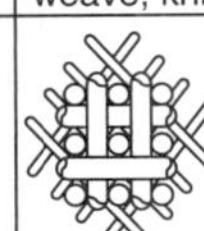 Multiaxial 3-D weave
	Plane element	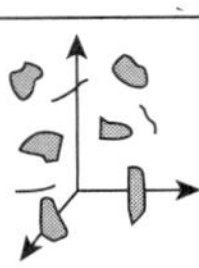	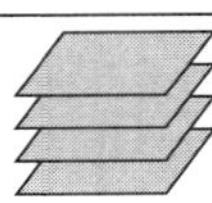 Laminate type	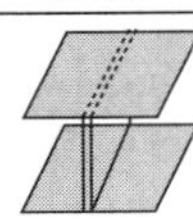 H or I Beam	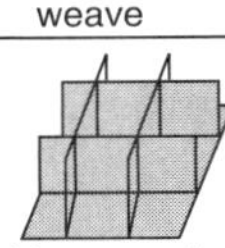 Honeycomb type	

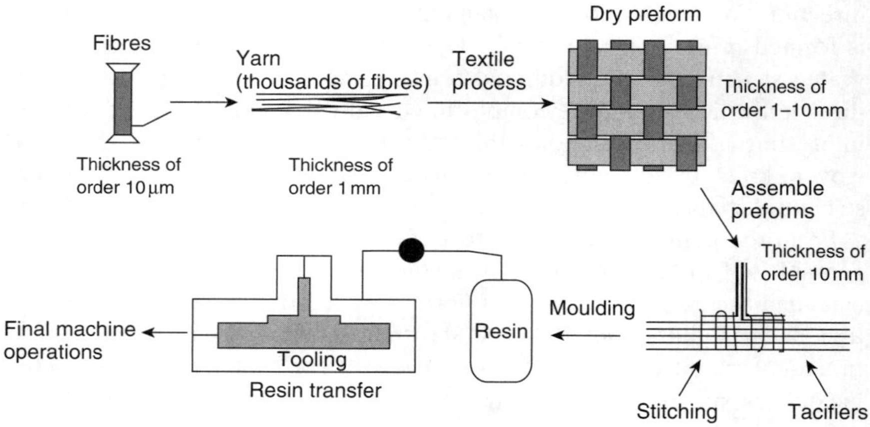

2.1 Production process of a textile composite.

formed skin/stiffener assembly. In the second step, the yarns are woven into plain woven cloth. The cloths are then laid up in the shape of the skin and stiffener and stitched together to create an integral preform. Finally, the composite part is consolidated by the infiltration of resin and curing in a mould.

There are many techniques available today for manufacturing thermoset composite parts. Some are still very low-tech and labour intensive, while some involve very sophisticated tooling and computer controls. However, all of these processes share some of the same challenges and requirements. They all consist of a tool to hold the fabric in the correct position while the resin is curing, and require some means of forcing the resin into the fabric. The major differences in the processes are the resulting part quality, limitations in size and geometry, cost of tooling, and process time.

The most basic and labour-intensive process is known as hand lay-up. In hand lay-up fabric is placed onto a tool where resin is applied by hand using rollers and squeegees. Each ply must be saturated as it is applied to the tool to ensure that no bubbles are left between plies. This makes hand lay-up very time consuming, but it does have its advantages. Carefully applying resin to each ply can ensure a part without dry spots. Unfortunately, the process is not performed under vacuum so micro-porosity is possible. Hand lay-up is very attractive due to the low cost of the tooling required. Since there is no pressure applied to the tool, it does not have to be very robust, and can be made out of a variety of materials. In many cases, the tool will have only one side to produce a nice finish on the outside of the part. Hand lay-up can also be used to produce very large parts. As long as there are enough people to apply the resin to the fabric before it cures, there are really no limitations on the size of the part. Hand lay-up is currently the most utilized method of manufacture for large wind turbine blades. Unfortunately, there are also many disadvantages to hand lay-up. The most obvious is the labour cost. In addition, the application of the resin in an open environment allows very volatile emissions to escape from the resin that can be harmful to humans and to the environment (Skramstad, 1999). It is anticipated that the use of hand lay-up for wind turbines will eventually be restricted due to the high volume of emissions. Other disadvantages are lower dimensional tolerances, poor fatigue performance, and less aerodynamic surfaces. Even with these considered, hand lay-up is still the fastest and cheapest way to produce a small number of composite parts with few defects, but the process is limited.

Beginning in the 1950s, more industrialized processes began to evolve for use on aircraft. These processes are generally referred to as resin transfer moulding processes, or RTM. In RTM the fabric is laid into a tool where the resin is forced into the fabric under pressure. These processes have several advantages over the hand lay-up process. The process has the potential to be more repeatable and consistent since the human involvement is reduced. This reduction in human involvement also reduces labour costs. In addition, the amount of volatile emissions is reduced. Much higher fibre contents can also be achieved, since the tool can clamp down on the fibre preform. Dimensional tolerances can also be increased if the tool is two-sided (Gebart, 1992). The disadvantages are the cost of the mould and the difficulty in forcing the resin through the fabric.

Modifications of the RTM process have been developed recently that reduce these disadvantages. Although there are many variants being used today, they all deal with these problems in a similar manner. Lower tool costs are achieved with the use of one-sided moulds. In these processes a vacuum is drawn on the fabric, while a flexible bagging is forced against the preform by atmospheric pressure. To deal with the problem of getting the resin to flow large distances through the fabric, a distribution network is used. This distribution network allows the resin to flow through high-permeability channels or layers to disperse it throughout the mould. The resin must then flow a much shorter distance in the plane or though the thickness of the part. Several variants of these processes are described in detail by Larson (2004) and will be discussed briefly here.

One process that has been successfully used on large structures is the Seemanns Composite Resin Infusion Molding Process (SCRIMP™). This process has been used since the 1980s and its use continues to increase. There are several variations of SCRIMP. One uses a series of channels above the fabric for resin distribution, and the resin is then forced to flow in the plane of the fabric between the channels. In other variants, a high-permeability layer may be placed over the fabric for resin distribution. The resin is then forced to flow though the thickness of the fabric. This layer is then peeled off after the process is complete. SCRIMP is capable of producing large parts very quickly, cheaply, and with high fibre volume fractions (Han, *et al.*, 2000).

A very similar process known as the Fast Remotely Actuated Channeling process (FASTRAC) is a more recent variation of this general principle. The main difference in the FASTRAC process compared to SCRIMP is a more refined distribution strategy. The distribution network is created by a 'FASTRAC layer' which is a flexible membrane with tightly spaced channels formed into it. The major difference is that these channels can be collapsed to force the extra resin though the fabric or out of the mould, rather than leaving them attached to the part as in SCRIMP. The FASTRAC layer also allows a positive pressure to be applied to the fabric to achieve even higher fibre volume fractions.

A process very similar to FASTRAC was developed by Larson which will be referred to as pressure bag moulding. In pressure bag moulding the distribution system is a channel that covers the whole surface of the fabric. Once the resin fills the channel, pressure is applied to a flexible film to force the resin into the fabric as in FASTRAC. In order to apply a positive pressure to the bagging, a second tool half is required. Although this adds an additional cost in the tooling, the second mould half would not require the surface finish and dimensional tolerance that the first half would. The mould for this process is illustrated in Figs 2.2 and 2.3. In these figures the flow channel is just empty space; however, it could also represent a highly permeable layer as in SCRIMP or FASTRAC.

Of the processes examined, the FASTRAC and pressure bag moulding process have been identified as having the largest injected volume per port. This is due to the fact that the distribution system covers the whole part.

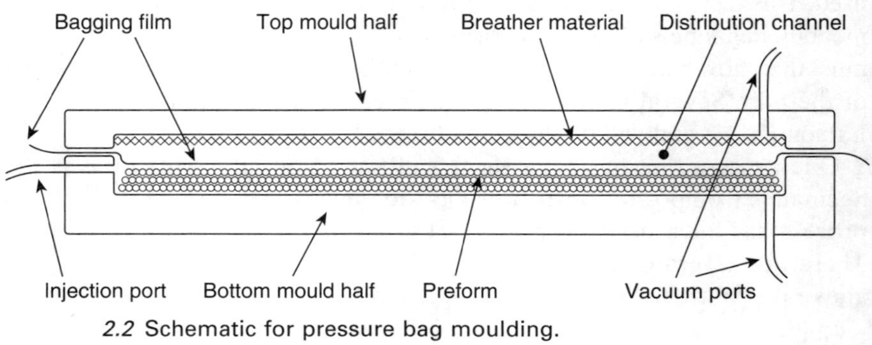

2.2 Schematic for pressure bag moulding.

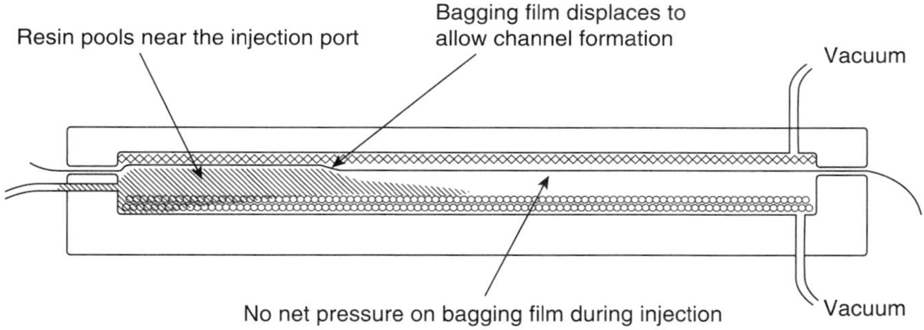

2.3 Pressure bag moulding during stage one.

For this reason, these processes are the most viable for large wind turbine blades, and will be the focus of this study. For future modelling this process will be described in two stages. Stage one consists of injecting the resin into the mould, and stage two is when pressure is applied to the bagging to force the resin through the thickness.

A summary of several of the processes described is presented in Table 2.3. Due to their similarity, the FASTRAC and pressure bag moulding processes are presented together.

2.2.3 Three-dimensional woven fabrics

Woven fabrics are probably by far the most commonly used form of textile composites in structural applications (Laroche and Vu-Khanh, 1994). Very good drapability, complex shape formation with no gap and reduced manufacturing cost are the main features of woven fabrics. They generally exhibit good dimensional stability in the warp and weft directions, offer highest cover or yarn packing density, and provide higher out-of-plane strength which can carry the secondary loads due to load path eccentricity, local buckling, etc. In addition, woven fabrics generally have a very low shear rigidity which gives a very good formability. However, they offer anisotropy, and relatively less extensibility for deep draw moulding compared to knitted and braided fabrics, and they are poor in resisting in-plane shear (Tan *et al.*, 1997). Most of the pure and hybrid woven fabrics used in textile composites are simple 2-D fundamental weaves, i.e., plain, twill and satin weaves.

The process of weaving is suited for the production of flat panels, and woven fabrics have been used for a number of years in 2-D laminated composites. However, these composites exhibited poor impact resistance, delamination strength and reduced in-plane shear properties, since typical

Table 2.3 Summary of manufacturing process details

Process	Basic principles	Advantages	Disadvantages
Hand lay-up	Open mould Manual infusion One-sided mould	Low cost Fastest implementation	Volatile emissions Health risks Inconsistent results Less efficient material usage
RTM	Closed mould In-plane resin flow Two-sided mould	Higher dimensional consistency Less volatile emissions Both sides finished	Higher mould cost Resin flow pattern critical Costly equipment required Lowest volume per port
VARTM	Closed mould In-plane resin flow Two-sided mould Evaluated mould	Higher dimensional consistency Less volatile emissions Both sides finished Higher quality products than RTM	Higher mould cost Resin flow behaviour critical Costly equipment required Complexity of vacuum porting
SCRIMP™	Closed mould In-plane resin flow One-sided mould Evaluated mould	Higher dimensional consistency Less volatile emissions Higher quality products than RTM	Proprietary process One side finished
FASTRAC + pressure bag	Closed mould Channel flow One side critical Evaluated mould	High quality High dimensional consistency Less volatile emissions Largest injection volume per port	Added cost of FASTRAC layer or top mould half Highest complexity Possible artefacts from bag Costly equipment required

2-D weaves possess fibres only in the 0° (warp) and 90° (weft) directions. To improve the impact and interlaminar properties, through-thickness reinforcement was required. This was accomplished by the use of multilayer 3-D weaving by angle-interlock weaves which use fibres to either weave together adjacent fabric layers (layer-to-layer interlock) or weave together all fabric layers (through-thickness interlock) (Kamiya *et al.*, 2000). These fabrics are produced principally by the multiple warp weaving method, which has long been used for the manufacture of double cloth and triple cloths for bags, webbings and carpets. The bending and friction resulting from the shedding motion and beat-up motion inherent to this weaving process tend to damage the high modulus yarns for the 3-D fabrics (Ko, 1989). The woven fibre architectures commonly used in 3-D woven composites are composed of several series of warp and weft yarns that form distinct layers, one above the other. The fabrics can be woven with a space between layers (core fabrics) or woven as thick, dense structures. The layers can be bound together by interlacing warp ends in the structure with the weft of adjacent layers (angle interlock) or by having the ends interlaced between the face and back layers (warp interlock). The binding yarns may also interlace vertically between the layers, producing an orthogonal weave (Mohamed, 1990).

The angle-interlock woven fabrics consist of three sets of yarns. The stuffers (warp yarns) and warp weavers are oriented along the longitudinal direction, i.e., along the loom feed direction. The fillers (weft yarns) are oriented transverse to the loom feed direction, and are inserted between layers of stuffers. The stuffers and fillers form an orthogonal array. The warp weavers traverse through the thickness of the weave, and interlock with filler layers. The warp weavers crisscross the weave thickness at off-axis angles. Different weave geometrical parameters are yarn size, yarn spacing, yarn distribution, interlock lengths and depths. There are two main types of angle-interlock preforms: through-thickness angle interlock weave (TTAW) and layer-to-layer angle interlock weave (LLAW). The TTAW is a multilayered preform in which warp weavers travel from one surface of the preform to the other, holding together all the layers of the preform. The LLAW is a multilayered preform in which warp weavers travel from one layer to the adjacent layer, and back. A set of warp weaves together hold all the layers of the preform (Naik *et al.*, 2002).

Angle-interlock or multilayer fabrics for flat panel reinforcement can be woven on traditional looms, mostly on shuttle looms. The warp yarns are usually taken directly from a creel. This allows mixing of different yarns in the warp direction. Other, more complex 3-D fabrics such as polar and orthogonal weaves require specialized weaving machines.

In polar weave structure, fibres or yarns are placed equally in circumferential, radial and axial directions. The fibre volume fraction is around 50%. Polar weaves are suitable for making cylindrical walls, cylinders, cones and convergent–divergent sections. To form such a shape, prepreg yarns are inserted into a mandrel in the radial direction. Circumferential yarns are wound in a helix and axial yarns are laid parallel to the mandrel axis. Since the preform lacks structural integrity, the rest of the yarns are impregnated with resin and the structure is cured on the mandrel. Polar weaves can be woven into near-net shapes. A near-net shape is a structure that does not

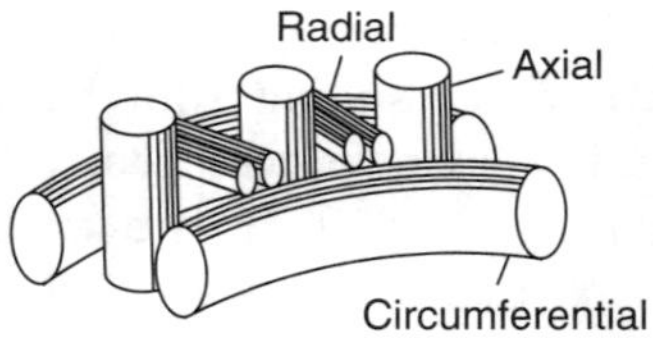

2.4 Polar weave.

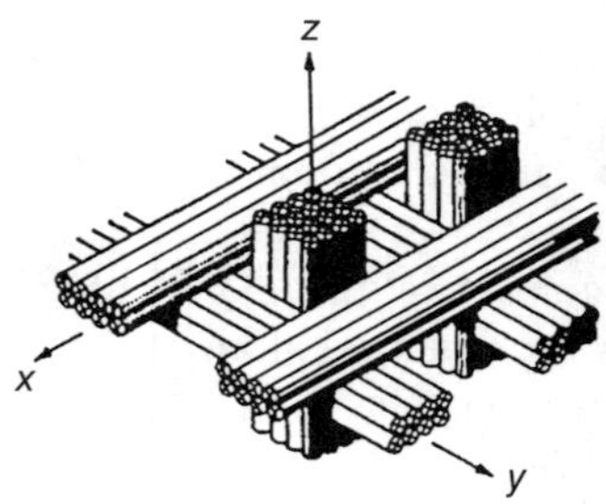

2.5 Orthogonal weave.

require much machining to reach the final product size and shape. A polar weave structure is shown in Fig. 2.4 (Adanur, 1995).

In orthogonal weave, reinforcement yarns are arranged perpendicular to each other in the *x*, *y* and *z* directions. No interlacing or crimp exists between yarns. The fibre volume fraction is between 45% and 55%. By arranging the amount of yarn in each direction, isotropic or anisotropic preforms can be obtained. Except for the components that are fundamentally Cartesian in nature, orthogonal weaves are usually less suitable for net shape manufacturing than the polar weaves. The unit cell size can be smaller than in polar weaves, which results in superior mechanical properties. Since no yarn interlacing takes place in polar and orthogonal structures, they are also referred to as 'non-woven' 3-D structures in the composites industry. However, it is more proper to label these structures as woven structures with zero level of crimp. A typical orthogonal weave structure is shown in Fig. 2.5 (Adanur 1995).

One of the most promising, recently developed textile processes is a new form of 3-D weaving (Brandt *et al.*, 1992; Dickinson *et al.*, 1999) being commercialized under the trademark 3WEAVE™ by 3TEX, Inc. With completely controlled and tailorable fibre orientations in the *x*, *y* and *z* directions, the ability to weave aramid, carbon, glass, polyethylene, steel fibres, etc. and any hybrid combination, at thicknesses up to one inch (2.54 cm) and widths up to 72 inches (183 cm), and the ability to make net shapes, an almost infinite number of 3WEAVE materials are possible with a tremendously wide range of performance (Mohamed *et al.*, 2001). Although these materials are typically more expensive than 2-D fabrics and mats, reduction of labour, higher performance and improved process efficiency result in overall cost savings in a variety of applications. When compared on a cost per square foot of finished composite structure, 3WEAVE reinforcements consistently outperform traditional 2-D materials.

Preforms made by the 3WEAVE process provide several important advantages in composites fabrication. The most obvious advantage of this material shows in manufacturing thick composites, owed to a dramatically reduced labour time, when multiple layers of 2-D fabric plies are replaced by one or a few 3WEAVE plies to obtain the required thickness in a composite structure. In many cases, a single 3WEAVE layer can replace multiple 2-D layers (Singletary and Bogdanovich, 2001).

It is natural to expect that the processing advantages of thick 3WEAVE preforms come at the expense of reduced conformability. In fact, it has been demonstrated that these preforms conform as well as or better than the most conformable 2-D fabrics. The absence of interlacing between warp and filling yarns allows the fabric to bend and internally shear rather easily, without buckling within the in-plane reinforcement (Mohamed *et al.*, 2003).

A fully automated 3-D weaving process with simultaneous multiple filling insertions has been developed at the North Carolina State University College of Textiles (Dickinson and Mohamed, 2000). This process is inherently 3-D from the onset, and does not involve the building up of layers one layer at a time. Rather, a single unit of thick fabric is formed during each weaving cycle. The essence of the innovation/patent centres around this simultaneous multiple insertion from one or both sides of the fabric.

2.2.4 Three-dimensional braided fabrics

Braiding was the first process used to manufacture a 3-D fibre preform for a composite. This process was developed in the late 1960s to produce 3-D carbon–carbon composites to replace high-temperature metal alloys in rocket motor components in order to achieve weight savings of 30–50%. Textile composites with braided fabric preforms are under consideration for aircraft applications that can be used at a lower cost and provide higher impact resistant/tolerant materials (Tan *et al.*, 1997).

The 3-D braiding technology is an extension of the well-established 2-D braiding technology wherein the intertwining or orthogonal interlacing of two or more yarn systems to form an integral structure constructs the fabric. It is one of the textile processes in which a wide variety of solid complex structural shapes can be produced integrally, resulting in a highly damage-resistant structural preform.

The 3-D braids are produced by a number of processes including the track and column (3-D circular loom) method (Brown and Ashton, 1989), the two-step braiding method (Popper and McConnell, 1987), and a variety of displacement braiding techniques. The basic braiding motion includes the alternate *x* and *y* displacement of yarn carriers followed by a compacting motion. The proper positioning of the carriers and the joining of various rectangular groups through selected carrier movements accomplish shape formation. Examples of the structural shapes successfully demonstrated as shown in Fig. 2.6.

3-D braided fabrics formed by the intertwining of yarn systems can be obtained in a variety of forms with laid-in yarns. At any time one half of the yarns are travelling in one direction at some angle to the axis down the fabric, while the other half are travelling in the opposite direction, passing over and under the strands of the first group. Figure 2.7(a)

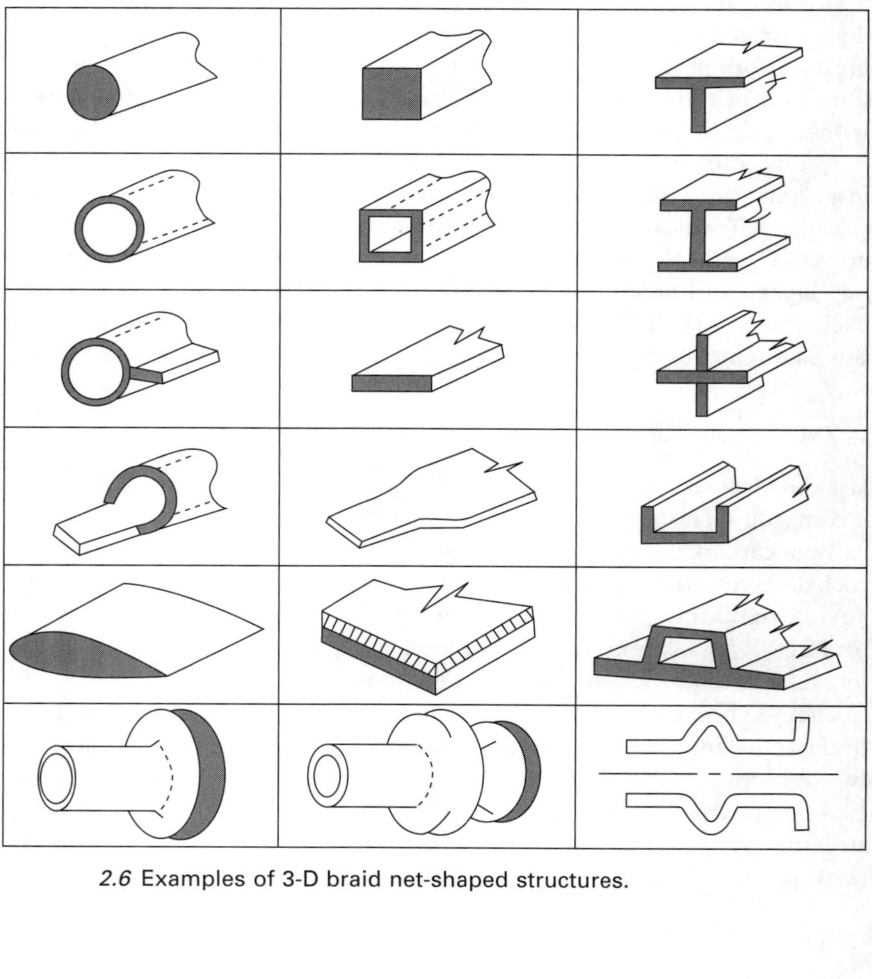

2.6 Examples of 3-D braid net-shaped structures.

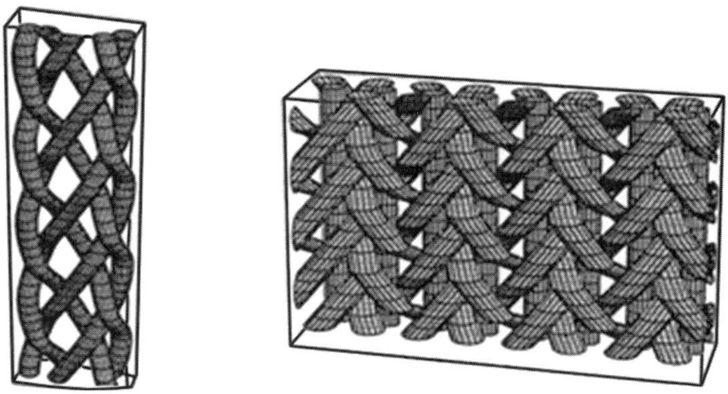

(a) Flat braided fabric (b) Circular braided fabric

2.7 3-D models for braided fabrics.

shows a braid fabric plaited by five strands. Braided fabrics have yarns interlacing at angles other than 0° and 90°. For a yarn orientation of ±45° interlacing is half that for the plain weave. This results in better adoption of fibre properties by the composite fabric due to reduced crimp. Therefore, braiding is one of the major manufacturing processes for textile composite preforms, although it is not a major manufacturing process for traditional textiles relative to weaving and knitting.

Figure 2.7(b) shows the local yarn structure in a circular braid consisting of three yarn systems. Two groups of yarns, one having an angle of 45° and the other having an angle of 90° to the mandrel axis, interlock to form a biaxial fabric. As an option, a third group of yarns mounted on the back side of the track ring of a braiding machine can be inserted through the centre of each horn gear. These yarns are deposited onto the mandrel in the axial direction. Clearly, the angle made by this group of yarns is 0° to the mandrel axis. All three groups, namely $\pm\theta$ and 0°, form a biaxially interlocked braid as shown in Fig. 2.7(b).

In the four-step braiding method, yarns are intertwined to form a multilayer 3-D structure (Fig. 2.8). Some of the braiding yarns traverse the internal layers and bind the two exterior layers together. Some sort of a beat-up is needed to push the yarns into the fabric structure after each round of braiding. In a two-step system to make 3-D braids, axial yarns and braider yarns are used. The braider yarns move around axial yarns which are fixed parallel to each other. The resultant tubular structure has good reinforcement in the axial direction but is weak in the circumferential direction. The multilayer interlock braiding method, which is similar to two-step braiding, allows greater circumferential reinforcement. In this method, yarns from adjacent layers are interlocked together.

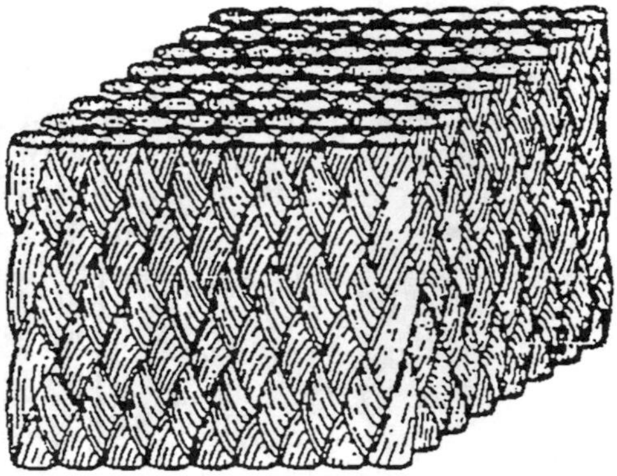

2.8 3-D braid produced by intertwining of multi-yarns.

In comparison to woven structures, due to the lack of beat-up during braid formation, braided fabric structure usually has low shear resistance and therefore is highly deferrable in the axial and radial directions. This characteristic of braided structures makes them particularly suitable to conform to surfaces of varying cross-sectional shapes such as cones and nozzles, and allows the production of near-net-shape structures. Hence, interest in braiding for composite manufacture has grown in recent years.

2.2.5 Three-dimensional knitted fabrics

Knitted fabrics are interlooped structures wherein the knitting loops are produced by the introduction of the knitting yarn either in the cross-machine direction (weft knit) or along the machine direction (warp knit). The most undesirable feature of weft-knitted structures is their bulkiness, which leads to the lowest packing density, or lowest level of maximum fibre volume fraction, compared to the other fabric preforms. While weft-knitted structures have applications in limited areas, multiaxial warp knit (MWK) 3-D structures are more promising and have undergone a great deal of development in recent years. Three-dimensional knitted fabrics are produced by either the weft-knitting or the warp-knitting process.

Multiaxial–multilayer structures are fabrics bonded by a loop system, consisting of one or several yarn layers stretched in parallel. Multiaxial–multilayer warp knits (MWK) are also termed non-crimp structures since the presence of knitted loops is to perform the function of holding layers of uncrimped inlay yarns. These yarn layers may have different orientation and different yarn densities of single ends. Multiaxial–multilayer fabrics are used to reinforce different matrices, since a combination of multidirectional fibre layers and matrices has proved capable of absorbing and distributing

2.9 Multiaxial–multilayer warp-knitted structures.

extraordinarily high strain forces. Among three basic types of these structures, Karl Mayer structures (Raz, 2000) are those in which, along with the ground tricot structure, inlaid yarns in warp, weft and both diagonal directions (30–60°) are incorporated in the fabric. Two other types, the LIBA and Malimo systems, along with layers of knits can also incorporate fibre/non-woven fleece between the layers to produce multiaxial–multilayer structures (Anand, 1996), which are predominantly applied to composite reinforcements.

Figure 2.9 illustrates tricot and chain-stitched multiaxial–multilayer knitted structures along with their cross-sectional view. Structural and mechanical studies on these knit preforms based on unit cell modelling have shown superior performance of these preforms for composite applications (Ko *et al.*, 1985; Dexter and Hasko, 1996; Du and Ko, 1996; Franzke *et al.*, 1997).

2.2.6 Three-dimensional non-woven fabrics

Use of 3-D non-woven surfaces in the composites industry is not so long established as that of woven and diagonal knit structures. Production of these structures is simple and quick. Needling and assembling by sewing (Malimo) are the most common methods. In addition to these, some uncurled structures can be defined as non-woven surfaces. Since there is no connection between the yarns in non-woven structures, they have weaker mechanical features compared to woven or knitted fabrics.

There is a long history of 3-D non-woven reinforcements, primarily in carbon–carbon composites. Orthogonal 3-D materials are fabricated by fixing a series of yarns in one direction (or rods which will later be withdrawn and replaced by yarns), and then inserting planar yarns in the two orthogonal directions around the fixed yarns.

Pioneered by aerospace companies such as General Electric, the non-woven 3-D fabric technology was developed further by Fiber Materials Incorporated. Recent progress in automation of the non-woven 3-D fabric manufacturing process has been achieved in France by Aérospatiale

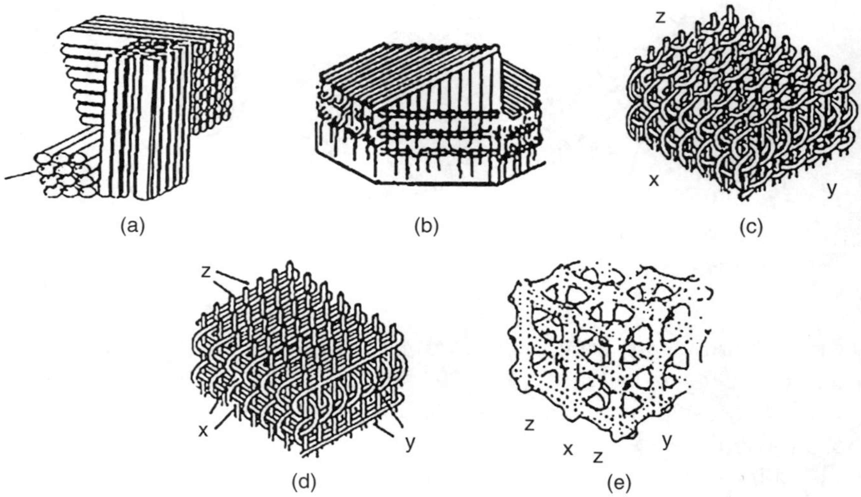

2.10 Orthogonal non-woven 3-D fabrics.

(Pastenbaugh, 1988; Geoghegan, 1988) and Bruno (Bruno *et al.*, 1986) and in Japan by Fukuta and co-workers (Fukuta *et al.*, 1982; Fukuta and Aoki, 1986).

The structural geometries resulting from the various processing techniques are shown in Fig. 2.10. Figure 2.10(a) and (b) show the single-bundle *xyz* fabrics in a rectangular and cylindrical shape. In Fig. 2.10(b), the multidirectional reinforcement in the plane of the 3-D structure is shown. Although most of the orthogonal non-woven 3-D structures consist of linear yarns in a non-linear manner, as shown in Fig. 2.10(c), (d) and (e) these structures can result in an open lattice or a flexible and conformable structure.

2.3 Application of three-dimensional fabrics to medical textiles

Textile materials in the medical field gradually have taken on more important roles. As more research has been completed, textiles have found their way into a variety of medical applications. In addition to protective medical apparel, textiles in fibre and fabric form are used for implants, blood filters and surgical dressings. Woven and knitted materials in both synthetic and natural form play a part in the biotextile field, but non-woven materials also have proven to be effective and cost-efficient. Textile structures in implantation are identified by structure, material composition, and behaviour of fibre surface and degradation. A major concern with artificial implants is the reaction that the body will have towards the implant. A biotextile in implantation must meet mechanical requirements and it must be biocompatible. Biocompatibility testing evaluates the response of the host system to the medical textile. Results of this testing must be viewed along with the risks and benefits of the device.

The use of 3-D fabrics in medical textiles is relatively new. Compared to 2-D woven fabrics, 3-D fabrics are dimensionally stable and find wide applications as vascular grafts, wound dressings, plasters, tissue engineering scaffolds, hospital bedding and uniforms and surgical gowns. Three-dimensional warp-knitted structures tend to be more stable and versatile and find applications as vascular implants, artificial tendons and ligaments, stents, compression bandages, surgical hosiery, etc. The applications of braided 3-D structures include artificial ligaments, tendons and absorbable and non-absorbable sutures. The non-woven 3-D fabrics have their applications as artificial implants and tissue engineering scaffolds.

In a study on woven prosthesis, a solid 3-D woven tubular prosthesis has been developed (Schmitt, 1999). This prosthesis has sufficient inherent wall stiffness so as to be radially self-supporting. This solid woven prosthesis (Fig. 2.11) is capable of being formed with a smooth, continuous inner wall that improves the hemodynamic flow compared with conventional 2-D woven structures, thereby facilitating the flow-through of fluid. The wall is believed to provide the prosthesis with sufficient radial stiffness to maintain an open lumen. The crimping additionally provides a degree of longitudinal compliance to the prosthesis. These devices may be used in a variety of locations in the body, such as in intraluminal applications in the vascular system, pulmonary system or gastrointestinal tract.

A biocompatible implant material made of a 3-D woven/knitted fabric has been reported by Yasuo Shikinami (Shikinami and Kawarada, 1998). This implant material has high mechanical strength and durability in all three orthogonal directions and synchronizes with the deformation characteristics of the surrounding biological tissues. It is developed using a

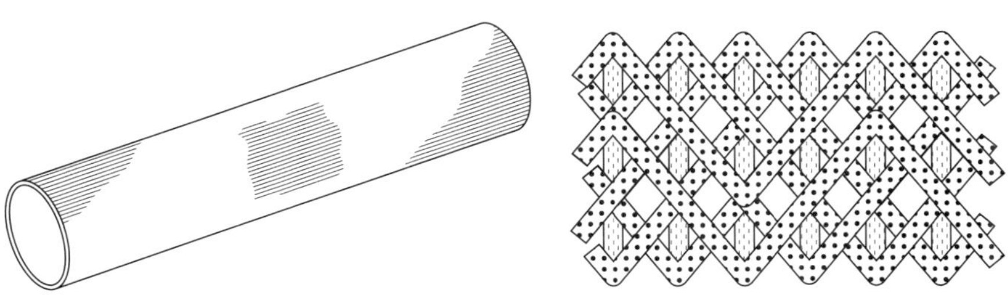

(a) (b)

2.11 Prosthesis made of 3-D woven fabric.

biocompatible bulk structure of a 3-D woven or knitted fabric of organic fibres. This implant material can be used to make artificial bones, cartilages, menisci and finger joints, bone fillers and materials for osteosynthesis and prosthesis.

Tissue engineering seeks to repair or regenerate tissues through combinations of implanted cells, biomaterial scaffolds and biologically active molecules. Three-dimensional woven, knitted and non-woven structures are finding application in tissue engineering for the manufacture of highly porous 3-D scaffolds, which can be produced by 3-D weaving techniques. A novel scaffold, mimicking the multidirectional biomechanical behaviour and the anisotropy of native cartilage, has been developed by Moutos *et al.* (2007). The porous structure was produced using 104-μm-diameter continuous multifilament polyglycolic acid yarn. The yarn was woven into two different 3-D structures containing a total of 11 in-plane fibre layers, five layers being oriented in the warp direction (0° or lengthwise in the loom) and six layers in the weft direction (90° to the lengthwise fibres) (Figs 2.12 and 2.13).

Non-woven fabrics are widely used as scaffolds in tissue engineering applications. However, non-woven fibrous matrices currently used in tissue engineering have a relatively large porosity, and a pore size in the range of several hundred micrometres, and have not been structurally optimized for specific applications. A non-woven 3-D scaffold produced by an electrospinning technique has been reported by Bhattarai *et al.* (2004). The study reports that the novel biodegradable block copolymer scaffold produced by the electrospinning process (Fig. 2.14) is a non-woven, 3-D, porous, nanoscale fibre-based matrix. The structure is suitable for soft tissue, such as skin and cartilage Fig. 2.15(a).

A tubular vascular prosthesis (Fig. 2.15(b)) with a complex triple-layered woven structure has been outlined by Latvian researchers (US Patent 6,863,696, 2005). The graft will find application in reconstructive surgery,

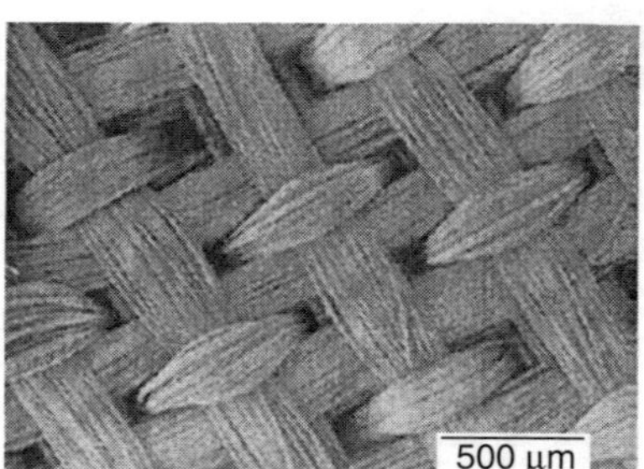

2.12 3-D woven structure for making porous scaffolds.

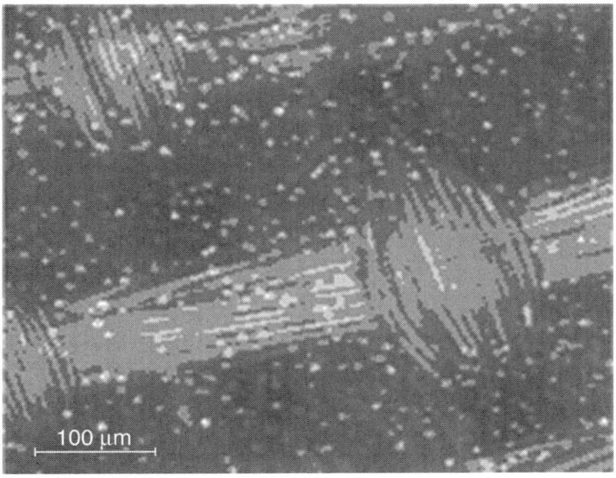

2.13 Cell growth on the freshly seeded scaffold.

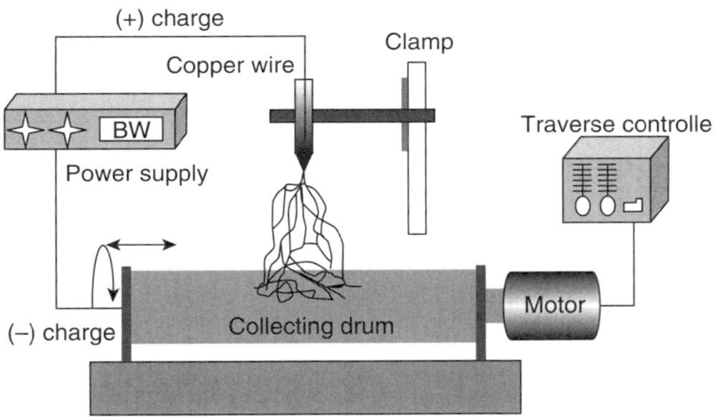

2.14 Electrospinning apparatus to produce 3-D porous non-woven structure.

such as for inborn vessel anomaly or for arterosclerotic damage or injury. The vascular prosthesis outlined in the patent is claimed to have good mechanical characteristics, non-ravelling ends, low permeability for blood, and good acceptance for ingrowth of tissue to seal the prosthesis walls.

2.4 Application of three-dimensional fabrics to sports

In the last decade tremendous progress has been made in the development of fibres and fabrics for sports end-uses. Textile materials are used in virtu-

(a) 3-D porous non-woven structure

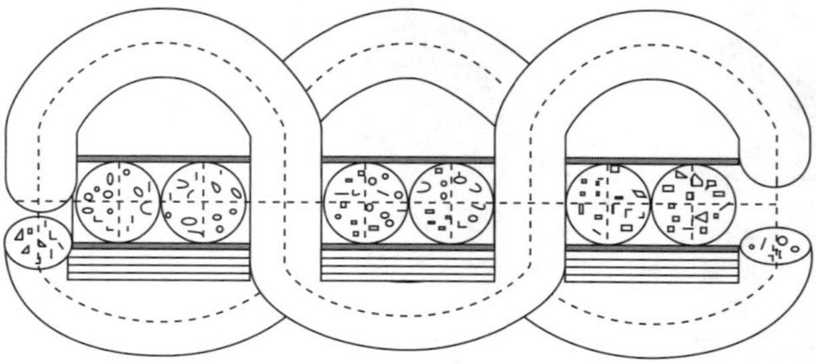

(b) Multilayer fabric for vascular prosthesis

2.15 3-D non-woven fabrics in medical application.

ally every sport from exercising and camping to football. High-performance textile fibres and fabrics are used in uniforms, equipment and sports facilities. The use of textile structural composites in sporting goods is increasing. Their applications include composite roller blades, bike frames, golf clubs, tennis rackets, ski and surf equipment, etc. Sports such as golf, baseball and tennis rely heavily on composites for their essential equipment. These sports would be drastically different if composite materials were not used. Due to their high strength and durability, composites have won favour in the sporting industry.

The application of 3-D textile fabrics is increasing in areas such as footwear, heavier fabrics such as for track suits and jogging suits, and shorts and shirts. Footwear, especially training/jogging shoes, is an area in which 3-D fabrics are now very widely used (Hill, 1985) to the almost total exclusion of leather in the construction of the main part of the upper. The fabrics in footwear are generally made of three layers: woven nylon outside, foam in the middle and warp-knit fabric inside (Adanur, 1995). The main advantages of 3-D fabrics over others are their consistency of quality and specification compared with the irregularities of leather, and their lightness, easy care properties and dimensional stability.

Garments for foul weather have received a great deal of attention recently as they are very much needed for such recreations as sailing, golf, climbing and running. The main direction has been to seek waterproof fabrics which 'breathe', that is, allow transportation of water vapour (perspiration) from the inside out. Multilayer fabrics have been found to be very successful in this direction. Very closely woven 3-D fabrics, multiaxial warp-knitted fabrics which have been calendered to close up the structure, and blended fabrics with coatings such as silicone–elastomer, are becoming popular.

Three-dimensional knits, also called spacer fabrics (Fig. 2.16(a)), are also attracting the attention of sportswear manufacturers for their capacity to trap air and offer extremely lightweight high-performance thermoregulation. Initially developed for industrial uses for cushioning and filtering, these fabrics are now popular in both sports shoes and garments. Sports shoes are one of the big initial success stories for spacer fabrics, making use of their light weight, high bulk and springiness. Climate properties, washability and a superior substitute for laminated foam are other advantages. Spacer fabrics are much like a sandwich and feature two complementary slabs of fabric with a third layer tucked in between. The inner layer can take a variety of shapes including tubes, pleats or other engineering forms, which gives the entire three-layer fabric a wide and ever-expanding range of potential applications (Bruer *et al.*, 2005) (Fig. 2.16(b)). This has led to successful experimentations with outerwear applications, where those properties found in sports shoes can be transferred to sports and leisure apparel.

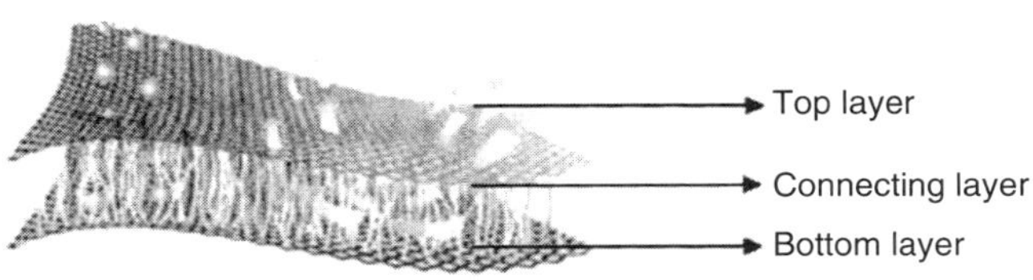

(a) 3-D knit spacer fabric

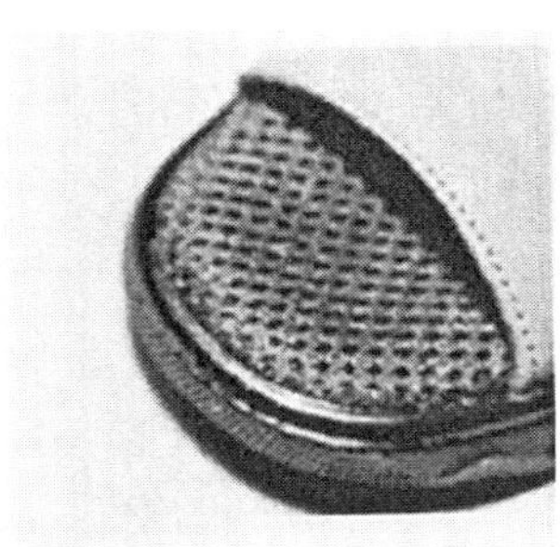

(b) Application in footwear

2.16 3-D knit fabric in sports. (Source: *Knitting International* (February 2002), Breathing Room, Knit Americas).

2.5 Application of three-dimensional fabrics to geotextiles

Geotextiles are permeable textile structures made of polymeric materials and are used mainly in civil engineering applications in conjunction with soil, rock or water. They are constructed as woven, non-woven, knitted, braided and combination materials. They are a member of a large family called geosynthetics. Other members of the family are geogrids, geonets, geomembranes and geocomposites. The first modern-day commercial geotextiles, also known as 'filter fabrics', were used for erosion control in the 1950s. The mechanical and hydraulic properties of the geotextile vary with the fabric type and can be adjusted to focus on five important performance functions: drainage, filtration, reinforcement, separation and armour. In addition, a geotextile composition must be selected to provide satisfactory placement and longevity for the design life of the structure. The first geotextiles were woven monofilament fabrics with a high percentage open area. Non-woven needle-punched fabrics were introduced in Europe in the late 1960s. The first non-woven (thermally bonded and needled) geotextiles were introduced in the United States in 1972 (Adanur, 1995).

There is a large family of textile structures available for geotextiles. Figure 2.17 illustrates examples of these structures. In the past two decades, apart from traditional woven fabrics, diversification into various forms including knits and speciality non-wovens has occurred. A particular class of textile structures that has been rediscovered and has undergone

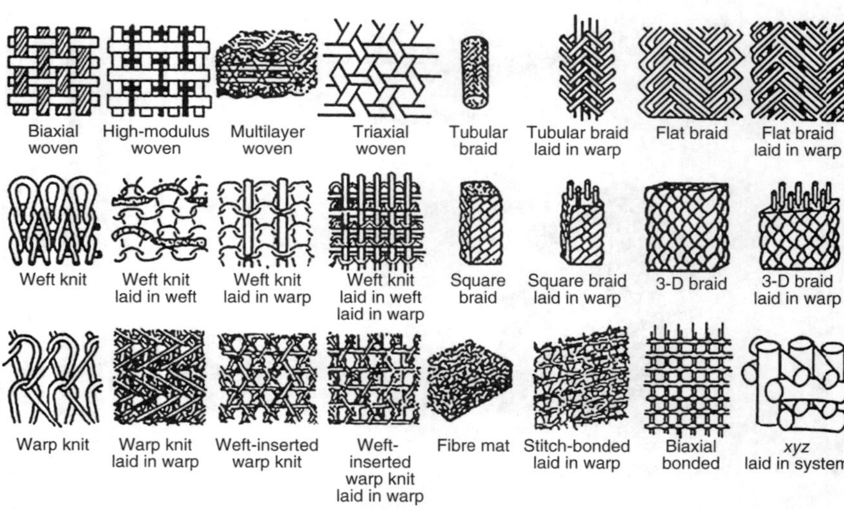

2.17 Textile structures for geotechnical applications.

extensive development for advanced composites and many other industrial applications is the 3-D textile structure.

Of the large family of textile structures, both woven and non-woven fabrics have found extensive applications as geotextiles because of their broad availability and low cost. Recently, 3-D woven, knitted, braided and non-woven fabrics have also been making inroads into the geotextiles area, although with limited application. Two traditional but less widely known technologies – braiding and warp knitting – have been rediscovered recently and have found many applications in marine, automotive and aircraft applications. Taking advantage of their multidirectional reinforcement capability, multiaxial warp knits (MWK) have been adopted extensively for large area coverage/reinforcement applications by boat builders as well as aircraft manufacturers, whereas braids have found applications in which linear, tubular and complex structural shapes are required, ranging from sporting goods to automotive components to concrete reinforcement. These structures also lend themselves to easy sensor incorporation, thus opening up new design opportunities for multifunctional geotextiles (Ko, 2004).

As produced by the Karl Mayer Malimo warp knitting system, the MWK fabric consists of warp (0°), weft (90°) and bias ($\pm\theta$) yarns plus the option of a non-woven backing held together by a chain or tricot stitch through the thickness of the fabric. Apart from the multidirectional and multicomponent nature of the fibre architecture, the MWK is characterized by its high productivity, at over one metre per minute and in widths as great as three metres. Since MWK fabrics possess a high level of tensile strength due to their non-crimped nature (and tear resistance as a result of the bias reinforcement and stitching yarn integration), they can be engineered for a wide range of geotechnical applications, including filtration and soil reinforcement.

MWK fabric composites incorporating a non-woven structure are ideally suited for many high-strength geotextile applications for which isotropic strength, resistance to tear and tear propagation, good water permeability, low creep and good fabric/soil interaction are required (Kaufmann, 1991).

A method for stabilizing soil and reinforcing vegetation using a 3-D woven fabric has been reported by Theisen (1996). In this method, a single-layered, three-dimensional, high-profile woven geotextile fabric (Fig. 2.18) was placed into the soil. The single-layered, homogeneous fabric was woven from monofilament yarns with different heat shrinkage characteristics such that, when heated, the fabric forms a thick 3-D cuspated profile. The monofilament yarns had a relatively high tensile strength and a relatively high modulus at 10% elongation so as to provide a fabric that is stronger and more dimensionally stable than other geotextile structures. The geotextile

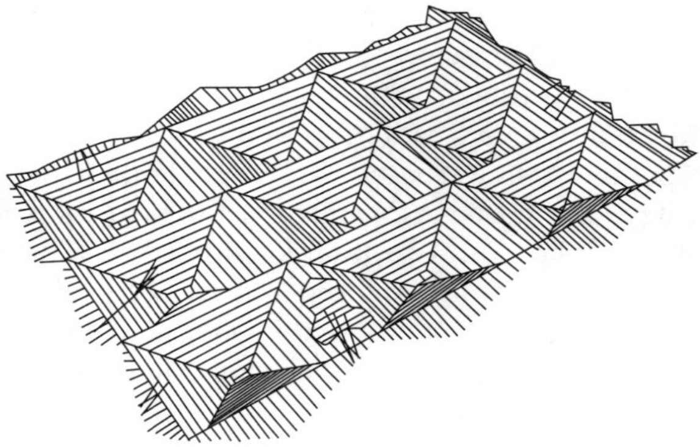

2.18 3-D woven fabric for soil reinforcement.

fabric thus produced was suitable for use on slopes, ditches and other embankments and surfaces where erosion control, soil stabilization and/or vegetative reinforcement may be necessary. The homogeneous, single-component nature of the fabric promotes easier handling and minimizes failure points, while offering a thick, strong and dimensionally stable product upon installation.

Braiding is a well-established technology which intertwines two or more systems of yarns to form a tubular structure. Longitudinal yarns can be laid in between the braiding yarns to form a triaxial braid and/or placed in the core of the braided tubular structure. Depending on the yarn diameter and the braiding angle, a continuous length of micron-diameter to metre-diameter structure can be produced. Taking advantage of the design flexibility and the wide availability of manufacturing capacity in the industry, braided structures can be employed as the foundation fibre architecture for the construction of ductile composite rebar systems as well as for seamless soil containment columns. By judicious selection of fibre materials and fibre architecture for the braid sleeve and the core structure, the load–deformation behaviour of the braided fibrous assembly can be tailored. The end product of this hybridization of material systems and fibre architecture is a composite rebar which has high initial resistance to tensile deformation followed by a graceful failure process manifested by a gradual reduction in the slope of the stress–strain curve before reaching a high level of ultimate strain (Ko, 2004).

Concrete structures (such as bridge decking) are frequently reinforced with steel bars (rebars) to prevent cracking during use. One of the major problems with steel rebars is corrosion due to salt from road deicing, environmental conditions and standing water on deck surfaces. Previously, attempts to replace steel with fibre-reinforced composites have been unsuccessful because of the brittle, sudden and catastrophic failure of these materials. In traditional metallic rebars, the steel will reach a yield point and lose modulus while extending. Current composite rebars (typically fibreglass-reinforced epoxy) rupture in two when the reinforcing element reaches a critical strain level. This type of catastrophic failure is unacceptable for civil constructions because there is no opportunity to avoid critical and catastrophic failure of the entire bridge structure. Replacement of steel rebars with non-corrosive braided hybrid composite rebars of equal strength and stiffness has been reported by Christopher (Pastore and Ko, 1999). This 3-D braided hybrid structure offers the opportunity for improved bridge durability and corrosion resistance. In addition, since composite rebars can be made to be lighter than steel, the designed capacity of the bridge increases and the labour cost of constructing and repairing the bridge decreases. The fibre-reinforced composite rebar consisting of a braided construction in a sheath–core combination has a core of high-modulus, low strain-to-failure material, whereas the sheath shows a high-extension, high-strength material. The structure uses a core of high-stiffness P-55 carbon fibre braided over by a sheath of Kevlar®49 aramid yarns, which provides high stiffness. The aramid fibres carry some load, resulting in a composite system that prevents structural system failure.

Non-woven geotextiles are complex 3-D structures formed by random arrangement of fibres. They are permeable and compressive textile materials and belong to the geosynthetic group which also includes geogrids, geonets, geomembranes and geocomposites. Three-dimensional needle-punched non-woven fabrics have been used as geotextiles for soil reinforcement, filtration and other civil engineering applications. Some of these applications require geotextiles to perform more than one function, including separation, drainage and filtration. The production processes for non-woven geotextiles involve fibre production, fibre preparation, web formation, web bonding and finishing. Thus, continuous filaments or short staple fibres are initially arranged in the form of a fibrous web in various orientations (random, cross, parallel or composite). Subsequently, these fibrous webs are bonded together by means of chemical, thermal or mechanical bonding processes. Mechanical bonding is generally carried out by a needle-punching process for producing geotextile structures. The needle-punched fabrics are produced by the penetrating action of barbed needles which reorientates and intermingles the fibres from a horizontal to a vertical direction. This forms a 3-D intermingled structure which fulfils the necessary requirements of geotextiles (Rawal and Anandjiwala, 2006).

2.6 Application of three-dimensional fabrics to automotives

The automotive industry is the largest user of technical textiles. Textiles provide a means of decoration and a warm soft touch to surfaces that are necessary features for human well-being and comfort, but textiles are also essential components of the more functional parts of all road vehicles, trains, aircraft and sea vessels.

Textiles in the automotive sector in the form of fibres, fabrics and composite structures are widely used because of the very high performance specifications and special properties required. Seat coverings, for example, are not easily removable for cleaning and indeed in automobiles they are fixed in place and must last for the lifetime of the car without ever being put in a washing machine. In trains, aircraft and passenger vessels they are exposed to much more rigorous use than domestic furniture. In addition they have to withstand much higher exposure to daylight and damaging ultraviolet radiation (UV) and because they are for public use they must satisfy stringent safety requirements such as flame retardancy.

In more functional applications, textiles are used in articles as diverse as tyres, heater hoses, battery separators, brake and clutch linings, air filters, parts of the suspension, gears, drive belts, gaskets and crash helmets. The most significant growth area in transportation textiles is expected in composites that straddle the textile and plastics industries.

The most familiar technical textile in transportation is car seat fabric, which is among the largest in volume and is growing rapidly. Car seat fabric requires considerable technical input to produce both the aesthetic and also the very demanding durability requirements. The processes developed for car seat fabric and the technical specifications provide some indication of the requirements for seat materials in other transport applications.

The application of 3-D textiles in the form of composites in the automotive sector is developing slowly. Because of their excellent dimensional stability and very high delamination resistance, these fabrics are finding their way into many applications in the automobile industry.

A multilayer fabric construction that offers greater air permeability and is intended as a seating material for use in automobiles has been described by Lohmann GmbH & Co KG (US Patent 5,747,393, 1999). The multilayered structure also wicks moisture away from the seat's occupant. This fabric is made up of multilayered non-woven polyester needlefelt fabric bonded to a viscose rayon felt (Fig. 2.19).

In a world patent, Mansour Mohamed and Kadir Bilisik (1998) describe a prototype 3-D woven fabric for reinforcement in automobile seats and other applications. The preform consists of multiple warp layers or axial yarns, multiple weft yarns, multiple z-yarns and ± bias yarns (Fig. 2.20). This

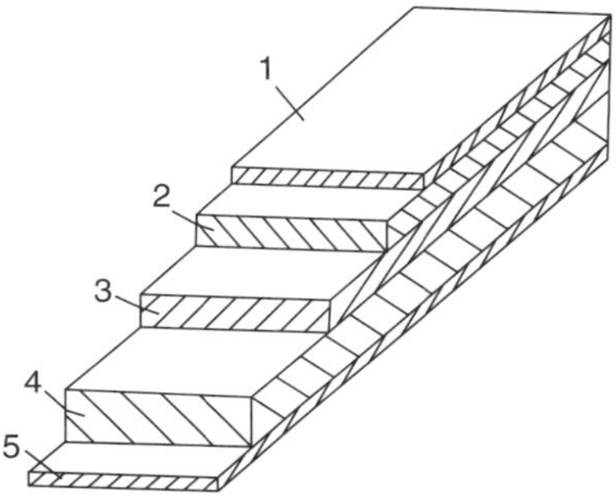

2.19 3-D multilayer fabric for automobile seats.

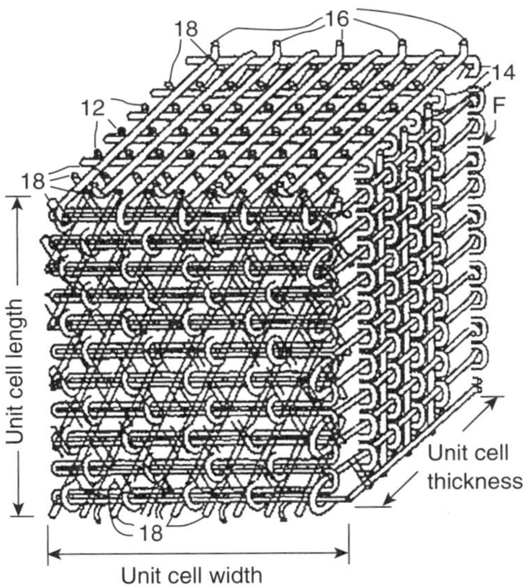

2.20 3-D woven fabric for automobile interiors.

3-D fabric could be used as a reinforcing fabric for automobile interiors such as seat coverings and other interiors.

Apart from the above applications, 3-D fabrics are used in floor coverings, while tufted and needle-punched 3-D non-wovens are widely used as pre-assembled interior components such as for bootliners, seatbacks, door panels and various types of filter. In addition, 3-D fabrics find application in roof linings, door kick-panels, parcel shelves, insulation materials (for heat, sound, vibration, etc.), wheelhouse covers, etc. Coated or laminated needle-punched 3-D non-wovens and 3-D warp-knitted fabrics are the main materials for these pre-assembled interior components (Mukhopadhyay, 2000).

2.7 Application of three-dimensional fabrics to protective clothing and the aerospace industry

Protective textiles refer to garments and other fabric-related items designed to protect the wearer from harsh environmental effects that may result in injury or death. There has been a large increase in the hazards to which humans are exposed as a result of developments in technology in the workplace and on the battlefield, for example. The need to protect against these agencies is paralleled by the desire to increase protection against natural forces and elements. The dangers are often so specialised that no single type of clothing will be adequate for work outside the normal routine. Previous studies have indicated that a series of protective clothing ensembles is required for a variety of potential hazards. Woven, knitted and non-woven fabrics have been designed to suit specific requirements. Blast-proof vests are most frequently made from aramid fibres, such as Kevlar® (DuPont) and Twaron® (AKZO) and Dyneema® (DSM) high tenacity polyethylene fibre. Different fabric constructions are required for protection against low-velocity and high-velocity ammunition. Most traditional ballistic armour used for bullet-resistant vests relies on multiple layers of woven fabric. The number of layers dictates the degree of protection. Three-dimensional fabric structures are being used extensively for ballistic protection. By adopting a 3-D multilayer weave construction, inventors in the USA claim that thinner, more comfortable protective vests can be produced without reducing the level of protection afforded. The 3-D fabric developed by the US inventors (US patent 5,456,974, 1996) is said to offer improved ballistic resistance but can be thinner than the 2-D fabrics that have been used previously, because it uses a 3-D weave and incorporates at least two planes of high-modulus warp yarns (Fig. 2.21).

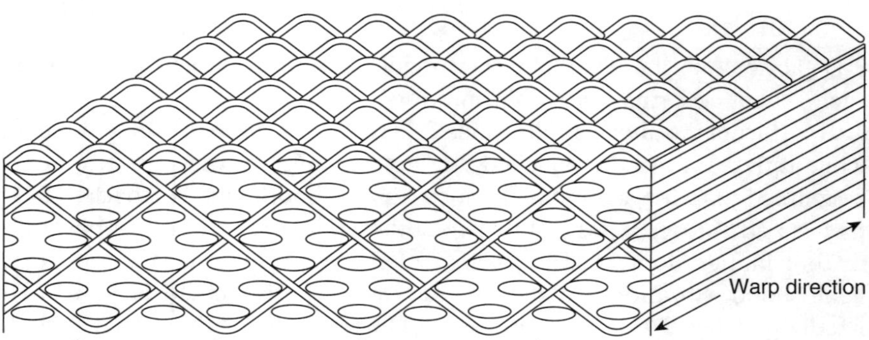

2.21 3-D woven fabric for improved ballistic protection.

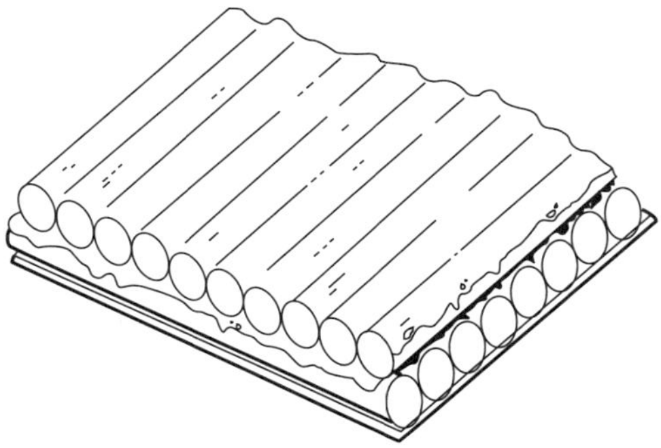

2.22 3-D non-woven fabric for increased ballistic protection.

Three-dimensional non-woven fabrics have also been reported as fabrics for ballistic protection. It is claimed by an inventor in the USA (US Patents 5,443,882 and 5,443,883, 1996) that a non-woven construction can be used for improved ballistic protection. The construction is based on an array of high-performance fibre bundles of Kevlar with multilayer concept. This 3-D construction (Fig. 2.22) is said to provide increased ballistic protection, and suitable body armours can be produced with the fabric.

Protective textiles, specially produced for ballistic applications, consist of a number of fabric layers stitched or quilted together. An alternative and cost-effective method would be to weave all the layers together. Relatively thick fabrics consisting of a number of warp and weft layers can be produced on conventional and specialized 3-D weaving machines. The warp and weft yarns are held together with interlacing z-yarns: orthogonal and angle-interlaced are the two prominent structures used. In these structures, most of the yarns remain non-crimped and hence these structures have high in-plane modulus and high longitudinal wave velocity (Potluri and Needham, 2005). 3TEX have developed a number of orthogonal weaves for the ballistic protection market (Singletary and Bogdanovich, 2000). They can typically weave up to 14 warp layers with a corresponding number of weft layers. Potentially, 3-D weaves have a number of advantages over broad cloth:

- Fabrics can be woven with much higher cover factors, since there are only small percentages of interlacing yarns.
- Warp and weft yarns have very little or no crimp at all in 3-D weaves. Hence, coarser yarns can be used as opposed to the fine yarns that are required in 2-D fabrics to minimize the effect of crimp.
- Labour cost can be reduced as a result of using coarser yarns and eliminating subsequent stitching processes.

Orthogonal 3-D woven fabrics from 3TEX have very low-crimp warp and weft yarns held in place by binding yarns. It is expected that these 3-D fabrics will eventually replace conventional multilayer constructions in body armour.

Aerospace structures require materials of high specific modulus, and composite materials are gaining increased popularity nowadays for use in aerospace structures because of their advantages such as high specific strength in certain preferred directions and material tailoring (Brandt *et al.*, 1990). Three-dimensional woven fabrics nowadays are extensively used in aerospace applications. The potential applications include space suits for astronauts, space shuttle components, aircraft seat cushions, bags for gas-filled aircraft, etc.

A multilayered 3-D fabric intended for making garments for astronauts has been developed by a German company. In the multilayer material patented by ERNO Raumfahrttechnik GmbH, the various layers of fabric are interconnected to create a 3-D structure. Different materials can be used in each layer to provide specific properties. The construction may be used to make space suits (Fig. 2.23).

A laminated multilayer fabric has been designed by Lockheed Corporation (World Patent WO 96/10666) for the bag of a gas-filled aircraft. The patent suggests that the fabric is woven from a strong fibre such as Vectran or Kevlar. The fabric (Fig. 2.24) is a 3-D multilayered structure and can be used to make bags for gas-filled aircraft.

McDonnell-Douglas Corporation has introduced a new multiaxial and multi-ply warp-knitted fabric for structural applications. In World Patent WO 98/10128 (1999), the company disclosed a 3-D warp-knitted fabric to be used in structural applications including the components of a space shuttle. The fabric (Fig. 2.25) contains one or more plies oriented at 0°, and

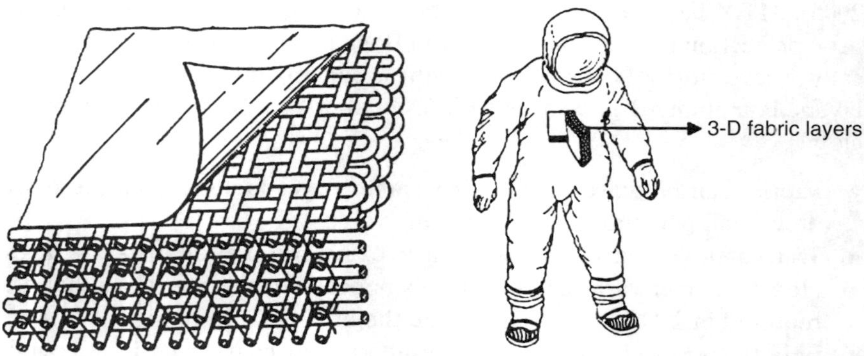

2.23 3-D woven fabric for space suits.

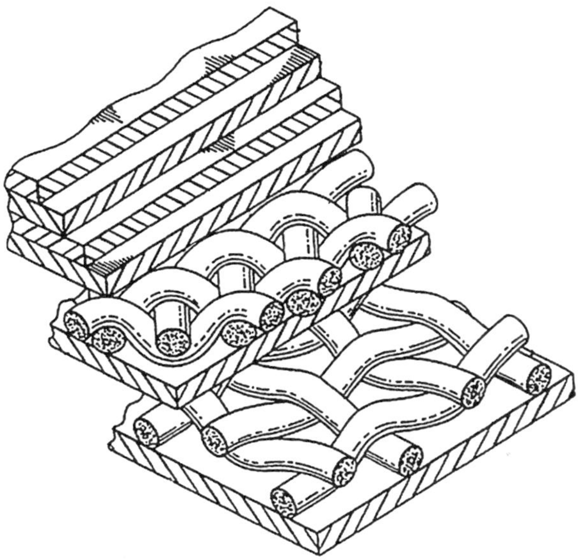

2.24 3-D woven fabric for the bag of a gas-filled aircraft.

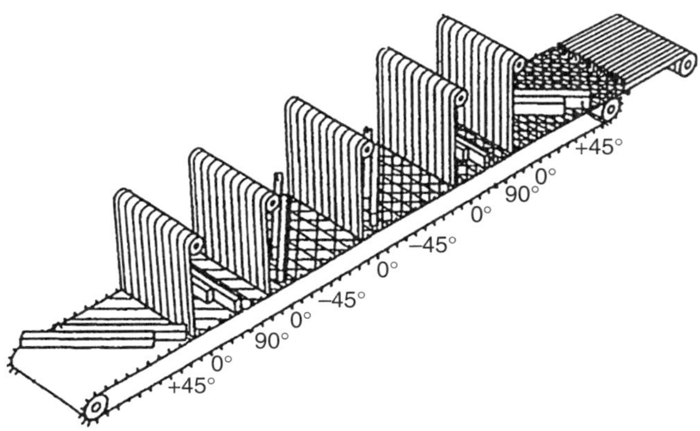

2.25 3-D multiaxial warp-knitted fabric for space shuttle.

provides localized reinforcement with better damage tolerance than 2-D structures. Using fibres of different moduli, it is possible to make fabrics in different weights and thicknesses for a variety of applications.

MWK fabric composites are being used more and more in the aerospace industry, which includes the military, aerospace and commercial aircraft industries. Relatively speaking, they are being used in this industry for many of the same reasons as in the marine industry, namely reduced weight and increased strength and integrity. Because of the flexibility and tailorability of mechanical and physical properties, MWK fabric composites can be customized for the application and specific properties can be emphasized to suit the particular need.

Currently, there are very few aircraft that use MWK fabric composites in critical structures such as the fuselage or wings. Most current applications centre around the skin of the aircraft. Other areas of use are in the top and side tail units, fuselage panelling, leading edges on side rudders, and engine panelling. MWK fabric composites are also being evaluated for rotor blades, outer skin and ballistic protection for helicopters. It is thought that the use of MWK composites is also being evaluated in the new military plane/helicopter, the V-22 Osprey, and the all-composite Beech Starship business plane (Kaufmann, 1991). The lower weight achieved through the use of MWK composite structures means that less fuel is consumed by the aircraft, which translates into significant energy savings for the user. Also, because of the improved structural integrity offered by the MWK fabric composite, it is believed that safety is enhanced.

2.8 References

Adanur S (1995), Textile structural composites, in *Wellington Sears Handbook of Industrial Textiles,* Technomic Publishing, Lancaster, PA, 231–271.

Alagirusamy R, Fangueiro R, Ogale V and Padaki N (2006), Hybrid yarns and textile preforming for thermoplastic composites, *Textile Progress*, **38**, 4, 1–71.

Anand S C (1996), Warp knitted structures in composites, *Proc. Seventh European Conf. on Composite Materials*, London, **2**, 407–413.

Bhattarai S R, Bhattarai N, Ho Keun Yi, Pyong Han Hwang, Dong Ji Cha and Hak Yong Kim (2004), Novel biodegradable electrospun membrane: scaffold for tissue engineering, *Biomaterials,* **25**, 2595–2602.

Brandt J, Drechsler K and Meistring R (1990), The application of three-dimensional fibre preforms for aerospace composite structures, *Proc. Int. Symp. organised by the European Space Agency*, ESTEC, Noordwijk, The Netherlands, 21–23 March, 71–77.

Brandt J, Drechsler K, Mohamed M and Gu P (1992), Manufacture and performance of carbon/epoxy 3D woven composites, *37th International SAMPE Symposium*, Anaheim, California, March, 9–12.

Brown R T and Ashton C H (1989), Automation of 3D braiding machines, paper presented at *4th Textile Structural Composites Symposium*, Philadelphia, PA, 24–26 July.

Bruer S M, Powell N and Smith G (2005), Three-dimensionally knit spacer fabrics: a review of production techniques and applications, *Journal of Textile and Apparel, Technology and Management,* **4**, 4, 1–31.

Bruno P S, Keith D O and Vicario A A Jr (1986), Automatically woven three dimensional composite structures, *SAMPE Quarterly*, **17**, 4, 10–16.

Chou T-W and Ko F K (eds) (1989), *Textile Structural Composites*, Elsevier, New York.

Dexter H B and Hasko G H (1996), Mechanical properties and damage tolerance of multiaxial warp-knit composites, *Composites Science and Technology*, **56**, 3, 367–380.

Dickinson L C and Mohamed M H (2000), Recent advances in 3D weaving for textile preforming, *Proc. ASME Aerospace Division*, Book no. H01214, 3–8.

Dickinson L C, Farley G L and Hinders M K (1999), Prediction of effective three-dimensional elastic constants of translaminar reinforced composites, *Journal of Composite Materials*, **33** (11), 1002–1029.

Du G-W and Ko F (1996), Analysis of multiaxial warp-knit preforms for composite reinforcement, *Composites Science and Technology*, **56**, 3, 253–260.

Franzke G, Offermann P, Bischoff T and Wulfhorst B (1997), Multi-axial warp knitted layers – a textile for reinforcing concrete, *Proc. 11th ICCM*, 14–18 July, Gold Coast, Australia, 870–880.

Fukuta K and Aoki E (1986), 3D fabrics for composites, paper presented at *15th Textile Research Symposium,* Philadephia, PA, September.

Fukuta K, Aoki E, Onooka R and Magatsuka Y (1982), Application of latticed structural composite materials with three dimensional fabrics to airtificial bones, *Bull. Res. Inst. Polymers Textiles*, 131, 159.

Fukuta K, Onooka R, Aoki E and Nagatsuka Y (1984), in S Kawabata (ed.), *15th Text. Res. Symp.*, The Textile Machinery Society of Japan, Osaka, 36–38.

Gebart B R (1992), Permeability of unidirectional reinforcements for RTM, *Journal of Composite Materials,* **26,** 8, 1100–1133.

Geoghegan P J (1988), Dupont ceramics for structural applications – the SEP Noveltex technology, *3rd Textile Structural Composites Symposium,* Philadelphia, PA, 1–2 June.

Han K, Jiang S, Zhang C and Wang B (2000), Flow modeling and simulation of SCRIMP™ for composites manufacturing, *Composites*, Part A, **31**, January, 79–86.

Hill R (1985), Textiles for sportswear, *Textiles*, **14**, 2, 30–36.

Kamiya R, Cheeseman B A, Popper P and Chou T-W (2000), Some Recent advances in the fabrication and design of three-dimensional textile preforms: a review, *Composites Science and Technology*, **60**, 33–47.

Kaufmann J R (1991), in *Proceedings of Fibre-Tex 1991, The Fifth Conference on Advanced Engineering Fibres and Textile Structures for Composites,* NASA Conference Publication 3176, Raleigh, NC, 15–17 October, 77–86.

Ko F K (1989), Three-dimensional fabrics for composites, in *Textile Structural Composites* (ed. Tsu-Wei Chou and Frank K. Ko), Elsevier, New York, 129–171.

Ko F K (2004), From textile to geotextiles, *Seminar in Honor of Professor Robert Koerner, September 13*, Department of Materials Science and Engineering, Drexel University, Philadelphia, PA.

Ko F K, Fang P and Pastore C (1985), Multilayer multidirectional warp knit fabrics for industrial applications, *Journal of Industrial Fabrics*, **4**, 106–113.

Laroche D and Vu-Khanh T (1994), Forming of woven fabric composites, *Journal of Composite Materials*, **28**, 18, 1825–1839.

Larson E (2004), Pressure bag molding: Manufacturing, mechanical testing, nondestructive evaluation, and analysis, Master's Thesis in Mechanical Engineering, Montana State University, Bozeman, MT, January.

Mohamed M H (1990), Three-dimensional textiles, *American Scientist*, **78,** 6, 530–541.

Mohamed M H and Bilisik K (1998), World Patent WO 95/12015, *High Performance Textiles,* October, 7.

Mohamed M H, Bogdanovich A E, Dickinson L C, Singletary J N and Lienhart R B (2001), A New generation of 3D woven fabric preforms and composites, *SAMPE Journal,* **37**, 3, May/June, 8–17.

Mohamed M H, Schartow R W and Knouff B J (2003), Light weight composites for automotive applications, *Proc. SAMPE 2003 Conference*, 11–15 May, Long Beach, CA.

Mouritz A P, Bannister M K, Falzon P J and Leong K H (1999), Review of applications for advanced three-dimensional fibre textile composites, *Composites*, Part A, **30,** 1445–1461.

Moutos F T, Freed L E and Guilak F (2007), A biomimetic three-dimensional woven composite scaffold for functional tissue engineering of cartilage, *Nature Materials*, **6**, February, 162–167.

Mukhopadhyay S K (2000), Automotive textiles – 1, *Textiles Magazine*, Issue 3, 5–7.

Naik N K, Azad N M and Durga Prasad P (2002), Stress and failure analysis of 3D angle interlock woven composites, *Journal of Composite Materials,* **36**, 93–123.

Pastenbaugh J (1988), Aérospatiale technology, paper presented at *3rd Textile Structural Composites Symposium,* Philadelphia, PA, 1–2 June.

Pastore C M and Ko F K (1999), Braided hybrid composites for bridge repair, *Research Briefs, National Textile Center*, USA, April, 23.

Popper P and McConnell R (1987), A new 3D braid for integrated parts manufacture and improved delamination resistance – the 2-step process, *32nd International SAMPE Symposium and Exhibition*, 6–9 April, 92–102.

Potluri P and Needham P (2005), Technical textiles for protection, in *Textiles for Protection* (ed. R A Scott), Woodhead, Cambridge, UK, 151–175.

Rawal A and Anandjiwala R (2006), Relationship between process parameters and properties of multifunctional needlepunched geotextiles, *Journal of Industrial Textiles*, **35**, 4, 271–285.

Raz S (2000), *The Karl Mayer Guide to Technical Textiles*, Karl Mayer Textilmaschinenfabrik, Obertshausen, Germany.

Scardino F L (1989), Introduction to textile structures, in *Textile Structural Composites* (ed. T-W Chou and F K Ko), Elsevier, Covina, CA.

Schmitt P J (1999), Solid woven tubular prosthesis, US Patent 5,913,894.

Shikinami Y and Kawarada H (1998), Potential application of a tri-axial three dimensional fabric as an implant, *Biomaterials*, **19**, 617–635.

Singletary J N and Bogdanovich A E (2000), Orthogonal weaving for ballistic protection, *Technical Usage Textiles*, **37**, 27–30.

Singletary J N and Bogdanovich A E (2001), Processing and characterization of novel 3-D woven composites, *Proc. SAMPE 2001 Conference*, 6–10 May, Long Beach, CA.

Skramstad J D (1999), Evaluation of hand lay-up and resin transfer molding in composite wind turbine blade manufacturing, Master's Thesis in Mechanical Engineering, Montana State University, *Bozeman*, MT, August.

Stobbe D and Mohamed M (2003), 3D woven composites: Cost and performance viability in commercial applications, *48th International SAMPE Symposium*, 11–15 May, Long Beach, CA.

Tan P, Tong L and Steven G P (1997), Modelling for predicting the mechanical properties of textile composites – a review, *Composites*, Part A, **28A**, 903–922.

Theisen M S (1996), Method of using high profile geotextile fabrics woven from filaments of differing heat shrinkage characteristics for soil stabilisation, US Patent 5,567,087.

US Patents 5,443,882 and 5,443,883 (1996), *High Performance Textiles*, June, 10–11.

US Patent 5,456,974 (1996), *High Performance Textiles*, June, 9–10.

US Patent 5,747,393 (1999), *High Performance Textiles*, January, 9.

US Patent 6,863,696 (2005), *Medical Textiles*, September, 2–3.

World Patent WO 96/10666 (1996), *High Performance Textiles*, August, 4.

World Patent WO 98/10128 (1999), *High Performance Textiles*, November, 4.

3 Multiaxial warp-knitted fabrics

Abstract: Over the past few years, multiaxial warp-knitted (MWK) fabrics have made significant inroads into the industrial composites arena. The use of MWK fabrics can lead to cost savings in the manufacture of composite components, and the uncrimped nature of the yarn can also produce improved mechanical performance when compared to traditional woven fabrics. MWK fabrics have a wide application scope ranging from geotextiles, pneumatic materials and construction to automobiles and aerospace-quality components as well as vessel-body parts due to their desired mechanical properties, and flexibility in design and low production cost. This chapter explains the structure, manufacturing methods, properties and advantages of multiaxial warp-knitted fabrics.

Key words: multiaxial warp-knitted (MWK) fabrics, manufacture of MWK fabrics, structure of MWK fabrics, shear of MWK fabrics, compression of MWK fabrics, applications of MWK fabrics.

3.1 Introduction to multiaxial warp-knitted fabrics

Multiaxial multi-ply fabrics constitute one of the most efficient predesigned reinforcement techniques for the production of composite components and also offer improved product performance. Compared with the most widely used woven fabrics, a multiaxial multi-ply fabric offers better mechanical properties (strength and Young's modulus), and in addition the degree of drape can be adjusted in a wide range by the design of the knitting system. The production costs of composites can be reduced by the use of multiaxial multi-ply fabrics. This can be demonstrated in various special applications such as city bus body and motorcycle wheel rim production.

Over the past few years, multiaxial warp-knitted (MWK) fabrics have made significant inroads into the industrial composites arena. They are a family of high-performance technical fabrics, which was developed in the early 1980s and entered into the field of structural composites in the 1990s. MWK fabrics have a wide variety of applications ranging from geotextiles, pneumatic materials and construction to automobiles, aerospace-quality components and vessel body parts, due to their desired mechanical properties, flexibility in design and low production cost (Kaufmann, 1991b; Dexter, 1992). The current trend in textile structural composites is to extend their applications from secondary non-load-bearing structures to primary load-bearing structures. This requires a significant improvement in their damage tolerance and reliability. For thin to medium-thickness structures, multiaxial warp-knitted (MWK) structures form an attractive family of textile preforms, and have the potential to meet the demand for structural composites. All layers of the insertion yarns in an MWK structure are placed in perfect order and reveal the uniformity of the non-crimped parallel yarns (Du and Ko, 1996). While the insertion yarns play a principal role in plane reinforcement, the stitch yarns provide through-thickness reinforcement, thus significantly increasing the damage tolerance, structural integrity and out-of-plane strength of the reinforced structure (Zhou Rongxing *et al.*, 2005). Being lightweight and resistant to corrosion and chemicals, multiaxial fabrics have opened up new developments in the fuselage and wings of aircraft, fast-moving machine parts, tennis rackets, skis and snowboards, and rotor blades for wind turbines (*Textile Month*, 2006).

Advanced composite structures comprising high-strength materials combined with a resin system are being used in the most diverse sectors of the industrial arena. MWK fabrics are ideally suited for this type of end-use because of their flexibility and engineerability. MWK fabrics, produced in a one-step process, have properties similar to those of quasi-isotropic lay-up structures. Besides good handleability, MWK fabric preforms are extremely conformable. The four axes of an MWK fabric are shown in Fig. 3.1.

In a nutshell, the use of MWK fabrics can lead to cost savings in the manufacture of composite components and the uncrimped nature of the yarn can also produce improved mechanical performance when compared to traditional woven fabrics. Their tear strength is also found to be higher than that of woven fabrics and this may be due to the shifting of yarn layers under force which bunch together to resist tearing. Warp inlay multiaxial structures show higher elastic modulus than woven fabrics. These fabrics have excellent dimensional stability and are outstanding in plane shear resistance in all directions.

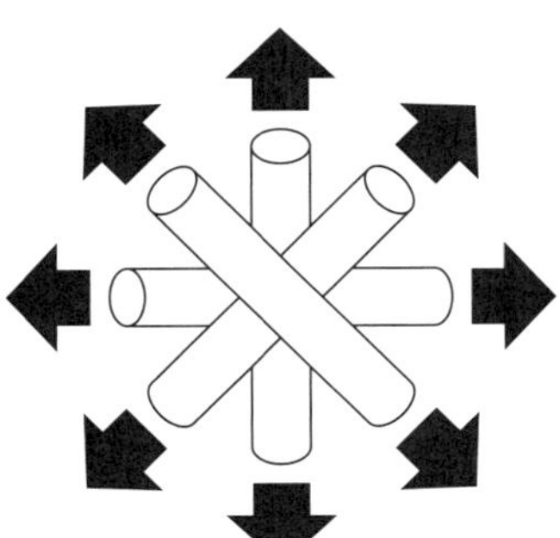

3.1 Four axes of an MWK fabric.

3.2 Advantages of multiaxial warp-knitted fabrics

The significant dominance of yarn layers against stitches in multiaxial structures leads to a fabric behaviour which has little to do with typical knitted structures.

- MWK fabrics generally possess up to four different load-bearing yarn systems arranged so that each can take on stress and strain in virtually all directions. Since these load-bearing yarns lie straight in the fabric, with no crimp, the physical parameters of the individual yarn system are fully utilized.
- Unlike the crimp inserted into the yarn in woven fabrics during weaving, the load-bearing yarn in MWK fabrics lies straight and parallel to other yarns in its yarn system. This characteristic of MWK fabrics allows for the yarn properties to be more fully utilized in withstanding in-plane forces.
- Fabric design is made easier because the designer can more accurately calculate the tensile load of an MWK fabric with a much higher degree of confidence than had been previously attainable with woven fabrics.
- MWK fabrics are capable of withstanding stresses and strains in an optimum fashion. This is due largely to the parallel and straight arrangement of the load-bearing yarns in the MWK fabric. Because the load-bearing yarns lie straight in the fabric, their tensile properties are fully utilized and they are able to absorb tension without the elasticity that occurs when the yarns are crimped or in a wavelike form, such as in a woven fabric.
- Due to the parallel nature of the load-bearing yarns in the MWK fabric, excellent tear propagation resistance is achieved. This resistance to tear propagation becomes increasingly important when an MWK structure is damaged while in use (such as in a sail, inflatable structure or aircraft skin). The damage is minimized by the resistance to further tearing.
- Because the yarn systems are not interwoven, but rather lie directly on top of each other and are held together by the fifth yarn system, conformability of the fabric is greatly improved. This allows the MWK fabric preforms to conform to many complex geometrical shapes and still maximize the translation of fibre mechanical properties to the composite structure. The conformability of the uncured MWK fabric also provides good shape retention during the laying up and curing process.
- By combining a web or non-woven fabric (usually nylon, polyester or fibreglass) to the MWK fabric during the knitting process, it is possible to control many other physical aspects of the fabric structure. Both the MWK fabric and the web can demonstrate their specific advantages. Because the web is fed into the knitting machine during production of

the MWK fabric, it is linked to the load-bearing yarns by the stitch yarn, rather than being rigidly bonded. Adding a web to the MWK structure allows the designer even greater design flexibility.

- Several advantages of multiaxial knits indicate clearly that a considerable reduction in yarn material (reaching up to around 30%) is justifiable. This is particularly important in lightweight composites, taking into account that multiaxial fabric production is more expensive than that of woven ones.
- One aim in composite production is to make maximum use of textile properties with a minimum of resin. Matrix ratio is the proportion of resin to fabric weight. In this connection multiaxial warp knits perform similar to woven fabrics. It is also claimed that wet resins are distributed better in a multiaxial fabric during impregnating, injecting or transfer moulding.
- Multiaxial knits could bear up to about 50% or more specific load capacity (breaking load per unit weight of fabric) in comparison to woven fabrics. This means that lighter warp knit constructions are possible for the same usage.

3.3 Manufacture of multiaxial warp-knitted fabrics

The MWK fabric system consists of warp (0°), weft (90°) and bias ($\pm\theta$) yarns held together by a chain or tricot stitch through the thickness of the fabric, as illustrated in Fig. 3.2. Theoretically, the MWK can be made to as many layers of multiaxial yarns as needed, but current commercially available machines allow four layers (the Mayer system, Fig. 3.3) of 0°, 90°, $\pm\theta$

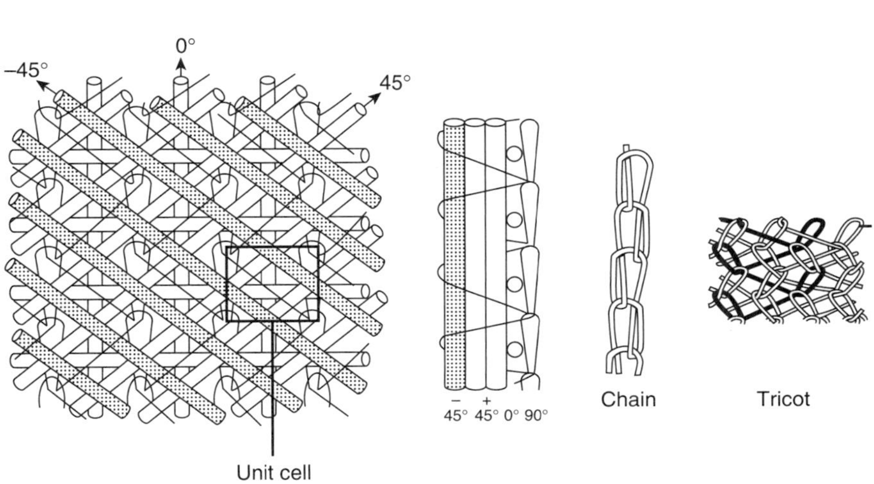

3.2 MWK fabric systems.

3.3 MWK fabric Karl Mayer system.

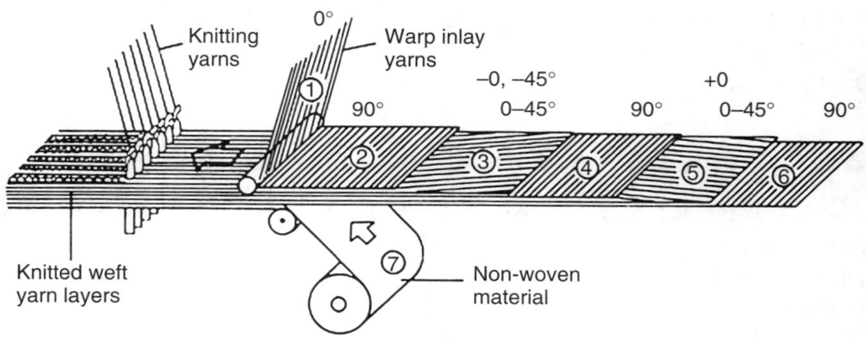

3.4 MWK fabric LIBA system.

for insertion yarns, or at most eight layers (the LIBA system, Figs 3.4 and 3.5) of 0°, 90°, three ($\pm\theta$) insertion yarns, to be stitched together. All layers of insertion yarns are placed in perfect order, each on top of the other, in the knitting process. Each layer shows the uniformity of the uncrimped parallel yarns. While the insertion yarns play a principal role in plane reinforcement, the stitch yarns provide through-thickness reinforcement, thus significantly increasing the damage tolerance, structural integrity and out-of-plane strength of the reinforced structure. To ensure the structural integrity, it is clear that the 0° yarns cannot be placed in either top or bottom layer. The insertion yarns usually possess a much higher linear density than the stitch yarns, and are, therefore, the major load-bearing components of the fabric (Du and Ko, 1996).

MWK fabrics are produced on a special Raschel machine and the structure consists of two diagonal weft layers, followed by a warp yarn and a horizontal weft yarn. The layers are held together by pillar stitches from

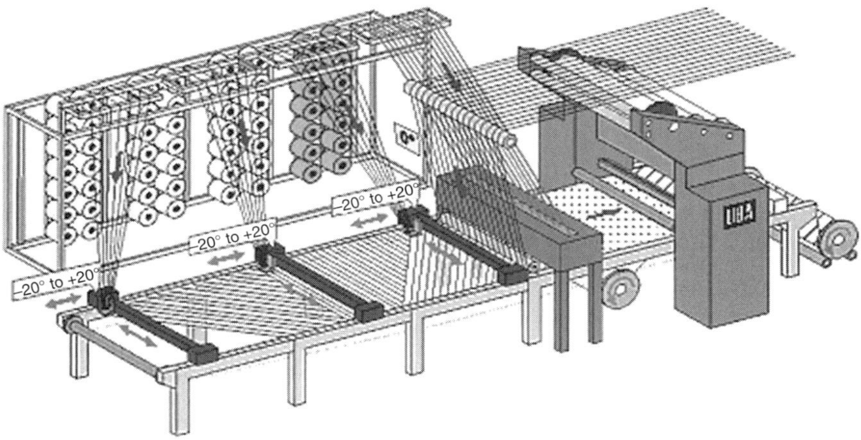

3.5 Schematic diagram of LIBA multiaxial warp-knitting machine.

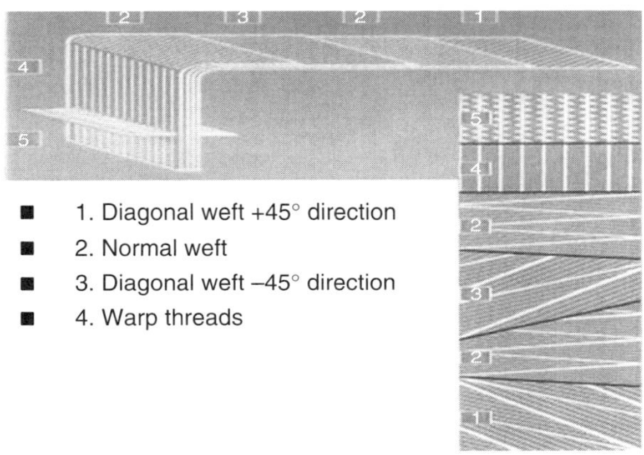

3.6 Production of MWK fabrics.

both sides. The diagonal weft insertions ensure parallel placement of yarns at constant distance. Weft insertion makes it possible to arrange that the least part of the fibre is in stretched form. The diagonal angle of weft can be varied in the range of 30° to 60° by suitably altering the course density. A schematic diagram showing the production of MWK fabrics is shown in Fig. 3.6.

3.3.1 The LIBA technique

By using the LIBA technique (LIBA Maschinenfabrik GmbH), MWK fabrics can be produced on a special tricot machine so that an undesirable bending of the fabric immediately after leaving the knitting zone is avoided. The machine is offered in gauges between E6 and E12 (needles per inch) with working widths of 50 or 68. The linear production rate ranges roughly between 30 and 90 metres per hour. Five to seven weft-inlaid yarn layers and a warp-inlaid yarn layer are possible. At each weft station the yarn can be laid horizontally in the range 30° to 60° in positive and negative directions, so that layer positions can be easily interchanged. The standard version has five independent segments; this can be increased to seven segments. All weft yarns are withdrawn from bobbins mounted on a creel. Production economy can be raised by working with a relatively small number of bobbins. In each segment the yarns can be laid horizontally or diagonally. The yarns are traversed by a weft carriage between two transport chains, fixed on them and continuously transported into the knitting zone. The insertion sequence is electronically controlled and can be programmed for the required 'packing density'. A change in diagonal angle is independent of the course density (Iyer, 1992).

The LIBA technique gives quadraxial structures, where yarns are arranged at 0°, +45°, –45° and 90°. This system uses latch needles with a rounded hook to reduce the possibility of damage to the reinforcing yarns, a significant advantage over other systems for stitching and knitting through multiple layers (Mohamed, 1990). On the other hand, in the LIBA system, six layers of linear yarns can be assembled in various stacking sequences and stitched together by knitting needles piercing through the yarn layers. While this piercing action unavoidably damages the reinforcing fibre, the penetration of the knitting needles also permits the incorporation of a non-woven as a surface layer for the composite.

A maximum of 73 variations have been calculated for the standard machine version with five weft insertion systems. As mentioned earlier, this can be extended to seven systems. Besides this, it is also possible to produce yarn-reinforced, directionally oriented fleeces. Variations can be summed up as follows:

- The fleece can be laid on the top or bottom of the yarn layers.
- The diagonal angle can be altered at each weft insertion system independently and without changing the course density.
- Layers can be omitted as required.
- Weft yarn density can be freely changed in each system.

It is possible to use any kind of yarn in the layers for directional orientation. Thus use can be made of high-modulus filament yarns like aramid, glass, carbon or high-tensile polyester and polypropylene for the production of high-performance composites. Yarn counts can lie anywhere between about 100 and 1200 dtex. Short fibre yarns can also be utilized for corresponding composites like mats and fleeces. Stitches are normally produced by using 'conventional' synthetic polymer filament yarns like polyester. While producing resin-based composites one should bear in mind the compatibility between resin, the fibre used for stitches and the fibres used in the yarn layers.

3.3.2 The Mayer technique

On a Raschel machine horizontal weft yarns are inserted by a magazine system. Guide bars are used for vertical warp yarns and stitches. With the help of special guides, rotating around the needle bar in one direction, diagonal weft yarns are laid exactly between two successive needles from one course to the next. Lateral guide movement and longitudinal fabric movement lead to a diagonal inclination of these yarns. The angle of inclination is determined by the course density. A multiaxial fabric contains two adjacent diagonal yarn layers, followed by vertical warp and horizontal weft yarns. This layer sequence cannot be altered. The production of yarn-reinforced fleeces is possible (Iyer, 1992).

The Malimo technique is another method to develop multiaxial stitch-bonded knitted structures by stitching several layers of yarns together at various angles or piles of skewed fabric on a modified stitch-bonding machine. This will improve mechanical properties and structural consistency (Iyer, 1994). In this machine, a parallel weft sheet is layered with the help of a weft layering apparatus. In the case of the weft threads the layering principle does not allow for a precise 90° orientation. The thread sheet is inserted into the transport system crossed at an angle of approximately 5° to the parallel weft layer due to the feed motion during the laying-up procedure. Finally the weft thread sheet and the warp threads (0° reinforcement) are fed into the knitting elements (Karlheinz and Burkhard, 1992).

In the production of MWK fabrics, the materials for inlays (knitting yarns) are normally high-modulus or high-temperature-resistant polymer filaments such as polyester, nylon and PEEK, whereas glass, aramid or carbon threads can be used as reinforcing fibre materials (Alagirusamy *et al.*, 2006).

The MWK fabric preforms, having four directional reinforcements similar to quasi-isotropic lay-up, can be produced in a single step. Besides good ease of handling and production economics, the MWK fabric preforms also provide the conformability to complex shapes, the flexibility in the principal yarn directions, and the improved through-thickness strength. The mechanical properties of MWK structural composites, especially out-of-plane strength and impact behaviour, are potentially superior to those of conventional woven laminated composites due to the elimination of fibre crimp in the insertion yarns and the presence of the through-thickness reinforcing stitch loops (Zhou Rongxing *et al.*, 2005). In addition to the above, Hearle

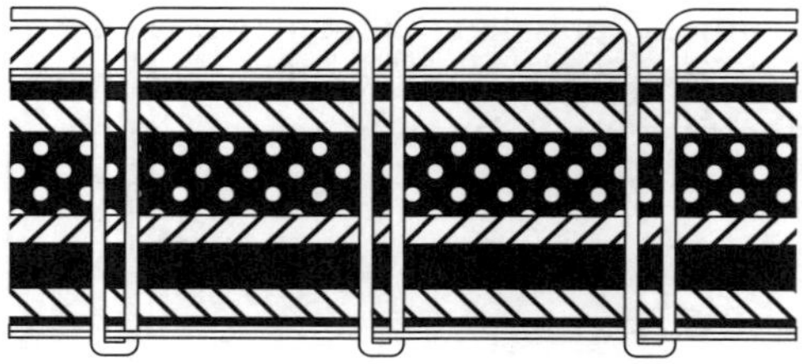

3.7 Knit-stitched structure.

(1995) notes that the Mayer machine is capable of producing four-layer fabrics up to 1.6 m wide, while six-layer fabrics up to 2.5 m wide can be achieved with the LIBA machine at a rate of 45 m/h. Besides the oriented fibres, this process also allows the incorporation of non-woven fabric layers. The Mayer machine utilizes a multiaxial magazine weft insertion mechanism; the attractive feature of this system is the precision of yarn placement with four layers of linear or non-linear bias yarns arranged in a wide range of orientations. Furthermore, stitches are formed at over one metre per minute. The system creates loops that hold the reinforcing yarns in place in two directions.

The stitching-through principle of making warp-knitted multiaxial–multilayer structures (Fig. 3.7) for fibre-reinforced composites ensures isotropic behaviour through the uniform distribution of the yarn ends in multiple directions (Dow and Dexter, 1997). Due to the non-crimped and parallel yarn sheets in them, they are particularly suitable for fibre-reinforced composites with special characteristics such as low specific weight, adjustable stiffness (between extremely stiff and extremely stretchable) and highest mechanical load resistance (Wagener, 1993).

Stitch-bonding is considered to be a special case of stitching with the warp-knitting technique, while the loop formation cycle is similar to warp knitting but deviates with respect to constructive design of the stitch-bonding area (horizontal needle arrangement, fixed retaining, backing rail), the needle types applied (stitching needle), the reference size for the machine gauge (number of needles per 25 mm), and conversion of unbonded fibrous webs to purely mechanically stitch-bonded non-wovens (Padaki *et al.*, 2006).

A simplified knitting cycle in a two-needle-bar Raschel machine (spacer or tubular fabrics) for producing 3-D multiaxial warp-knitted (MWK) fabrics is represented in Fig. 3.8 and the knitting options are shown in Fig. 3.9. The use of two needle bars opened new horizons in warp knitting and different machine types are built for production of a wide variety of products ranging from packing sacks to artificial blood vessels.

1. The guide bars are positioned at the back of the machine, above the back needle bar. The front sinker bar is placed forward to secure the

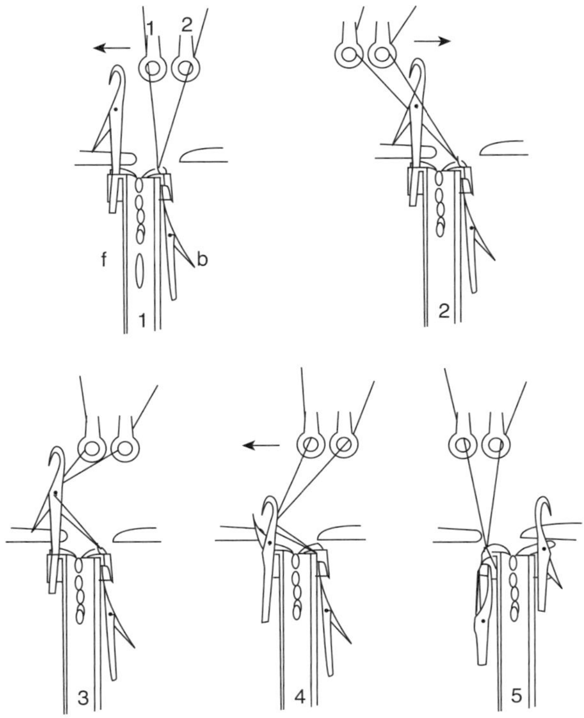

3.8 Knitting action of a double-needle-bar Raschel machine (for description of shapes 1–5 see text).

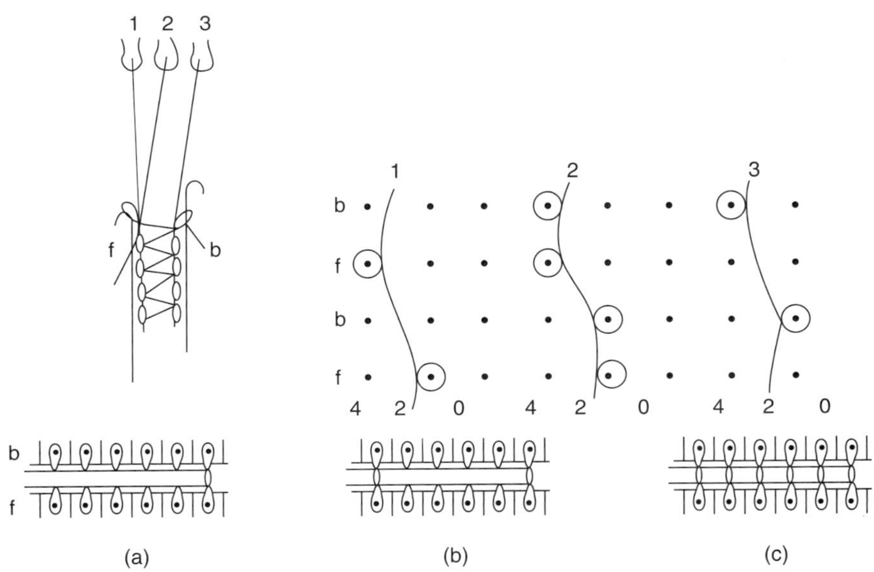

3.9 Knitting options with two needle bars and more than two guide bars (for description of shapes a–c see text).

fabric while the front needle bar ascends to the clearing position. The guide bars perform the underlap shogging movement for the needle bar and then swing to the hook-side.

2. On the hook-side of the front needle bar, the guide bars shog an overlap according to the pattern mechanisms and then swing back.
3. The swing back is completed; the yarns are wrapped within the needle hooks so that the front needle bar can start to descend. The front sinker bar retreats while the back one moves forward.
4. The front needle bar descends the previously formed loops resting on the needle stems close to the latches. The front sinker bar continues to retreat and the back one is now above its needle bar. The guide bar swings for the third time, this time to the front in order to clear the way above the back needle bar. The underlap shogging movement for the back needle bar can start.
5. The front needle bar is at knockover position and the needles form new loops. The back needle bar, now with its fabric secured by the sinker bar, ascends to the clearing position.
   a. The third guide bar is only threaded through one guide finger on one side, resulting in two separate fabrics, produced by the fully threaded guide bars.
   b. By threading the middle guide bar through two guide fingers, one on each side, a tubular fabric is formed.
   c. A fully threaded middle bar is used to produce a sandwich of two fabrics connected by yarns.

Typical MWK structures produced by the LIBA system are shown in Fig. 3.10.

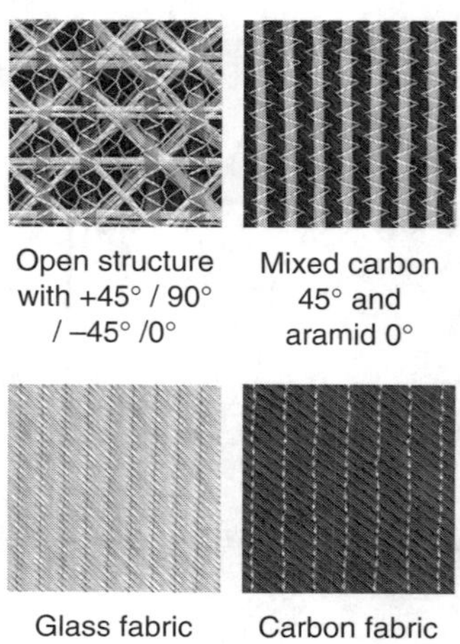

3.10 Types of MWK fabrics.

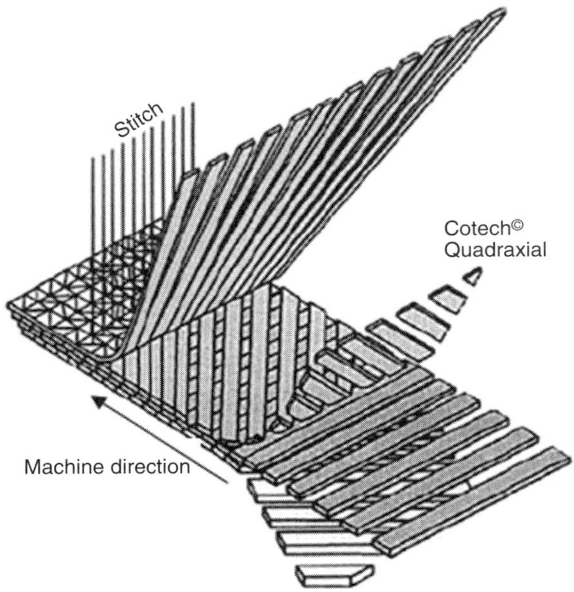

3.11 Yarn arrangement in MWK fabric.

3.4 General structure and behaviour of multiaxial warp-knitted fabrics

3.4.1 Structure

MWK fabrics are unique structures which are produced by warp-knitting techniques. With these techniques, straight ends of parallel and uncrimped yarns are inlaid into the knitted structure to give the ideal combination of mechanical properties at a favourable production cost. This produces MWK fabrics, with the combined advantages of design flexibility, performance, productivity and availability and the potential to become a major preform for industrial composites.

MWK fabrics have a structure where one or several yarn systems are held together by a binding yarn system. The interlacing yarn structure characterizing woven fabrics is not found in MWK fabrics. Instead, one or more layers of fibre tows are stacked on top of each other and held together by the binding yarn system (Fig. 3.11). Each layer consists of parallel fibre tows. Several layers are stacked in different directions to create the desired fabric properties. The reinforcing fibre systems commonly consist of tows from glass or carbon, but other fibres are available as well. The binding yarn system is usually made from polyester fibres (Edgren, 2006).

There are several basic structures used for the binding yarn system. A knitted structure is characterized by the interlacing loops created by the knitting yarn or thread (Byun and Chou, 2000). In a warp-knitted structure each thread forming the loops runs in the warp direction (Chou, 1992). To create a fabric suited for reinforcement in composite laminates, warp and weft yarns can be inserted into the warp-knitted structure, creating a multiaxial fabric. A schematic of a warp-knitted tricot structure with inserted warp and weft yarns is presented in Fig. 3.12. Reinforcing yarns can be inserted also in other directions than the 0° and 90° directions, e.g. ±45° (Cox and Flanagan, 1997), and a triaxial fabric structure is shown in Fig. 3.13.

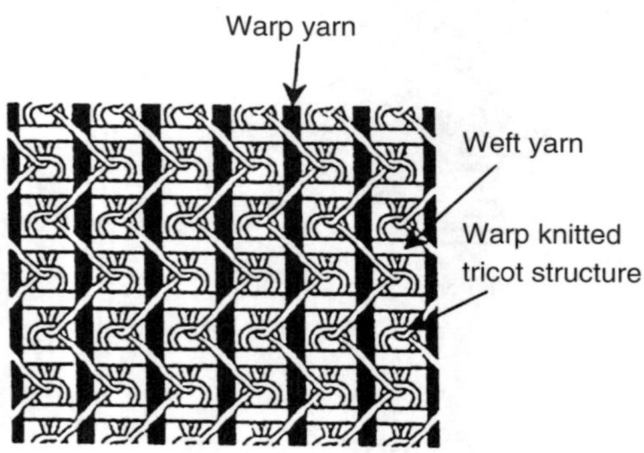

3.12 Structure of biaxial MWK fabric.

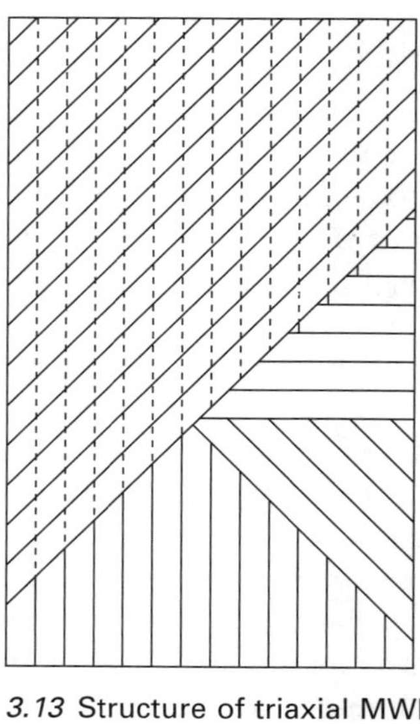

3.13 Structure of triaxial MWK fabric.

MWK fabrics are often characterized through the number of reinforcement yarn directions inserted into the binding structure. Biaxial, triaxial and quadraxial fabrics are the most common fabrics commercially available. It is also common to give information on the surface weight of reinforcing fibres in each direction. These weights can then, together with a known fibre density, be transformed into the fibre volume fraction if the ply thickness is known.

MWK fabrics can be produced on a multiaxial knitting machine, the principle of which is shown in Fig. 3.14. The typical structure of an MWK fabric is illustrated in Fig. 3.15. There are many kinds of MWK fabrics as variations of the typical structure, e.g. as shown in Fig. 3.16 in which the structure is less complicated than the typical one, e.g. structure (c) contains only two inserting yarn systems – warps and wefts – and according to the arrangement of inserted yarns, the MWK fabrics could be classified into five types, as shown in Fig. 3.16.

From Figs 3.15 and 3.16, it can be seen that there are four groups of yarns in the fabric structure. The vertical strength and formability depend on the inserted warp yarns while the horizontal strength and formability depend on the inserted weft yarns. The angled strength and formability are dependent on the oblique yarns with alterable angle arrangement of 30°–90°. The yarns in each plane group are unbent and parallel to each other. The four

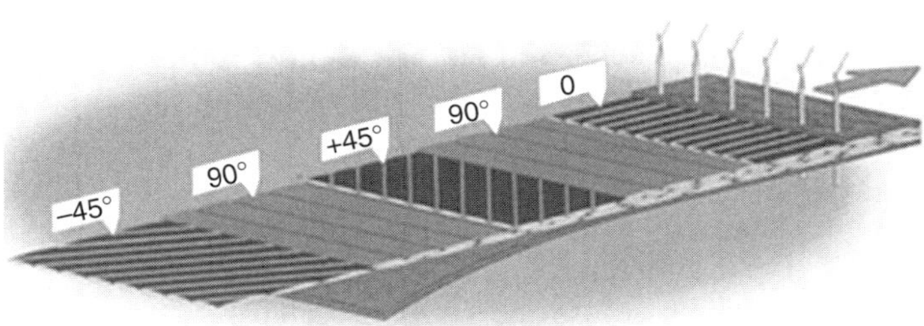

3.14 Principle of MWK production.

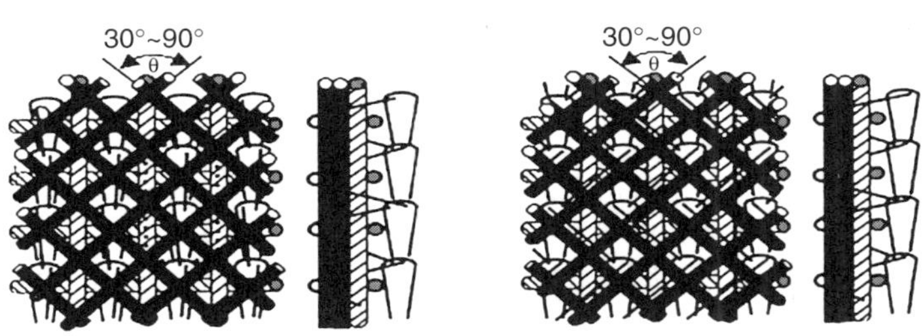

3.15 Typical structure of MWK fabrics.

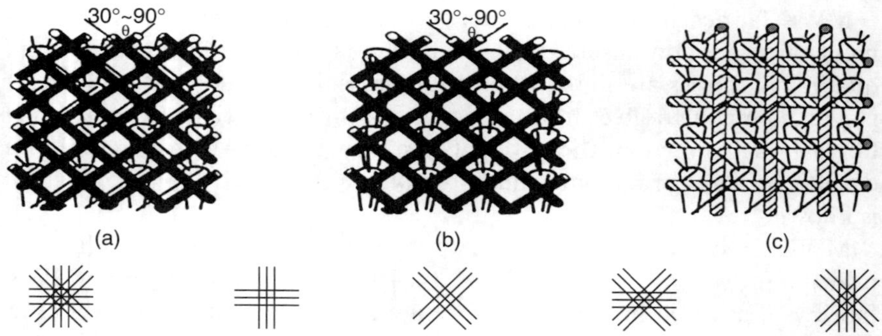

3.16 Variations in the MWK structure.

groups of yarns are bonded together by a group of yarns called warp-knitting ground yarns.

From a structural geometry viewpoint (Fig. 3.15), fabrics consist of warp (0°), weft (90°) and bias (± various degrees) yarns that are stitched together by a chain or tricot stitch during the warp-knitting process by a fifth yarn system through the thickness of the fabric structure. The warp and weft yarns stabilize the fabric in the machine and cross-machine directions while the diagonally arranged or biased yarns absorb tension from any required angle. The fifth yarn precisely binds together all of the load-bearing yarn systems. The bias or diagonal yarns in the fabric can be inlaid at any angle along the plane of the machine direction, the most common of which is ±45 degrees. It should be noted that all four load-bearing yarn systems do not have to be used in an MWK fabric construction. Also, different yarn types and counts can be used in each of the yarn systems (Kaufmann, 1991a).

By using sewing methods, more complicated structures of MWK fabrics can be made. For example, in Fig. 3.17, the part on the left is the MWK fabric with four groups of inserted yarns. More layers of MWK fabrics can be combined together, and the part on the right shows four layers of fabric piled up together using the single needle sewing method to make a 16-layer MWK fabric.

Unit cell geometry

The key geometric parameters of the MWK fabric preforms, which affect the reinforcement capability and the composite processability, include the number of yarn axes, the orientation of bias yarns, total fibre volume fraction, pore size and pore distribution, and percentage of stitch fibres to total fibre volume fraction. The process variables adjustable to control the MWK microstructure include the type of knit stitch, the ratio of stitch-to-insertion yarn linear density, the orientation angle of bias yarns, and the thread count.

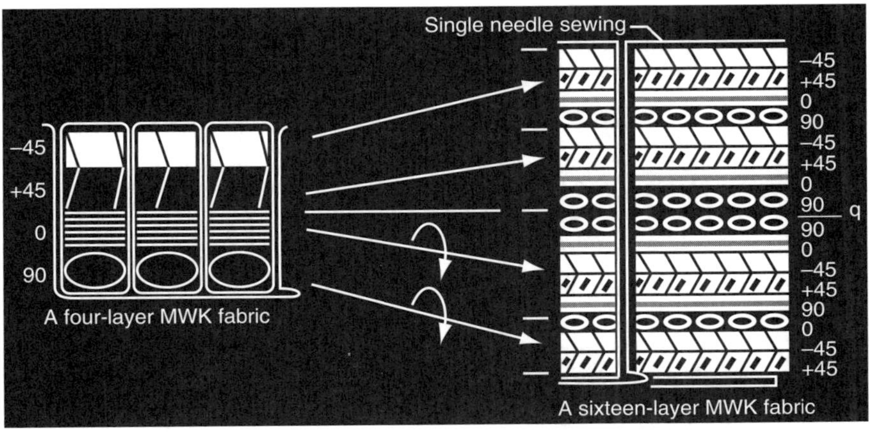

3.17 A 16-layer MWK fabric.

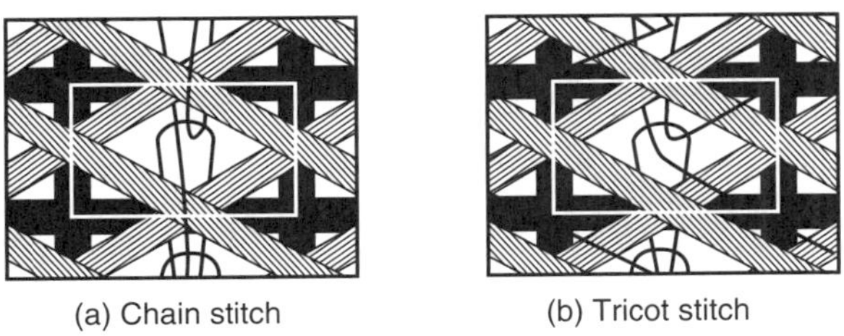

3.18 Unit cell of MWK fabric.

The concept of a unit cell is used to establish the relationship between the geometric parameters and process variables.

Normally, yarn bundles consisting of numerous continuous filaments are used for fabric preforms; thus, the fabric has three microstructure levels: geometry of interfibre packing in the yarn bundle (fibre level), cross-section of yarn bundles in the fabric (yarn level), and orientation and distribution of fibres in the 3-D network (fabric level). The first step in the unit cell-based modelling is to determine the unit cell dimensions, so that it is the smallest repeating unit of the structure. The second step is to assume some idealized cross-sectional shapes of the yarn bundles, based on experimental observations. The final and most important step is to identify the overall unit cell geometry, from which expressions for the key geometric parameters can be derived, and the geometric limits applied to the structure can be defined. This unit cell-based modelling technique has been demonstrated successfully for 3-D braided structures.

As can be seen in Fig. 3.18, the unit cell for the MWK fabric preform can be defined in many ways to meet the definition for a unit cell. However, it

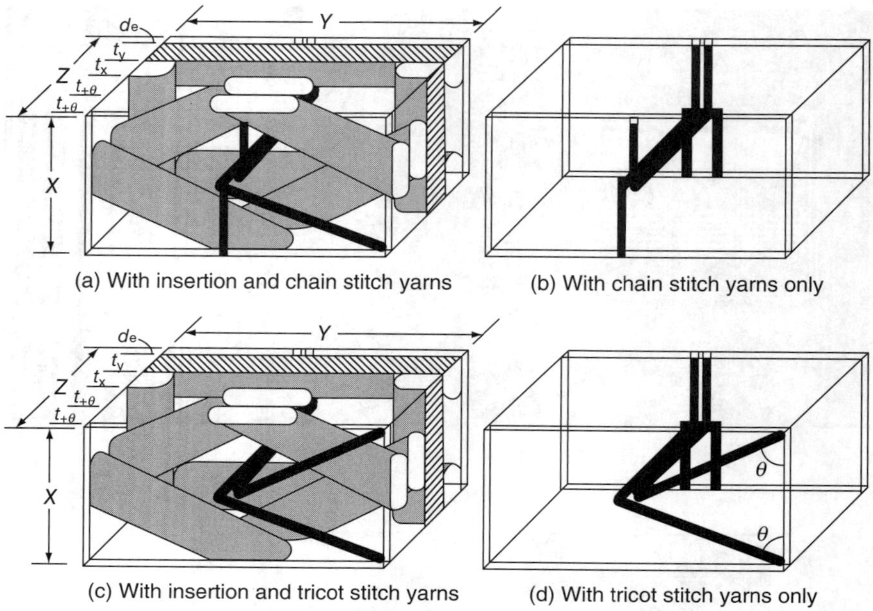

3.19 Unit cell geometry of MWK structure.

would be most reasonable to have a unit cell which consists of a complete knitting stitch, and the insertion yarns in the unit cell are all symmetrical. Figure 3.18(a) and (b) show the unit cells for the MWK structure with chain and tricot stitch, respectively. Within the outline in the figure, the unit cell consists of one each of 0° and 90° yarns, two each of $\pm\theta$ and $-\theta$ yarns as well as a knit stitch (Du and Ko, 1996).

Figure 3.19 shows the idealized unit cell geometry for the MWK structure, including the shape, dimension, orientation and position of all the insertion and stitch yarns within the fibre 3-D network. The knit stitch is assumed to have the tightest loop construction, and the curved loop is idealized to a rectangular shape, as illustrated in Fig. 3.19(b) and (d). The dimensions of the unit cell are X, Y and Z, corresponding, respectively, to the 0° axis, the 90° axis, and the thickness axis vertical to the 0°–90° plane, as shown in Fig. 3.19(a) and (c).

The unit cell geometric analysis of a four-layer system is used as an example to generate the $V_f - \theta$ functions for the MWK fabric, where V_f denotes the volume fraction. This analysis can be generalized to include other MWK systems with six or more layers of insertion yarns. The fibre volume fraction relation in Fig. 3.19 shows that for the fixed parameters selected, only a limited window exists for the MWK fabric construction. The window is bounded by two factors: yarn jamming and the point of 90° bias

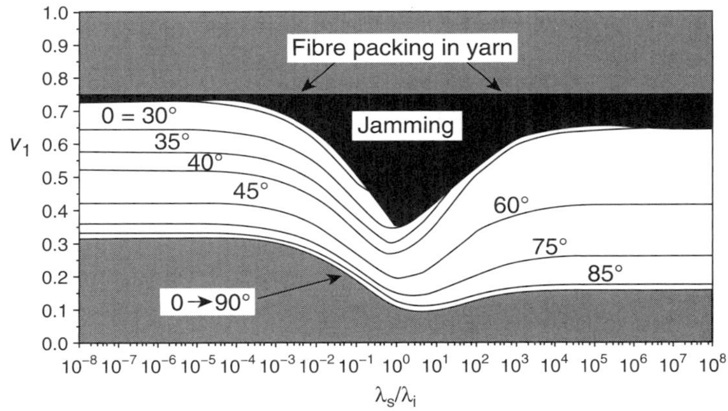

3.20 Fibre volume fraction versus ratio of stitch-to-insertion yarn linear density (tricot stitch, $\kappa = 0.75$, $\rho = 2.5$ kg/m^3, $f_i = 5$, $\eta = 0.5$).

yarn angle. Fabric constructions corresponding to the curve marked 'jamming' are at their tightest allowable point, and constructions at the $\theta \to 90°$ curve have the most open structure. When $\theta < 30°$, jamming occurs in the whole range of yarn linear density ratio from zero to infinity.

The fibre volume fraction relation is shown in Fig. 3.20. When θ is in the range of 30° to 40°, the fibre volume fraction decreases with an increase in yarn linear density ratio until jamming occurs. When $\theta = 45°$, the fibre volume fraction decreases with an increase in yarn linear density ratio to a minimum at about $\lambda_s/\lambda_i = 1$ where λ_s = stitch yarn diameter and λ_i = insert yarn diameter, and starts to increase until jamming occurs. When $\theta \geq 60°$, the fibre volume fraction has the same trend as when $\theta = 45°$, but yarn jamming never occurs. The fibre packing in the yarns, taken as 0.75, limits the maximum fibre volume fraction in the fabric (Ko, 1999).

3.4.2 Mechanical behaviour

MWK fabrics have good dimensional stability that allows them to be handled easily in the composites manufacturing processes. The stitches allow relative fibre movement in the fabric while at the same time maintaining uniform fibre spacing. The fabrics' excellent conformability makes them suitable for making composite parts of complicated shapes (e.g. parts with double curvatures) without excessive cutting, joining and post-consolidation machining. Since multiple fibre layers are handled in a single step, the composites manufacturing process is significantly simplified. The mechanical properties of the MWK composites, especially in compression, may be superior to those of conventional woven fabric composites due to the elimination of fibre crimp. A thin, textured polyester yarn is often used as the stitching yarn. A fabric's dimensional stability and conformability can be altered by controlling the stitching yarn density or the stitching pattern. A high performance aramid or glass yarn may be used as the stitching yarn to improve the interlaminar properties and damage tolerance of the composites.

The literature on the mechanical properties of MWK fabrics and their composites is very limited. Due to their close fibre packing and dense structures, a reasonably high fibre volume fraction (about 60%) in the final composite part can be obtained even from the wet manual lay-up process. This can result in good mechanical properties as well as significant savings in resin consumption.

In an attempt to understand the mechanical behaviour of MWK fabrics and their composites, some researchers have studied the properties of fabrics such as tearing, shear, compression, impact strength, etc. The effect of reinforcement of MWK fabrics on the performance of composites has also been studied. This section summarizes the works of various researchers on the mechanical behaviour and properties of MWK fabrics and their composites.

The basic mechanical properties of MWK fabrics are somewhat superior to the equivalent volume fraction of woven roving-reinforced material. For example, Hogg *et al.* (1993) determined the Young's modulus and tensile strength of a biaxial fabric with glass-reinforced polyester, volume fraction 33%, to be 21 GPa and 264 MPa, respectively, which are 13% and 20% higher than the values found for an equivalent volume fraction of plain woven-reinforced composite. Quadraxial reinforcement of the same fibre volume fraction gave similar results (24 GPa and 286 MPa, respectively).

The improvement in properties compared to woven-reinforced composites is emphasized by the work of Godbehere *et al.* (1994) in tests on a carbon fibre reinforced MWK fabric epoxy resin and equivalent unidirectional (UD) laminates. All the composites had 0°/±45° orientations. Although the MWK laminates had poorer properties than the UD laminates, the reduction was small (e.g. less than 7%) in the 0° direction. For example, the UD equivalent laminate gave values of Young's modulus and tensile strength of 58 GPa and 756 MPa, respectively, compared to MWK fabric values of 56 GPa and 748 MPa (for fibre volume fractions of 56%).

The increases in through-thickness reinforcement achieved by MWK fabrics have been demonstrated by a number of authors. For example, Backhouse *et al.* (1995) compared the ease of delaminating polyester stitched 0°/±45° carbon fibre MWK fabric with equivalent carbon fibre/epoxy UD laminates. There were large increases, some of 140%, in the measured parameters used to quantify resistance to delamination (the mode I and mode II toughness values) for the MWK fabrics compared to the UD material.

Shear behaviour

Traditional fabrics, such as woven, non-woven and knitted fabrics, tend to form a large deformation in the direction of the shearing stress which is parallel to the diagonal direction of the fabric. Consequently, they cannot perform well as reinforcement material in composites when there is a parallel shearing stress on the composites. Typical MWK fabrics have yarns inserted in the diagonal directions called binding yarns, which retard the shearing deformation. Compared with biaxial warp-knitted fabrics and other woven fabrics, MWK fabrics have better shearing properties (Zhou Rongxing and Chen Mingzhen, 1999).

Typical shear compliance curves obtained for MWK fabrics with both tricot and pillar warp-knitted stitching threads are presented in Fig. 3.21

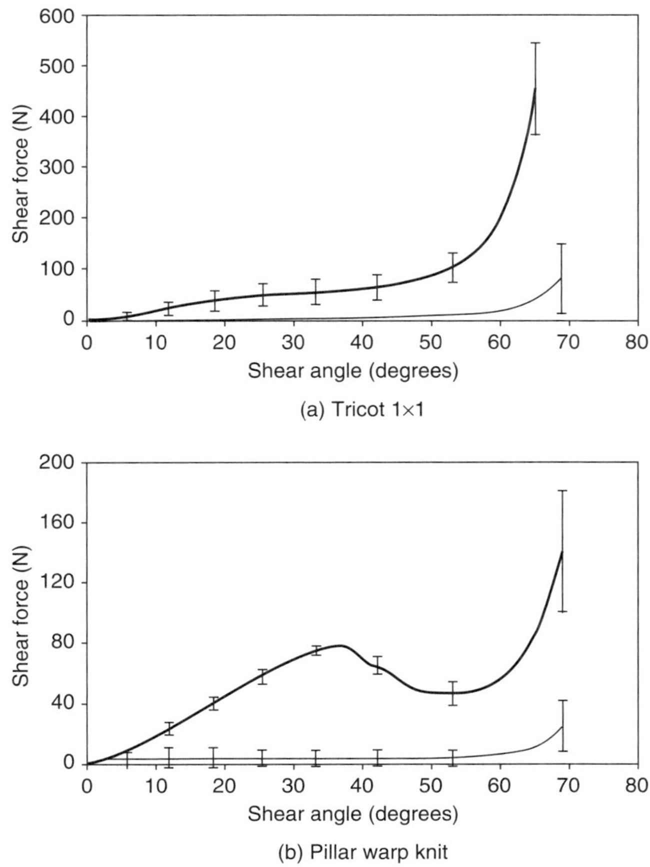

3.21 Shear compliance curves for MWK fabrics (adapted from Wang, 2002).

(Long *et al.*, 2002). A zigzag stitching pattern was observed for tricot warp knit, whereas the pillar warp knit was similar to a chain stitch. In both cases it was apparent that the compliance was lower when the fabric was sheared parallel to the stitching direction. The authors observed that testing in this direction resulted in a tensile strain within the stitch and caused an increase in shear force. In the case of pillar warp stitch, as the majority of the stitching thread was aligned with the applied force, the effect was more pronounced. Testing parallel to the stitch resulted in a linear increase in force until the stitching thread snapped. The force reduces after this point until interrow compaction occurs. The authors opined that the directionality exhibited by non-crimp fabrics during shear could result in non-symmetric fibre patterns during forming.

Compression

Compressive behaviour of MWK fabric composites has been reported by Wang (2002). In this study different varieties of biaxial, triaxial and quadraxial fabrics were manufactured and incorporated into composites as reinforcing fabrics. The mechanical testing of these fabrics included tensile, flexural and compressive tests. The compression test was carried out to determine the compressive strength and modulus using Surfalloy-faced hydraulic grips without tabs. The sample dimensions were 100 mm by 25.4 mm and the specimen gauge length (between grips) was about 25.4 mm. The test procedure for the compressive behaviour of fabrics is illustrated in Fig. 3.22.

The test results for the compression test are summarized in Table 3.1. The compressive moduli measured were lower by about 30% than their respective tensile values, as observed from the experiments. This could partially be due to experimental errors, as the small specimen gauge length in compression hinders mounting the transducer directly on the specimen. The compressive strengths are generally similar to or higher than their tensile values. Typical compressive test curves are shown in Fig. 3.23. Ductile failure is observed for laminates with all the fibres oriented at ±45° to the loading

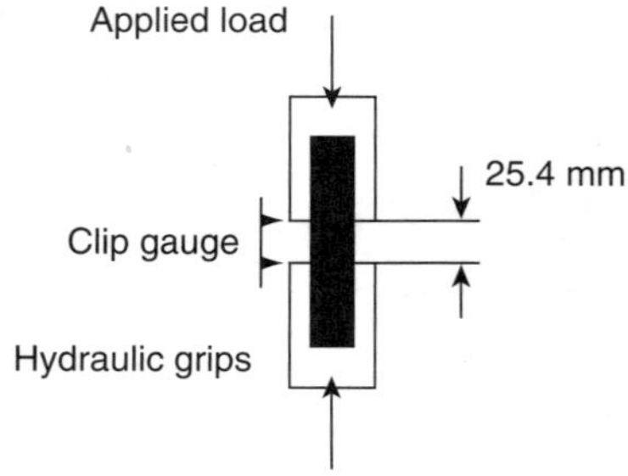

3.22 Compression test for MWK fabrics.

Table 3.1 Compression and flexural test results of MWK fabrics

Fabric	Volume fraction (V_f) (%)	Direction	Compression*		Flexural*	
			E (GPa)	σ (MPa)	E (GPa)	σ (MPa)
Biaxial	47.8	0°	7.5	110	10.9	251
		90°	7.7	114	11.2	257
		45°	15.7	377	19.0	380
Biaxial	51.4	0°	6.5	84	7.2	187
		90°	7.3	90	8.1	201
		45°	16.1	254	17.0	531
Biaxial	38.3	0°	13.1	330	15.4	359
		90°	13.7	336	16.8	414
		45°	7.1	110	9.1	211
Triaxial	49.7	0°	11.6	270	16.4	423
		90°	12.2	256	14.9	440
		45°	14.3	353	17.2	503
Triaxial	52.1	0°	6.0	198	18.3	570
		90°	7.5	127	9.1	221
		45°	12.2	272	11.1	352
Quadraxial	41.5	0°	12.1	313	15.0	356
		90°	11.4	306	14.8	368
		45°	12.0	323	12.9	384

* E = Young's modulus, σ = strength.
Source: adapted from Wang (2002).

direction. Delamination associated with in-plane shear deformation is evident and the ×-shaped failure bands are along the ±45° fibre directions as seen in Fig. 3.23.

A higher compressive strength is generally observed when tested along a fibre direction. In such cases the laminates showed a brittle failure with a sudden load drop after reaching the maximum load. Examination of the failed specimens reveals that the dominant mode is delamination, and the single failure band is perpendicular to the loading direction. Figure 3.24 shows the compressive failure pattern typical of all the laminates tested along a fibre direction. In contrast, for a woven composite with yarn undulations, it is easier for a ply to buckle under compressive load. It has been observed that compressive failure in woven composites is initiated in the crimped yarn along the loading direction (Wang and Li, 1995).

Flexural behaviour

The flexural modulus is similar to the respective tensile modulus, as reported in Table 3.1. The flexural strengths of all the laminates tested are

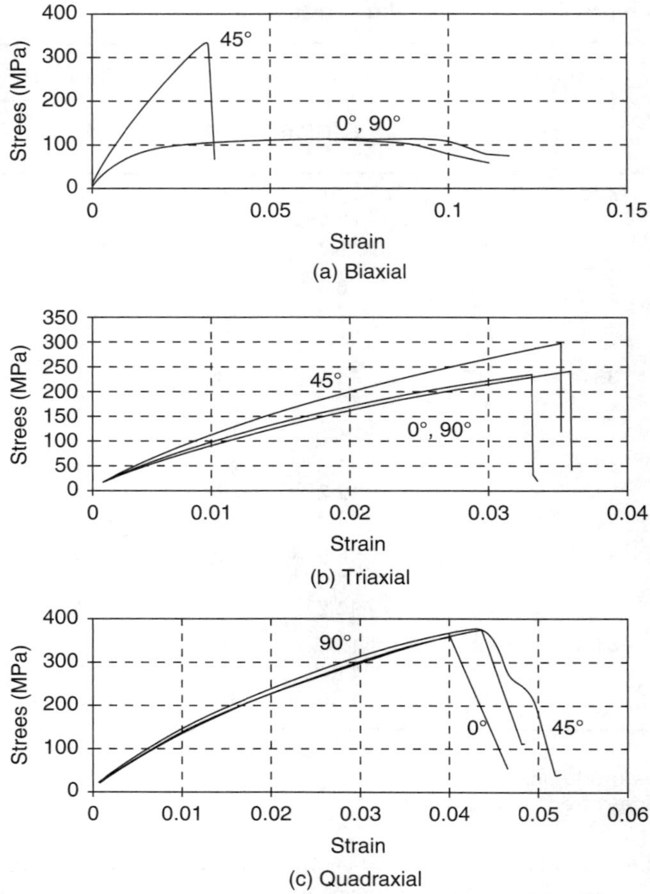

3.23 Typical compressive test curves for MWK fabric laminates (adapted from Wang, 2002).

significantly higher than their tensile strengths, and are also higher than or similar to their compressive strengths. The flexural load–deflection responses, shown in Fig. 3.25, exhibit less non-linearity than the tensile and compressive responses. All the laminates tested along a fibre direction show brittle failure, generally by outer ply delamination on a tensile surface. The delamination zone starts at the middle of the specimen where the bending moment is maximum, then propagates outwards until significant fibre rupture occurs at the middle section on the tensile surface. This typical failure mode can be observed in Fig. 3.26 for a biaxial specimen tested along a fibre direction. In contrast, laminates with all fibres oriented at ±45° to the specimen length direction show ductile failure, as in compression. Delamination along the fibre strands can be observed on both the tensile surface (Fig. 3.26(b)) and

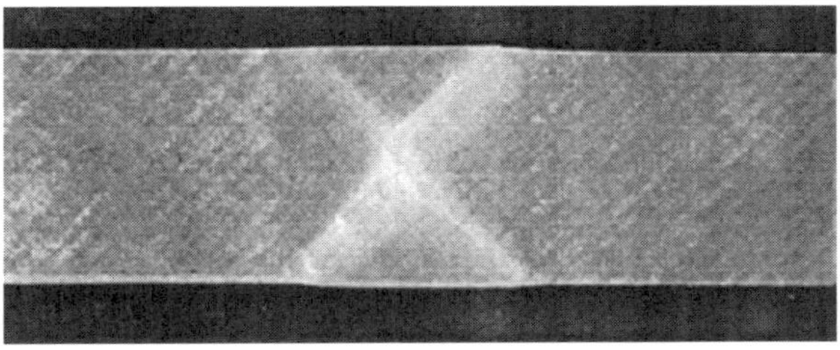

(a) Biaxial, along 0° direction (fibres are at ±45° to the loading direction)

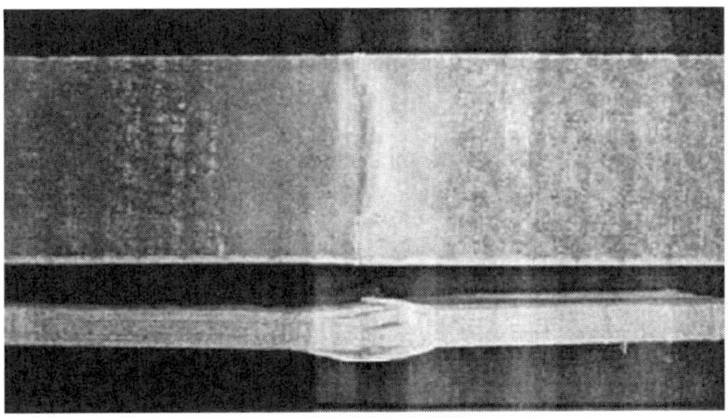

(b) Quadraxial, along 0° direction

3.24 Typical failure pattern of compressive specimens (adapted from Wang, 2002).

the compressive surface (Fig. 3.26(c)). Large deflection is reached before the final failure when fibre rupture occurs.

Tearing behaviour

Fabrics are often subjected to the effects of external forces. When the external force works sharply on the fabrics, the fabrics could become torn or even pulled apart into pieces by the force. The tearing property is a very important mechanical criterion for fabric testing. There are many tearing test methods to evaluate the tearing properties of fabrics, such as tearing strength by the falling pendulum apparatus, by the tongue method, by the trapezoid method, etc.

Many researchers have focused their work on the tearing properties of traditional fabrics in the past. Because of the complicated structure and multilayer knitting combination of MWK fabrics, it would be expected that the tearing properties of MWK fabrics are superior to those of traditional woven and knitted fabrics. Actually, many researchers have compared the tearing resistance of woven fabrics and MWK fabrics and their results show that the MWK fabrics have a higher initial Young's modulus, lower rupture

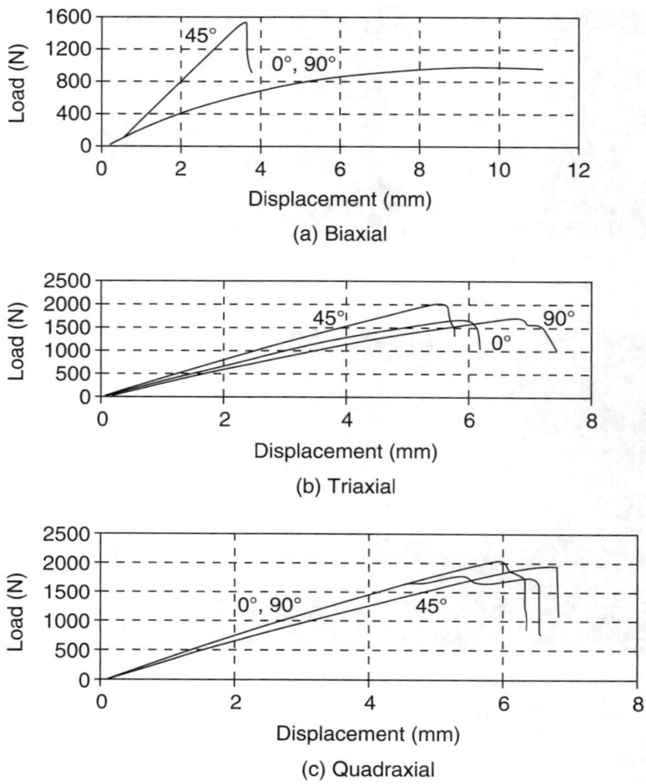

3.25 Typical flexural test curves for MWK laminates (adapted from Wang, 2002).

elongation, higher tear resistance and a higher percentage of usable potential. Further applications of MWK fabrics put the requirements on how exactly the best tearing resistance property can be achieved and what parameters of the MWK fabrics affect the entire tearing properties. The tearing property of MWK fabrics has been one of the primary research areas within the published literature. As with many of the 3-D fibrous assemblies described here, the tearing properties and mechanisms of MWK fabrics are very complex and not well understood. However, the available research results allow some general behaviour to be understood. Figure 3.27 compares the appearance of tearing in woven fabric and biaxial knitted fabric.

The tearing performance of biaxial warp-knitted fabric has been studied by Li Lvye and Shen Wei (2005). They attempted to find the relationship between the inserted warp and weft yarn density on the direction of suffering load and the tear resistance. The tearing characteristics of the biaxial warp-knitted fabric were also analyzed in their paper. They pointed out that the tearing strength has a general linear relationship with the density of the inserted warp yarn and weft yarn. With increase in the density, the tearing strength improves to a much higher extent. The knitting insertion method of the warp yarn also influences the tearing strength. From their test results, two samples with the same background weave (Denbigh plain tricot) and different warp insertion methods show that the tearing strength of the fabric knitted by using the single needle distance inserting method is poorer than the fabric knitted by using the zero needle distance inserting method. They also concluded that the background knitted weave plays an insignificant role in tearing propagation.

(a) Tested along 45° direction, tensile surface (fibres are at 0° and 90° to the loading direction)

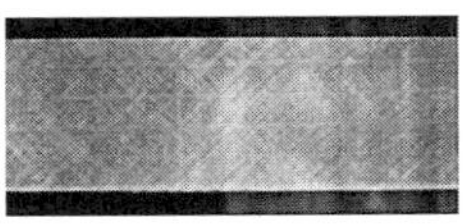

(b) Tested along 0°, tensile surface (fibres are at ±45° to the loading direction)

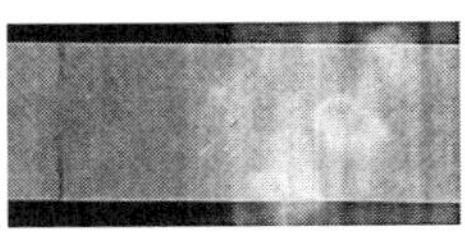

(c) Tested along 0°, compressive surface

3.26 Typical failure pattern of flexural specimens (adapted from Wang, 2002).

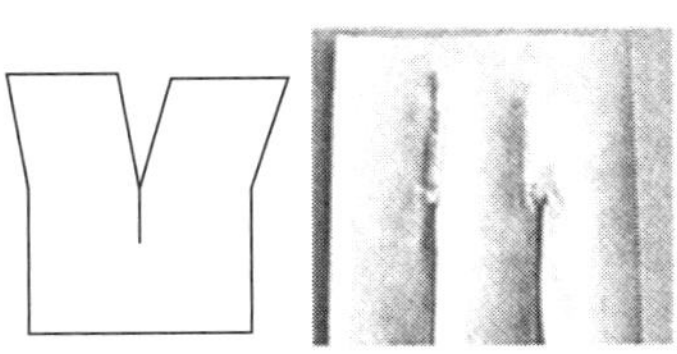

(a) Tearing of woven fabric

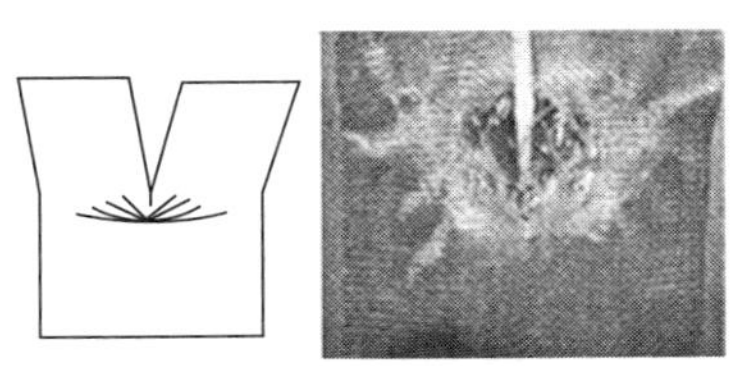

(b) Tearing of biaxial warp-knitted fabric

3.27 Comparison of tearing behaviour of woven and knitted fabrics.

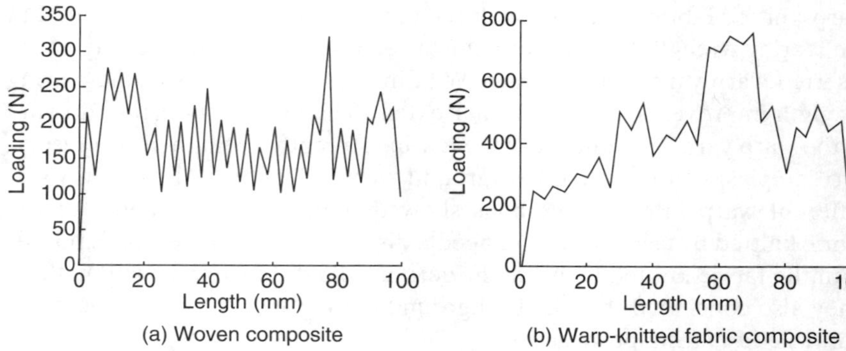

3.28 Tearing behaviour of woven and warp-knitted fabric composites.

Research on the tearing behaviour of reinforced plastic film with biaxial warp-knitted fabric skeleton by Shen Wei *et al.* (1998) further shows that the specification of laid-in and warp yarn has the greatest influence on the tearing strength of the reinforced plastic film and there is an exponential relation between the tearing strength of the reinforced plastic film and the number of yarns in the direction of the force.

Chen Nanliang and co-workers (Chen Nanliang, 1999; Jiao Wei-hong and Chen Nanliang, 2004; Xi Shiping *et al.*, 2005) studied warp-knitted reinforced composites and reported that the tearing resistance is much better than its counterpart in the woven plain weave, as shown in Fig. 3.28. The testing was conducted according to the UK industry standard (BS3424 Method 7A). In Fig. 3.28, the length means the tearing length, and the loading was recorded with the tearing length. It is clear that the tearing resistance of the biaxial knitted fabric reinforced composite is better than that of the woven plain fabric reinforced composite. The biaxial warp-knitted fabric shows high tearing strength because of the yarn clustering effect. The authors pointed out that the tearing resistance was largely dependent on the binding yarn's tearing strength and the higher density of the fabric.

From the literature above, it may be summarized that the tearing strength of MWK fabrics mainly depends on the following factors: the density of inserted yarns, the strength of the material, the insertion method of the warp and weft yarns, the knit structures of the MWK fabrics, the friction between the yarns and layers, the number of layers, etc. In addition, the tearing properties of MWK fabrics show different resistance under the same value of tearing force but with different values of tearing speed. If the speed is higher, the tearing resistance of the MWK fabrics will be poorer. All the previous results show that the MWK fabrics exhibit better tearing resistance than woven fabrics on the whole, because of the higher number of interlacing yarns in their structures.

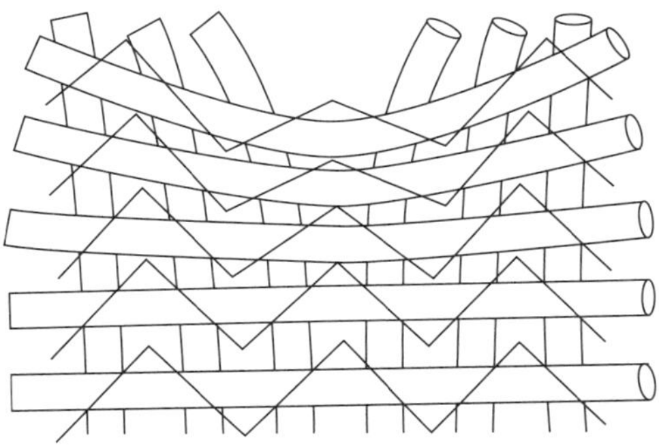

3.29 Binding effect in MWK fabrics.

As far as the tearing mechanism of MWK fabrics is concerned, when MWK fabrics are subjected to the tearing force, slippage and deformation will occur in the structure. With increase of the tearing stress, the deformation becomes larger, and once it increases to the rupture point, the yarns break first and then the entire fabric. As long as the breakage heading direction comes across the yarns that are perpendicular to the breakage direction, the stress will be shared by the perpendicular yarns, preventing the breakage from advancing. If the tearing force acts continuously upon the MWK fabric, the yarns will break one by one. Hence it is recommended to consider two indices, i.e. the yarn's breakage strength and the maximum tearing resistance, when evaluating the tearing properties of MWK fabrics. The mechanisms of tearing resistance for MWK fabrics and traditional woven fabrics show an obvious difference. In the case of a biaxial warp-knitted fabric, for example, the inserted warp yarns and inserted weft yarns are not in the same plane and the MWK fabric will not easily be destroyed by the tearing force as in the traditional woven fabric because of the additional binding effect, as shown in Fig. 3.29. When the MWK fabric is subjected to the tearing stress, nearby yarns are inclined to slip towards the breakage. There will be more yarns to share the tearing stress, and that is another characteristic of the MWK fabric to prevent the tearing breakage from advancing.

3.5 Applications of multiaxial warp-knitted fabrics

The advantages of using MWK fabrics in composite structures are clearly the flexibility and freedom of choice in the desired properties in all directions which can be matched to individual needs. As a result, MWK fabric composites are uniquely suited to a wide range of industrial applications. The majority of current end-uses for industrial composites made from MWK fabrics can be separated into two different industries, marine and

Table 3.2 Current end-uses of MWK fabrics in composites

Industry	Marine	Aerospace	Other
Estimated % of MWK fabric composite market	65%	20%	15%
Applications	• Hulls • Decks, superstructure, substructure • Support beams • Motor bays • Sails • Racing shells	• Aircraft skin • Tail units • Fuselage panelling • Leading edges on wings and rudders • Engine panelling • Rotor blades • Ballistic protection	• Flooring • Geotextiles • Wall panels • Automotives • Protective helmets • Industrial belting • Inflatables

aerospace. Probably 65% of all MWK fabrics currently made are used in marine composite applications, while another 20% are used in the aerospace industry. The remaining 15% encompass all of the varied end-use applications being evaluated with MWK fabric composites. Table 3.2 shows many of the current end-uses of MWK fabrics.

MWK fabrics are becoming the fabric preform of choice in the marine industry, especially in yachts, sailboats and high-speed racing boats. The MWK fabric composite is generally used in these vessels for the hulls, deck superstructures and substructures, and motor bays. Because of the isotropic properties of the MWK fabric structure, boat designers are finding that they can use less MWK fabric in the composite structure and still maintain, or often improve upon, the structural integrity and torsional stiffness of the boat. This also means that boat hulls made of MWK fabric composites can withstand greater stresses and strains with less overall weight. Less overall weight obviously requires less energy to power the boat, which translates into fuel savings and/or faster boats. The improved structural integrity makes the boat safer at higher speeds.

MWK fabric composites are being used in most of the fastest ocean-going racing boats and yachts because of their increased stability and weight savings. The hulls and masts of several of the sailboats used in the America's Cup competition were made of MWK fabric composites because of the performance edge experienced by using these composite structures. Partially as a result of using MWK fabric composite structures, speedboats and racing boats are achieving speeds previously thought to be unreachable with any degree of safety.

Another example of improved performance as a result of using MWK composite structures in marine applications was seen in a new generation of racing shells used by some of the top rowing teams in the country. The shells (long narrow row boats, usually powered by eight rowers) were found to give a greater translation of power into speed because of the improved torsional stiffness. This allowed energy to be translated more directly into speed rather than being absorbed by the shell when flexing.

MWK fabrics are also being looked at for applications in sails. In this case, however, the composite structure is the MWK fabric combined with a plastic film, usually through laminating. Because of the lack of crimp in the yarn in the MWK composite structure, the force of the wind is immediately translated into power and not absorbed at all by the crimp deformation associated with woven fabric structures.

Fabric composites of all types are being used in the marine industry because of their inherent resistance to corrosion. This saves on the manufacturing cost because expensive metal treatments and repeated paintings are not needed to protect the craft from the corrosive nature of salt water.

MWK fabric composites are being used more and more in the aerospace industry, which includes the military, aerospace and commercial aircraft industries. Relatively speaking, they are being used in this industry for many of the same reasons as in the marine industry, namely reduced weight and increased strength and integrity. Because of the flexibility and tailorability of mechanical and physical properties, MWK fabric composites can be customized for the application and specific properties can be emphasized to suit the particular need.

Currently, there are very few aircraft that use MWK fabric composites in critical structures such as the fuselage or wings. Most current applications centre around the skin of the aircraft. Other areas of use are in the top and side tail units, fuselage panelling, leading edges on side rudders, and engine panelling. MWK fabric composites are also being evaluated for rotor blades, outer skin and ballistic protection for helicopters. It is thought that the use of MWK composites is also being evaluated in the new military plane/helicopter, the V-22 Osprey, and the all-composite Beech Starship business plane (Kaufmann, 1991b). The lower weight achieved through the use of MWK composite structures means that less fuel is consumed by the aircraft, which translates into significant energy savings for the user. Also, because of the improved structural integrity offered by the MWK fabric composite, it is believed that safety is enhanced.

Other various applications for MWK fabric composites can be found in the industrial composites arena. In Europe, MWK fabric composites are being used for flooring in sports halls where the combination of multidirectional force distribution and excellent tear resistance are beneficial. The MWK fabric composite also helps to improve the sound damping characteristics of the flooring. MWK fabrics coated with rubber are also being used in the industrial roofing industry. They can be used not only to replace traditional materials, but also to improve the performance of many new industrial composites which seem to have reached their performance limit. MWK fabric composites can be used to replace traditional structural materials such as concrete, wood and steel, thus creating new possibilities in various industries and end-uses.

MWK fabric composites incorporating a non-woven structure are ideally suited for many high-strength geotextile applications, where isotropic strength, resistance to tear and tear propagation, good water permeability, low creep and good fabric/soil interaction are required.

With the flexibility of fibre placement and potentially high productivity, MWK fabric composites are ideally suited for many structural load-bearing applications in the automotive and aerospace industries. MWK fabrics, because of their structural make-up, have good flexibility, which allows them to be formed during moulding into virtually any desired shape. The through-thickness reinforcement provided by the stitching process helps to reduce the possibility of delamination of layers in the composite structure.

Other applications for MWK fabrics include protective helmets and armoured protection of vehicles, buildings and people. Various drive belts, V-belts, fan belts and conveyor belts will benefit from the availability of diagonal load-bearing yarns in the composite structure. Inflatable rafts, cushions, balloons and fuel cells are ideal applications for MWK fabric composites because of their good isotropic strength and tear resistance.

3.6 Summary

This chapter presented a comprehensive study of the manufacture, structure and properties of multiaxial warp-knitted (MWK) fabrics. These newly developed fabrics offer an excellent means of reinforcement in high-quality composite structures. They hold a great deal of potential for the manufacture of specific types of composite components. Three-dimensional knitted fabric products have a number of important advantages over conventional 2-D fabrics, particularly very high drape properties and superior impact damage resistance. Their excellent impact performance makes them ideal for service conditions where energy absorption or damage tolerance are critical. Therefore, MWK fabrics are widely used as reinforcement in automotive, marine, transportation and construction applications, wind turbines, sports equipment, electronics and so on.

MWK fabric structures consist typically of two to four layers of straight fibre strands held together by a chain or tricot stitch through the thickness. The process involves arrangement of fibre layers followed by stitching. The fibres in each layer can be oriented in the warp (0°), filling (90°) or a bias direction (typically between 30° and 60°). Unlike a woven fabric in which yarns are crimped due to interlacing, MWK fabrics preserve the unidirectional characteristics of each fibre layer. They are densely packed structures, and a reasonably high fibre volume fraction (about 50%) in the final composite part can be obtained from the wet manual lay-up process. This can result in good mechanical properties as well as significant savings in resin consumption.

MWK fabrics exhibit excellent conformability and therefore are well suited for composite parts ranging from flat panels to complicated shapes. A single layer of MWK fabric may contain several plies of straight fibres oriented in different directions, and therefore the behaviour of an MWK composite can be easily tailored to meet specific needs.

Recently, MWK fabric composites have been developed to overcome the damage tolerance problem and weak strength in the through-thickness direction by improving the properties in that direction. The stitched MWK fabric composites are constructed with multiple stacks of multidirectional warp-knitted fabrics stitched together in the through-thickness direction. The reinforcement in the through-thickness direction by stitching can provide significant improvements in resistance to delamination, resistance to impact and strength in the through-thickness direction. The stitched MWK fabrics also have some excellent advantages in the fabrication aspect such as their ability to conform to complicated contours, low fabrication cost and mass productivity conjunction with resin transfer moulding to produce high-quality composite materials.

3.7 References

Alagirusamy R, Fangueiro R, Ogale V and Padaki N (2006), Hybrid yarns and textile preforming for thermoplastic composites, *Textile Progress*, **38**, 4, 1–71.

Backhouse R, Blakeman C and Irving P E (1995), Mechanisms of toughness enhancement in carbon-fibre non-crimp fabrics, *Proc. 3rd Int. Conf. on Deformation and Fracture of Composites*, University of Surrey, Guildford, UK, published by Institute of Materials, 307–316.

Byun J H and Chou T-W (2000), Mechanics of textile composites, in *Comprehensive Composite Materials, Volume 1: Fibre Reinforcement and General Theory of Composites* (ed. T-W Chou), Elsevier, Oxford.

Chen Nanliang (1999), Mechanical property research of the agricultural plastic film reinforced with the warp-knitted net fabric (in Chinese), *Journal of China Textile University*, **25**, 4, August, 20–23.

Chou T-W (1992), *Microstructural Design of Fibre Composites*, Cambridge University Press, Cambridge, UK.

Cox B N and Flanagan G (1997), *Handbook of Analytical Methods for Textile Composites*, NASA contractor report, 4750.

Dexter H B (1992), An overview of the NASA textile composites program, *Sixth Conference on Advanced Engineering Fibres and Textile Structure for Composites*, Fibre-Tex, 1–31.

Dow M B and Dexter H B (1997), Development of stitched, braided and woven composite structures, NASA/TP-97–206234 (web based), USA.

Du G-W and Ko F (1996), Analysis of multiaxial warp-knit preforms for composite reinforcement, *Composites Science and Technology*, **56**, 3, 253–260.

Edgren F (2006), Physically based engineering models for NCF composites, Ph.D. thesis, Department of Aeronautical and Vehicle Engineering, School of Engineering Sciences, Kungliga Tekniska Högskolan, SE-100 44 Stockholm, Sweden.

Godbehere A P, Mills A R and Irving P (1994), Non-crimped fabrics versus prepreg CFRP composites – a comparison of mechanical performance, *Proc. Sixth Int. Conf. on Fibre Reinforced Composites, FRC '94*, University of Newcastle upon Tyne, Institute of Materials Conference, 6/1–6/9.

Hearle J W S (1995), Textiles for composites, *Textile Horizons*, **15**, 11–15.

Hogg P J, Ahmadnia A and Guild F J (1993), The mechanical properties of non-crimped fabric-based composites, *Composites*, **24**, 423–432.

Iyer C (1992), Manufacture and application of biaxial and multiaxial knits as textile reinforcements in composite materials, *Proceedings of VTT Symposium 133, Textiles and Composites '92* (ed. H Meinander), Tampere, Finland, 15–18 June, 138–146.

Iyer C (1994), Directionally oriented inlay warp knits – some aspects of production and application, *Indian Journal of Fibre of Textile Research*, **19**, 195–202.

Jiao Wei-hong and Chen Nanliang (2004), Performance advantage of warp knitting bi-axial fabrics used as coating substrates (in Chinese), *Journal of Donghua University*, **30**, 6 December, 91–95.

Karlheinz H and Burkhard W (1992), New types of textile fabrics for fibre composites, *Proceedings of VTT Symposium 133, Textiles and Composites '92* (ed. H Meinander), Tampere, Finland, 15–18 June, 147–153.

Kaufmann J R (1991a), in *Proceedings of Fibre-Tex 1991, The Fifth Conference on Advanced Engineering Fibres and Textile Structures for Composites*, NASA Conference Publication 3176, Raleigh, NC, 15–17 October, 77–86.

Kaufmann J R (1991b), Industrial applications of multiaxial warp-knit composites, Chapter 5 in *High-Tech Fibrous Materials* (ed. Tyrone L Vigo and Albin F Turbak), American Chemical Society, Washington, DC, 81–89.

Ko F K (1999), 3-D textile reinforcements in composite materials, in *3-D Textile Reinforcements in Composite Materials* (ed. A Miravete), Woodhead, Cambridge, UK, 9–42.

Li Lvye and Shen Wei (2005), The study of warp-knitted biaxial fabric tearing property (in Chinese), *Chan Ye Yong Fang Zhi Pin*, No. 4, 20–22.

Long A C, Souter B J, Robitaille F and Rudd C D (2002), Effects of fibre architecture on reinforcement fabric deformation, *Plastics, Rubber and Composites*, **31**, 2, 87.

Mohamed M H (1990), Three-dimensional textiles, *American Scientist*, **78**, 6, 530–541.

Padaki N V, Alagirusamy R and Sugun B S (2006), Knitted preforms for composite applications, *Journal of Industrial Textiles*, **35**, 4, 295–321.

Shen Wei, Chen Jigang and Feng Xunwei (1998), Tearing property of reinforced plastic film with bi-axial warp-knitted fabric skeleton (in Chinese), *Journal of China Textile University*, **24**, 2 April, 9–12.

Textile Month (2006) Issue 6, 51. Technical Feature Article.

Wagener G (1993), Stitch-bonded non-crimp products – New possibilities for the reinforcement of composites, *Proceedings of International Techtextil Symposium*, **3**, 2, 1–7.

Wang Y (2002), Mechanical properties of stitched multiaxial fabric reinforced composites from manual lay-up process, *Applied Composite Materials*, **9**, 81–97.

Wang Y and Li J (1995), Properties of composites reinforced with E-glass nonwoven fabrics, *Journal of Advanced Materials*, **26**, 3, 28–34.

Xi Shiping, Cui Guide and Chen Nanliang (2005), Tearing property study on the calendered flexible composites (in Chinese), *Chan Ye Yong Fang Zhi Pin*, No. 12, 17–19.

Zhou Rongxing and Chen Mingzhen (1999), MWK technology and its application in composites (in Chinese), *Journal of Wuhan Textile S&T Institute*, **12**, 3 September, 82.

Zhou Rongxing, Hu Hong, Chen Nanliang and Feng Xunwei (2005), An improved MWK structure for composite reinforcement, *Textile Research Journal*, **75**, 4, 342–345.

4 Multilayer woven fabrics

Abstract: Three-dimensional multilayer woven (MLW) fabrics are textile structures having fibres oriented along the three directions of a unit cell. These fabrics are becoming increasingly important owing to their excellent performance characteristics such as permeability, compressibility, drapeability, ease of handling and ability to conform to complex shapes. The fabrics can be woven with a space between layers (core fabrics) or woven as thick, dense structures. The development of multilayer woven fabrics has shown that these structures need not be interlaced throughout the fabric to have all the advantages of traditional weaving. An attempt is made in this chapter to describe the various structures, manufacturing principles and advantages of multilayer woven fabrics.

Key words: multilayer woven (MLW) fabrics, 3-D woven fabrics, 3-D weaving, structure of MLW fabrics, tensile behaviour of MLW fabrics, shear behaviour of MLW fabrics.

4.1 Introduction to multilayer woven fabrics

Traditional woven fabrics are produced by interlacing two sets of yarn, warp and weft. Plain, twill and satin are the most commonly used structures for composites, which can be produced on either tappet or dobby looms. A number of weft-insertion technologies, including shuttle, rapier, projectile and air-jet, can be used for producing woven preforms. Rapier looms are the most popular type as they can handle a variety of yarns, from very delicate to thick yarns, even metal wires.

Woven fabrics have a number of advantages over unidirectional material. Because of the interlacement, woven composites have better structural integrity and can be handled in the dry form, whereas unidirectional materials cannot be handled before impregnation. Unlike unidirectional prepregs, woven preforms have no shelf-life problems. Woven preforms are highly suitable for RTM processes and also highly formable and hence suitable for drape onto complex shapes.

Although woven fabrics are superior to unidirectional prepregs, there are a number of limitations, including the time involved in preparing the number of layers needed to build up the desired thickness, and possible delamination of the composite due to lack of through-thickness reinforcement. To improve the through-thickness strength, multilayer woven fabrics have been developed, both on conventional looms and on specialized machines (Potluri *et al.*, 2000).

Three-dimensional multilayer woven fabrics are becoming increasingly important owing to their excellent performance characteristics such as permeability, compressibility, drapeability, ease of handling and ability to conform to complex shapes. They are becoming one of the important forms of reinforcements for composite materials because of their good resistance to delamination over the laminated reinforcements (Chen *et al.*, 1992).

Three-dimensional multilayer woven fabrics are textile structures having fibres oriented along the three directions of a unit cell. A three-dimensional fabric should have three or more yarns in the thickness direction in order to distinguish itself from a planar fibre assembly. These fabrics are composed of several series of warp and filling yarns that form distinct layers, one above the other (Watson, 1955). The fabrics can be woven with a space between layers (core fabrics) or woven as thick, dense structures. In addition to the 3-D features of the macrogeometry, the internal structure of the 3-D multilayer woven fabric is characterized by continuous fibres oriented and integrated in three or more directions. In other words, these fabrics will have more than three yarns in the through-thickness direction and are manufactured with an inherent through-thickness yarn component (Ko, 1989).

The development of multilayer woven fabrics has shown that these structures need not be interlaced throughout the fabric to have all the advantages of traditional weaving. It turns out that the addition of vertical yarns interlaced with the top and bottom horizontal yarns provides the same kind of reinforcement in a 3-D structure that is provided by the over-and-under interlacing of yarns in flat weaving (Mohamed, 1990).

4.2 Advantages of multilayer woven fabrics

Multilayer woven fabrics offer a number of advantages over 2-D structures. Some of the important advantages can be summarized below:

- 3-D woven fabrics exhibit higher through-thickness and interlaminar properties because of their integrated structure in the presence of orthogonal and/or angle interlock constructions.
- These fabrics have complicated geometry and can be made to near-net-shape structures.
- 3-D multilayer woven fabrics can serve as non-crimp reinforcement preforms with the highest fibre volume leading to stronger and lighter structures.
- The ability of 3-D weaving to produce near-net-shape preforms can greatly reduce the cost of a component by reducing material wastage, the need for machining and joining, and the amount of material handled during lay-up.

- 3-D multilayer woven fabrics demonstrate high ballistic impact damage resistance and low-velocity impact damage, which have been a major problem with 2-D laminates in military aircraft structures. These fabrics have *z*-fibres through the part thickness and this feature dramatically improves impact damage tolerance by suppressing delamination.
- High-performance fibres such as carbon, ceramic, aramid, quartz and metal fibres can be woven into multilayer fabrics which can be made up to thicknesses up to one inch (2.54 cm) and widths up to 72 inches (183 cm) conforming to customer-specific complex shapes.
- Multilayer woven fabrics can be given additional strength by insertion in each layer of stuffing yarns, which remain straight and contribute their full strength to that direction. Yarns that interlace between layers as binding yarns contribute partially to the strength of their direction in orthogonal woven fabrics and contribute immensely to the strength in the thickness direction.
- Multilayer woven fabrics are highly conformable to complex mould shapes, providing uniform part thickness and eliminating complicated darting and pin-wheeling. This is because of the absence of interlacing between warp and filling yarns which allows the fabric to bend and internally shear rather easily, without buckling within the in-plane reinforcement.
- 3-D woven fabrics wet out faster in both open and closed moulding, improving quality, reducing moulding times and facilitating migration to vacuum infusion. The structural regularity and internal openness of the fibre architecture, which is strictly defined by the *z*-yarn placement, can explain this effect. Moreover, the *z*-yarns act as capillary channels to transfer resin into the preform interior from the outer surface. The increased permeability translates directly into reduced cycle time, due to faster and easier wet-out.
- Although these structures are typically more expensive than 2-D fabrics and mats, reduction of labour, higher performance and improved process efficiency result in overall cost savings in a variety of applications. When compared on a cost per square metre of finished composite structure, 3-D woven fabric reinforcements consistently outperform traditional 2-D materials.

4.3 Manufacture of multilayer woven fabrics

The conventional 2-D weaving process is carried out employing the mono-directional shedding operation. This enables either a single or a multiple layer warp to be displaced to form only one row-wise shed. Subsequent picking of the weft in the produced single shed results in the interlacing of the corresponding warp type with only one weft. In general it is characterized by one filling insertion (pick) per weave cycle or loom rpm, unavoidable crimp, limited thickness and high production speed, whereas 3-D weaving is characterized by multiple filling insertions (picks) per weave cycle or machine rpm, no internal crimp, better performance and greater thickness but lower production speed. A comparison of the traditional 2-D weaving and the 3-D weaving systems is presented in Fig. 4.1.

The principle of the 3-D weaving process is represented in Fig. 4.2. The heart of the 3-D weaving process is the dual-directional shedding operation. Through this operation the multiple layer warp can be displaced to form alternately multiple column-wise and row-wise sheds (Fig. 4.3). Subsequent picking of wefts in the corresponding sheds of the two directions results in the complete interlacing of the multiple layer warp (Z) with the two mutually perpendicular sets of wefts (X and Y).

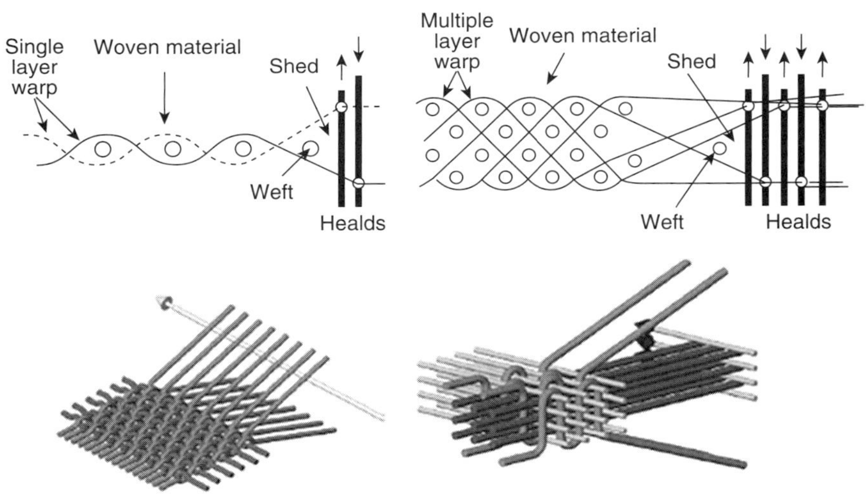

(a) 2-D weaving (b) 3-D weaving

4.1 Comparison of 2-D and 3-D weaving.

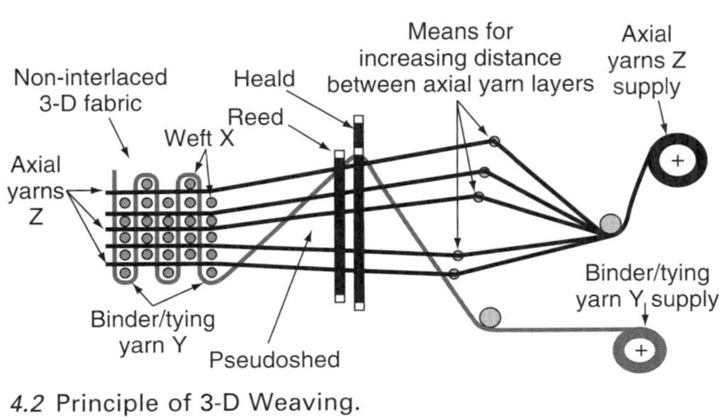

4.2 Principle of 3-D Weaving.

Another point to be noticed in Fig. 4.3 is the alternate formation of the column-wise and row-wise sheds. Accordingly, two mutually perpendicular sets of wefts, the vertical and horizontal sets, will be required. The number of wefts in each set will thus correspond with the number of sheds of the respective directions. As shown, first multiple column-wise sheds are formed into which a corresponding number of vertical wefts are picked. Next, multiple row-wise sheds are formed and a corresponding number of horizontal wefts are picked. Such shedding and picking will result in the interlacing of the multiple layer warp with two orthogonal sets of wefts as shown in Fig. 4.4 and hence an interlaced 3-D fabric, referred to as 3-D woven 3d fabric, will be obtained (Khokar and Peterson, 1998).

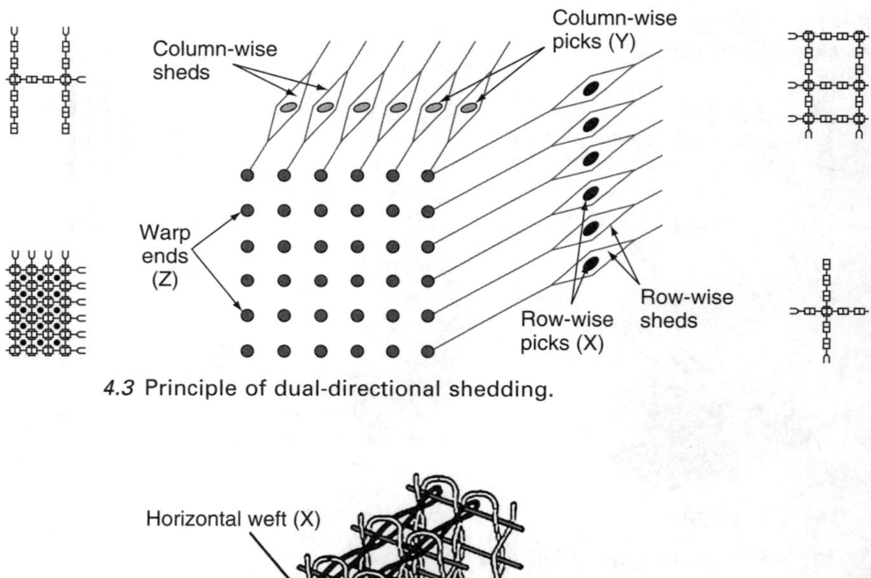

4.3 Principle of dual-directional shedding.

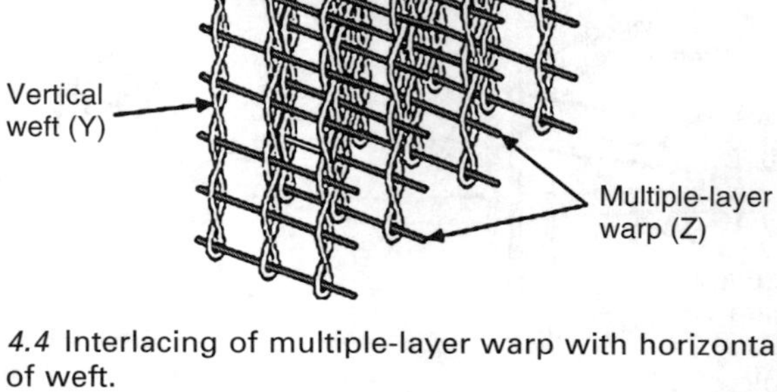

4.4 Interlacing of multiple-layer warp with horizontal and vertical sets of weft.

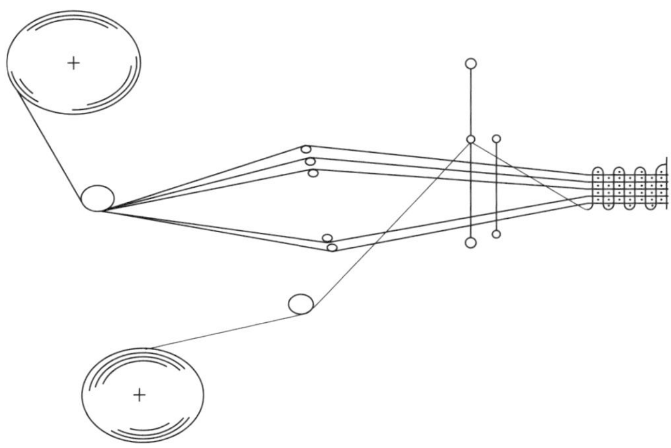

4.5 Multiwarp weaving loom.

Three-dimensional multilayer woven fabrics are produced principally by the multiple warp weaving method, which has long been used for the manufacture of double cloth and triple cloths for bags, webbings and carpets. A typical setup of a multiwarp weaving loom is shown in Fig. 4.5 (Ko, 1989). The number of layers of yarns in the fabric is governed by a special shedding mechanism which controls the height through which the harnesses are lifted. Fabrics with as many as 17 layers have been woven successfully with this method. By the weaving method, various fibre architectures can be produced including solid orthogonal panels, variable thickness solid panels and core structures simulating a box beam, or truss-like structures. Furthermore, by proper manipulation of the warp yarns, as exemplified by the angle interlock structure, the through-thickness yarns can be organized into a diagonal pattern. One limitation of the multiwarp weaving method is the difficulty of introducing yarns in the bias direction as in the triaxial weaving or circular weaving process.

Traditional weaving machines can be adapted well to make multilayer woven fabrics. Mohamed (1990) has described a method of weaving these fabrics with different shapes using a variety of fibres. The special feature of the process is that the multiple weft yarns are inserted simultaneously. The machine was capable of producing continuous thick panel and structural elements with different cross-sections such as T, double-T, I and double-I, with fairly thick walls in all directions. These shapes were made in one step using variety of fibres. A schematic diagram of a loom to produce 3-D multilayer woven fabrics is shown in Fig. 4.6.

Weaving a 3-D woven fabric requires several steps. The steps involved in the actual weaving process are described in Fig. 4.7 (from (a) to (h)). Before weaving begins, two sets of yarns are fed from bobbins and placed in the

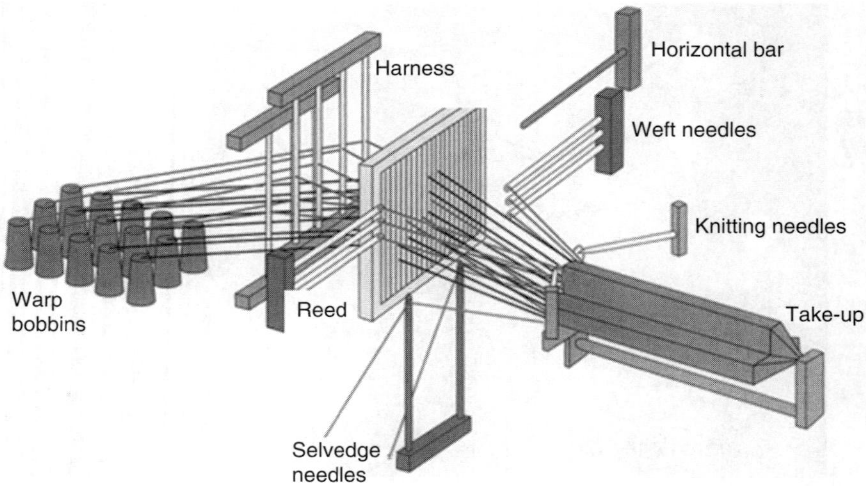

4.6 3-D multilayer weaving system (Mohamed, 1990).

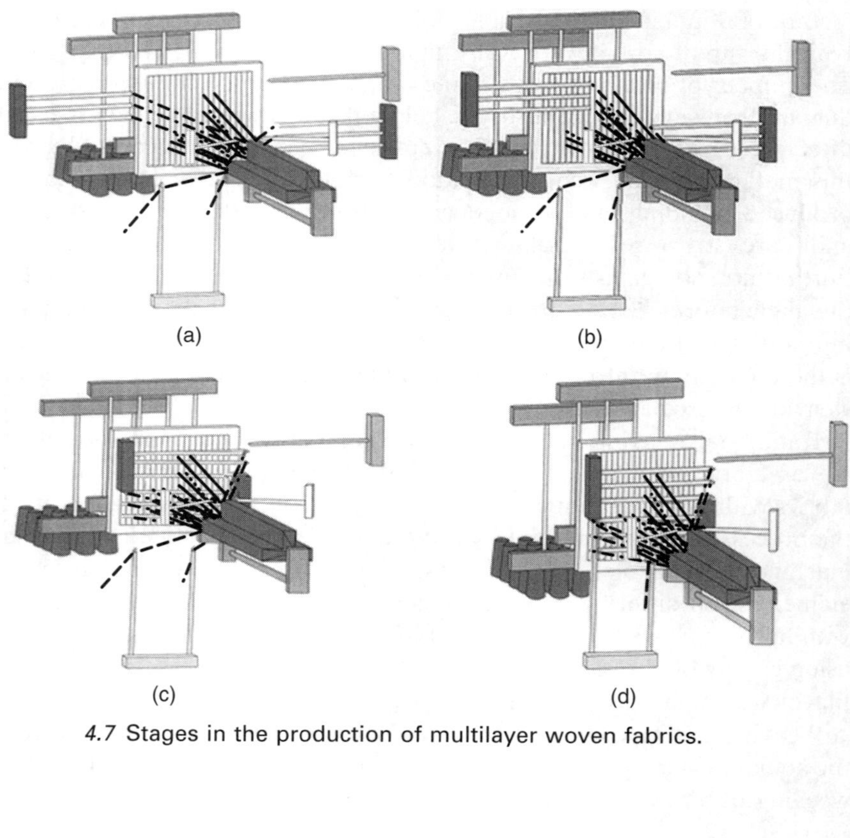

4.7 Stages in the production of multilayer woven fabrics.

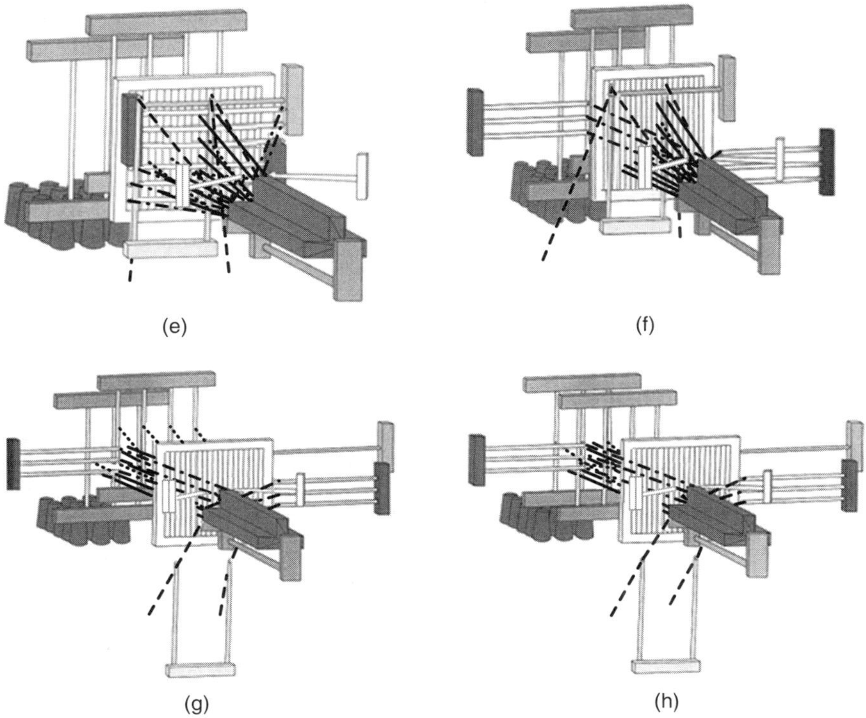

4.7 Continued

lengthwise direction. The first set, the warp yarns (—), will remain stationary during weaving and will become the lengthwise yarns in the finished structure. Harnesses suspend the *z*-yarns (. . . .), the set that will become the vertical yarns, at oblique angles – some from above, some from below. Two additional yarns are used: filling yarns (–·–·), which are inserted from the side by two horizontal sets of needles, and selvedge yarns (----), inserted from below by a pair of vertical needles. Two knitting needles are positioned so that they can knit loops of selvedge yarn together at the corners of the woven beam. Weaving begins as the filling needles move between layers of warp and vertical yarns to insert the filling in a crosswise direction (b, c). Before these needles retract, the vertical selvedge needles move up to catch the filling; a horizontal bar in turn catches the selvedge yarn at the top (d, e). The filling needles retract (f). The pair of knitting needles clasps the two selvedge yarns, allowing the selvedge yarn and crossbar to retract also. The reed – the comb-like device just in front of the harness – moves horizontally to pack the yarns into their finished configurations; during this step, the *z*-

yarns are pushed from their diagonal position into a vertical alignment (g). At the end of the cycle the knitting needles pass the new loop of selvedge yarn through the previous one, and the harnesses switch to reserve the position of the vertical yarns for the next cycle (h).

As in any loom, the 3-D multilayer weaving system begins with a supply arrangement for the lengthwise yarns. This can be either a collection of bobbins on a creel or a set of beams. Two sets of yarns are arranged separately along the length of the loom: the warp (x) yarns and the vertical (z) yarns. The filling (weft) (y) yarn and the yarn that will secure the vertical selvedge edge must be dispensed separately. One of the lengthwise groups of yarns, the warp (x), does not move during the operation of the machine. To form a multilayer structure, the warp yarns are held lengthwise in the machine in a grid pattern matching the cross-sectional arrangement of layers. In a 3-D structure, there are many layers of filling (y) yarns, which must be inserted between the layers of warp yarns in multiples. This is accomplished by inserting the filling across the warp with a vertical row of needles, each needle being perpendicular to the warp and threaded with a length of yarn. To form a simple rectangular block structure, a single set of needles carries filling yarns across all the warp yarns.

A key step in assuring the strength of the finished structure occurs each time the needles cross the warp. A vertical needle, threaded with the yarn that will secure the selvedge, the loops on the vertical edge, is inserted from below the weaving area, coming up to catch the filling yarn that has just been brought across the warp. This selvedge needle holds a loop of each filling yarn at the edge of the warp as the filling needles return to their original positions. Thus a double length of filling is inserted with each cycle. The selvedge needles retract for the next step, but the selvedge yarn – which now holds the filling loops – is clasped and held in its vertical position by a knitting needle. To form a finished corner edge, the loops of selvedge yarns are knitted together as the process continues.

The step that is called beat-up in traditional weaving is used in 3-D weaving to position the vertical (z) yarns. These yarns are threaded through heddles suspended from harness frames (similar to those used in the traditional loom) and are passed through the vertical openings in a reed at angles, crossing on the opposite side of the filling needle. Once every cycle, as the z-yarns are suspended in diagonal positions, the reed moves horizontally to push the filling against the already woven length of fabric. This action pushes the crossed z-yarns into a vertical arrangement. The harnesses holding the z-yarns then move up and down to reverse the positions of the yarns before the process begins again. In this way the vertical yarn that has just been stretched from the bottom to the top of the textile is passed over the topmost filling yarn and held in position to be placed in the opposite direction at the next stage of the weaving.

In summary, 3-D multilayer woven fabrics are fabricated by modifying the conventional weaving mechanisms. Harnesses with multi-eye heddles are used to arrange the warps into three sections in plane form for weaving convenience. The mainframe and flanges are interlaced by a set of warps moving to and fro as a joint. Weft passes through the clear warp sheds separated by multi-eye heddles to form the 3-D woven fabrics in plane form. The differential feeding length between the warp yarns gives rise to extra friction, and therefore hairiness may occur. In order to reduce this friction the warps are passed through the tensioner and weight with ceramic eyes individually between the creel and weaving loom. The thickness of the central portion of the flattened fabrics is different from that of the side portions. Therefore the cloth roller cannot be used to take up the flattened fabrics. The fabric is clipped and pulled by a pair of rollers set in front of the loom as a take-up device.

4.4 General structure and behaviour of multilayer woven fabrics

4.4.1 Structure

In 2-D structures, yarns are laid in a plane and the thickness of the fabric is small compared to its in-plane dimensions. Single layer designs include plain, basket, twill and satin weaves which are used in laminates. Two-dimensional woven fabrics are generally anisotropic, have poor in-plane shear resistance and have less modulus than the fibre materials due to the existence of crimp and crimp interchange. Reducing yarn crimp in the loading direction or using high-modulus yarns improves fabric modulus. To increase isotropy, in-plane shear rigidity and other properties in the bias or diagonal direction, triaxial woven fabrics have been developed in which three yarn systems interlace at 60° angles.

Unlike in 2-D fabrics, in 3-D woven fabric structures the thickness or z-direction dimension is considerable relative to the x and y dimensions. Fibres or yarns are intertwined, interlaced or intermeshed in the x (longitudinal), y (cross) and z (vertical) directions. For 3-D structures, there may be an endless number of possibilities for yarn spacing in a 3-D space (Adanur, 1995). In practice, there are enormous variations in multiple layer textile structures. A structure can be of different layers, each layer may have different weaves, the yarns from different layers may be arranged in different orders, and so on (Chen *et al.*, 1992).

As illustrated in Fig. 4.8 (Pastore and Cai, 1990), there are four basic components to a generalized 3-D woven fabric geometry: warp, web, fill and surface weave yarns. Warp yarns are the system of yarns which run in the machine direction and have no crimp. These are also called 'stuffer' yarns

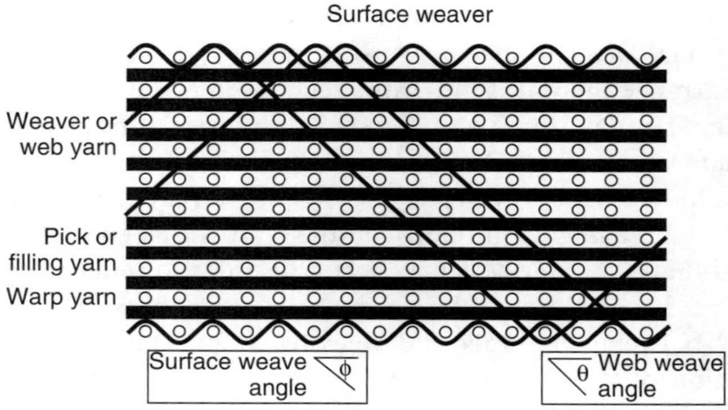

4.8 Schematic illustration of generalized 3-D woven fabric projected to the *x–z* (fabric length–thickness) plane.

or 'longitudinals'. Because of their very low crimp, these yarns provide the primary strength and stiffness in the longitudinal (x) direction of the material. Web yarns run in the machine direction and provide the interlacing necessary for fabric integrity. These yarns contain crimp in the through-thickness direction, providing the z-directional properties of the system. These yarns are sometimes called 'weavers'. The 'weave angle' of the web yarns refers to the angle of orientation of the web yarn with respect to the warp direction. Fill yarns are perpendicular to the machine direction and interlace with the web yarns. These yarns are sometimes called 'picks'. These yarns also possess crimp in the through-thickness direction, but this crimp is negligible compared to that of the warp yarns for these fabric systems. These yarns provide the transverse (y) directional properties of the composite system. Surface weave yarns run in the machine direction and form what is essentially a 2-D weave on the surface of the fabric. Surface weave yarns are incorporated into the structure when the web yarns are insufficient to provide a smooth surface on the face and back of the cloth. These yarns experience crimp in the through-thickness direction. When surface weave yarns are employed in the fabric, there are two yarns for every warp plane of the fabric. This system of yarns contributes the least to the mechanical properties of the composite.

Pattern design of three-dimensional multilayer woven fabrics

Three-dimensional woven structures can be classified into angle interlock and orthogonal interlock binding according to the orientation of binders, or through-thickness and layer-to-layer binding if the penetration depth of binders is concerned. Combination of the definitions can evolve four basic binding possibilities, i.e., angle interlock/through-thickness binding (A/T),

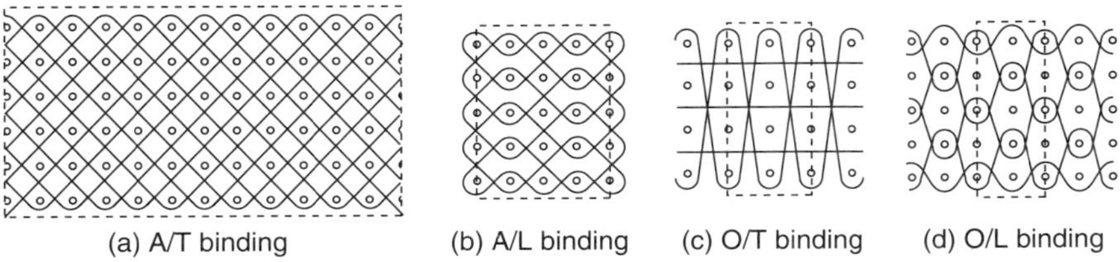

4.9 3-D woven structures with various binding patterns.

angle interlock/layer-to-layer binding (A/L), orthogonal interlock/through-thickness binding (O/T) and orthogonal interlock/layer-to-layer binding (O/L). These four binding patterns, for which the unit cell is enclosed by a dotted rectangle, are shown in Fig. 4.9.

Binding patterns are of prime importance in determining the fibre architecture of 3-D woven preforms and in analysing the performance of woven composites. Other conditions being the same, an orthogonal interlock binding can provide a greater fibre volume fraction than an angle interlock binding, particularly in the thickness direction. This provides a possible way to modify the performance of woven composites by selecting a suitable binding pattern with respect to the orientation of binders. On the other hand, woven preforms with an angle interlock binding possess better pliability and distortion capability than those with an orthogonal binding. It is therefore appropriate to select angle interlock binding for producing woven composites with complex configurations (Yi and Ding, 2004). In addition, a layer-to-layer binding can also offer a greater fibre volume fraction than its counterpart of a through-thickness binding because of the overlapping arrangement of binders in the thickness direction. Therefore, the design of penetration depth of binders could provide another way to improve the performance of woven composites.

Weave instruction in cross-sections of weft

After the binding pattern of a 3-D woven structure is determined, a method of instructions to drive the weaving machine should be given in order to clearly illustrate the geometrical arrangement of the constituent yarns and consequently to guide the drawing and denting processes. To demonstrate the method, a typical 3-D woven pattern as shown in Fig. 4.10(a) is used. No doubt, the 3-D woven structure can be presented by means of the traditional design method for a compound structure (Goerner, 1989). However, this seems to be over-complex for the 3-D woven structures, particularly when the number of fabric layers is beyond the common range.

The three lengthwise-arranged yarns including binder, stuffer and surface warp are jointly called warp yarns for convenience. They are numbered

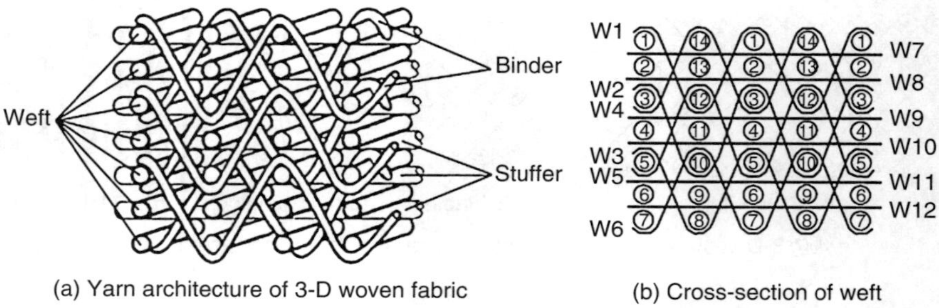

4.10 Schematic diagram of a typical 3-D woven structure.

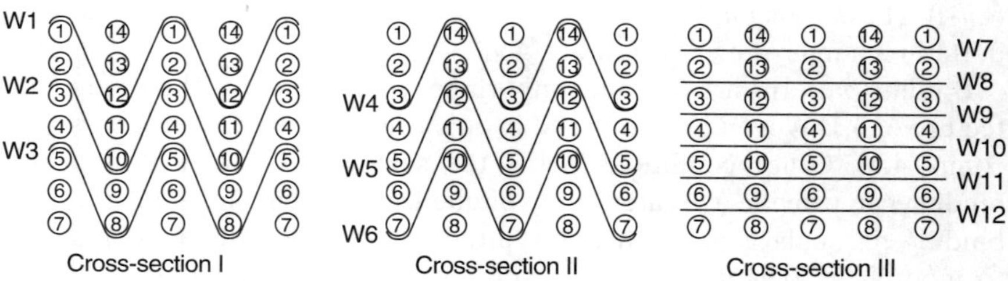

4.11 Binder and stuffer in a dent.

consecutively corresponding to the harness frames. Similarly, the wefts are numbered to illustrate the picking sequence, as shown in Fig. 4.10(b). To ensure a steady working condition of harness frames, wefts should be numbered in descending order from the top to the bottom along a weft column, then in ascending order from the bottom to the top along the next column, and so on.

To avoid severe yarn-to-yarn friction within a dent of reed, the warp yarns that are not interlaced on a cross-section of weft are grouped and drawn in a dent. From this point of view, for the woven structure shown in Fig. 4.10, three cross-sections with corresponding warp yarns are identified, as shown in Fig. 4.11.

It can be seen from Fig. 4.11 that the warp yarns in each cross-section are arranged in parallel and do not exchange their positions in the vertical direction during the weaving process. Once they are drawn in a dent, no friction between them will occur. For this particular structure, there are 12 warp yarns within a weave repeat, including six binders divided into two groups and six stuffers. Consequently, three dents are required with the denting program 3–3–6. Binders W1–W3 are inserted in the first dent, binders W4–W6 in the second dent and stuffers W7–W12 in the third one. It follows that if the straight draw method is used, the weave instruction can easily be obtained. For 3-D structures other than the one described, the proposed method can still be used.

Two of the most common architectures of 3-D woven fabrics used in most applications are the orthogonal and layer interlock weaves, which are illustrated in Fig. 4.12(a) and (b). The important difference between these two weaves is the weave pattern of the through-thickness binder yarn.

The angle interlock woven fabrics consist of three sets of yarns. The stuffers (warp yarns) and warp weavers are oriented along the longitudinal direction, i.e., along the loom feed direction. The fillers (weft yarns) are oriented transverse to the loom feed direction, and are inserted between layers of stuffers. The stuffers and fillers form an orthogonal array. The warp weavers traverse through the thickness of the weave, and interlock with filler layers. The warp weavers crisscross the weave thickness at off-axis

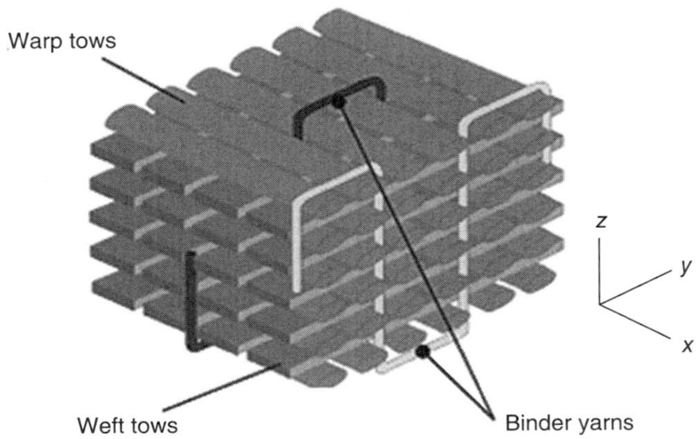

(a) Orthogonal

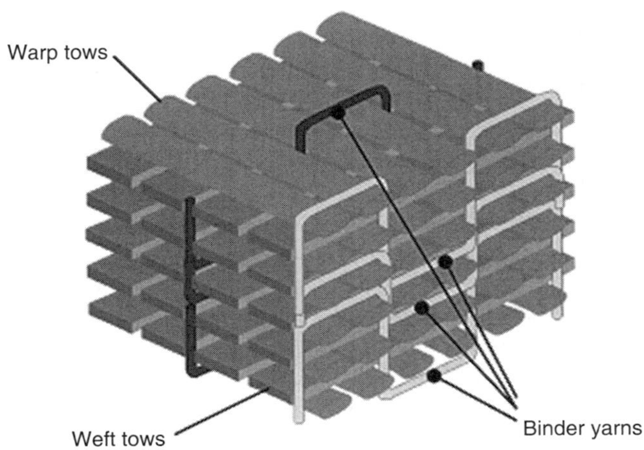

(b) Layer-interlock

4.12 3-D woven structures.

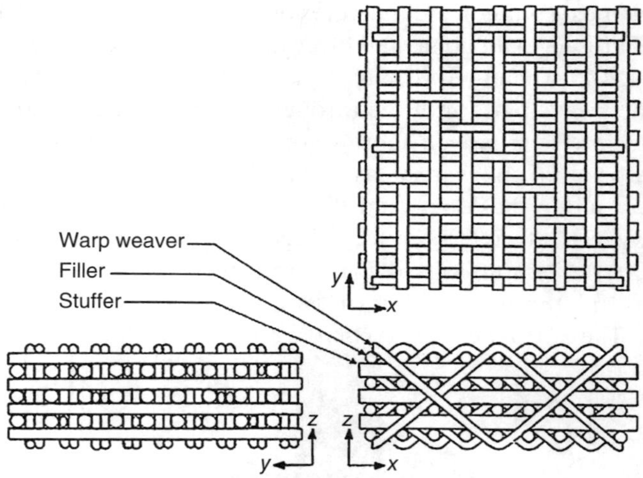

(c) Schematic representation of a 3-D through-thickness angle interlock woven structure

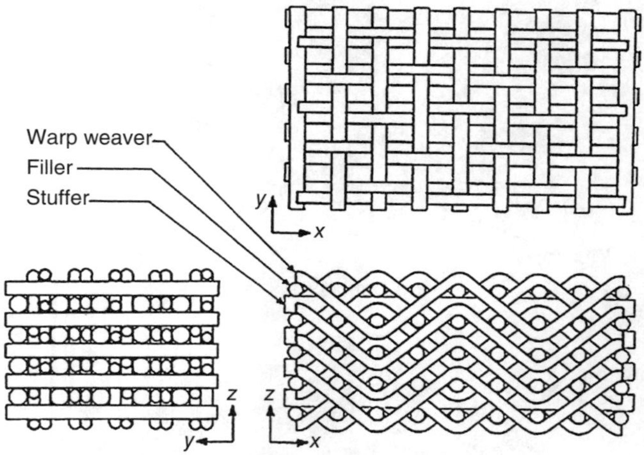

(d) Schematic representation of a 3-D layer-to-layer angle interlock woven structure

4.12 Continued

angles. Different weave geometrical parameters are yarn size, yarn spacing, yarn distribution, interlock lengths and depths. There are two main types of angle interlock preforms: through-thickness angle interlock weave (TTAW) and layer-to-layer angle interlock weave (LLAW).

The TTAW is a multilayered preform in which warp weavers travel from one surface of the preform to the other, holding together all the layers of the preform. The LLAW is a multilayered preform in which warp weavers travel from one layer to the adjacent layer, and back. A set of warp weaves together hold all the layers of the preform.

The schematic arrangements of TTAW and LLAW with ideal geometry are shown in Fig. 4.12(c) and (d). Here, the in-plane yarns, i.e., stuffers and fillers, are straight and the warp weavers are segmentwise straight. Once the preform has been formed, it is densified to form the final composite structure. Densification is the process of surrounding all the fibres/yarns within the preform structure with a resin, leading to a solid composite structure consisting of preform reinforcement and resin. Normally, using the 3-D preforms and resin transfer moulding, 3-D composites are made. Typically, the preform is formed thicker than the mould opening to ensure pressure on the preform during densification. However, such a compressive force applied to the preform necessarily results in some distortion of the yarns within the composite. The thickness of the composite would be less than the nominal thickness of the preform. Even before densification, the yarns within a preform would be wavy (Naik *et al.*, 2002).

4.4.2 Mechanical behaviour of multilayer woven fabrics and composites

Multilayer woven (MLW) fabrics are composed of warp, weft and sometimes binding yarns. The yarn stiffness in one set during weaving will influence the bending behaviour of the yarn in the other sets. When two yarns from two sets meet to interlace, it is the flexible one that bends more. Furthermore, the path of the yarn in the fabric is another factor which controls the yarn bending character. There is a variety of choices of 3-D fabric construction. It is important to understand the mechanical properties of MLW fabrics in terms of tensile, compressive and shear behaviour as these properties will have a direct bearing on the final characteristics of the composite produced. Although a very limited literature exists on the mechanical behaviour of MLW preforms, many researchers have attempted to study the effect of the mechanical behaviour of different varieties of MLW fabrics on the final performance of the composites.

Tensile behaviour

In a study on the tensile behaviour of 3-D woven composites by using different fabric structures, Gu and Zhili (2002) reported the tensile properties of four different types of 3-D woven fabrics and their effect on the tensile properties of composites. The 3-D woven structures analyzed for their tensile behaviour are presented in Fig. 4.13.

The tensile strength results for MLW preforms are presented in Table 4.1. The tensile strength values of orthogonal structures with and without layer-to-layer binding are comparable, and those of angle interlock and modified angle interlock are also comparable, although the difference between the

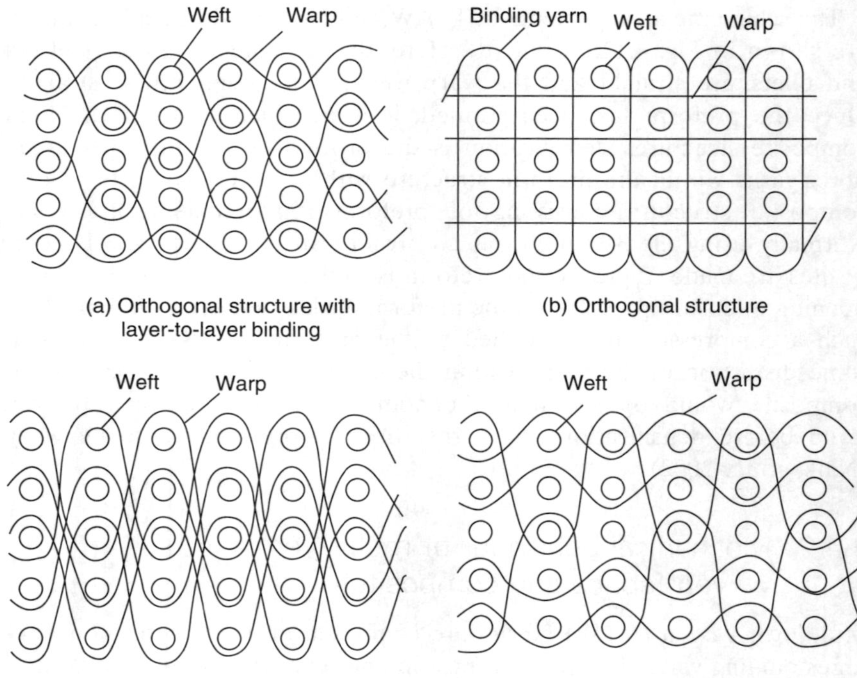

(a) Orthogonal structure with layer-to-layer binding

(b) Orthogonal structure

(c) Angle interlock structure

(d) Modified angle interlock structure

4.13 3-D woven structures.

Table 4.1 Tensile strength results for MLW preforms

Type of fabric	Tensile strength (kN)			Average (kN)
Orthogonal structure with layer-to-layer binding	2.20	1.90	2.00	2.03
Orthogonal structure	2.75	2.70	2.80	2.75
Angle interlock structure	2.50	2.60	2.60	2.57
Modified angle interlock structure	2.95	2.80	3.00	2.92

Source: adapted from Gu and Zhili, 2002.

first two samples is significantly larger. The difference may be attributed to the totally different warp passages in these two samples. All the warp ends in sample 1 are in a binding state, those on the top and bottom being bent slightly, the others to a large extent. The strength values of samples 3 and 4 also exhibit a large difference and this may again be attributed to the bending behaviour of the yarn in the fabric. These differences in the strength values of different fabrics do affect the performance of the final composite for a specific application.

The high strain rate (HSR) stress–strain behaviour of 3-D Kevlar woven hybrid composites using an epoxy resin system under fill, warp and through-thickness loading is presented in Figs 4.14 to 4.16. A representative quasi-static stress–strain curve is plotted in the same figures. The quasi-static behaviour is almost linear up to the ultimate stress level for this loading.

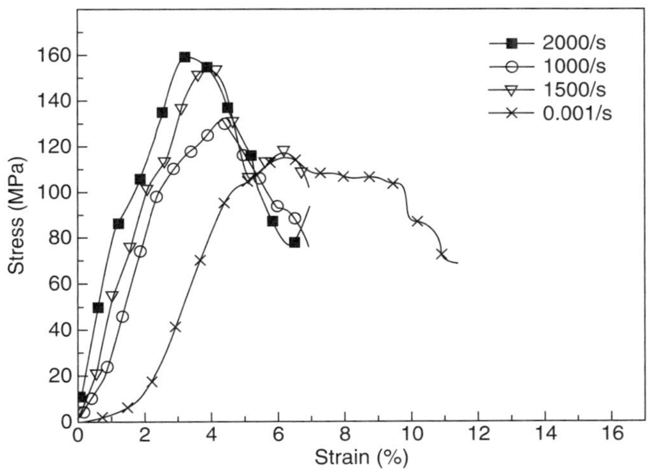

4.14 Stress–strain curves of 3-D woven composites at various strain rates in the in-plane warp direction (adapted from Guo *et al.*, 2007).

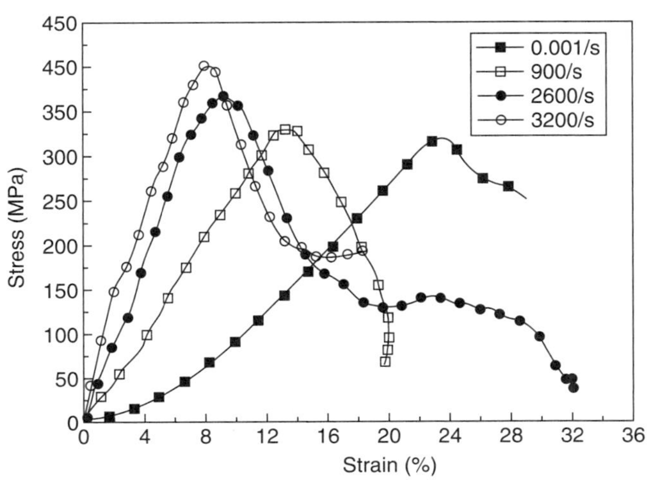

4.15 Stress–strain curves of 3-D woven composites at various strain rates in the in-plane fill direction (adapted from Guo *et al.*, 2007).

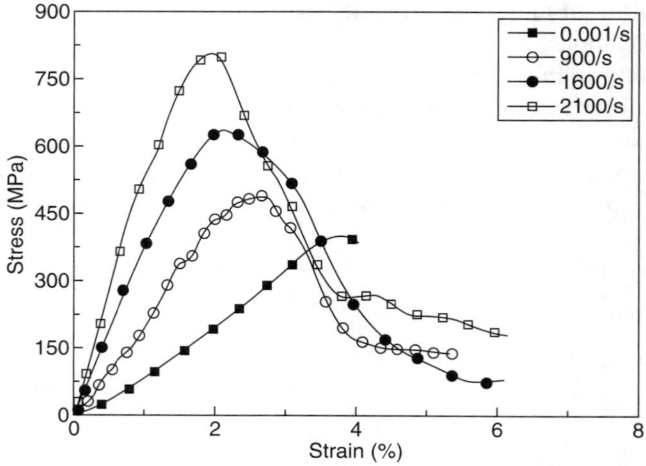

4.16 Stress–strain curves of 3-D woven composites at various strain rates in the through-thickness direction (adapted from Guo *et al.*, 2007).

Similar to quasi-static tests, all the HSR specimens showed some linear behaviour at lower loads (Guo *et al.*, 2007). As the propagating stress pulses further load the specimens, they show progressive non-linear behaviour in the stress–strain response. The non-linear behaviour of the material is dominated by matrix cracks and confirmed by visual inspection and microscopy after the test. In the fill or warp direction, the first region in the stress–strain curve was the elastic region, which extended from the beginning. The elastic region was related to the strain rate. It was found that with the increase of strain rate the elastic region decreased. The first significantly non-linear behaviour occurred after the elastic region as a result of microbuckling and some matrix microcracking. Around peak load, delaminations and formation of kink bands lead to a sequence of load drops. In the out-of-plane direction, there existed an elastic region and non-linear behaviour, but it was different from that in the in-plane direction because the loading pressed in the longitudinal direction, not in the transverse direction. Because there are some *z*-yarns in the out-of-plane direction, the elastic behaviour was affected by the *z*-yarns and the other yarns. In any region, there existed some relative damage modes.

Shear behaviour

Three-dimensional orthogonal woven fabric composites have high in-plane tensile stiffness and strength in the warp and weft directions because there is no crimp in the warp and weft yarns in the in-plane direction. Furthermore, the composites exhibit higher interlaminar shear strength than the laminates because the z-direction tows exist in the through-thickness direction. Some researchers have studied the high-strain-rate response of 3-D textile structure composites under compressive and tensile loading (Hosur *et al.*, 2003; Sun *et al.*, 2005a, b, c). The shear stress–strain curve, shear failure stress and shear failure strain in quasi-static and high-strain-rate loading conditions were compared. The shear deformation and failure mode of the composite in quasi-static conditions and at high strain rates were observed to demonstrate the failure mode at different strain rates.

The shear stress–strain curves of 3-D orthogonal woven composites at various strain rates in the warp and weft directions are respectively shown in Figs 4.17 and 4.18. Figure 4.17 shows that the shear failure stress is nearly invariable with increasing shear strain rate. However, the failure strain decreased with increasing shear strain rate. The shear stiffness at different shear strain rates is increased with increasing strain rate. It can be seen from Fig. 4.18 that the shear failure stress of 3-D orthogonal woven composites increased with increasing shear strain rate, and the shear failure strain decreased with increasing shear strain rate. Further, the shear stiffness of 3-D orthogonal woven composites increased with increasing strain rate. The shear failure stress, shear failure strain and shear stiffness are listed in Tables 4.2 and 4.3 for warp and weft direction shear, respectively (Baozhong Sun and Bohong Gu, 2006).

Figure 4.19 depicts the shear failure stress of 3-D orthogonal woven composites at various strain rates in the warp and weft directions of loading.

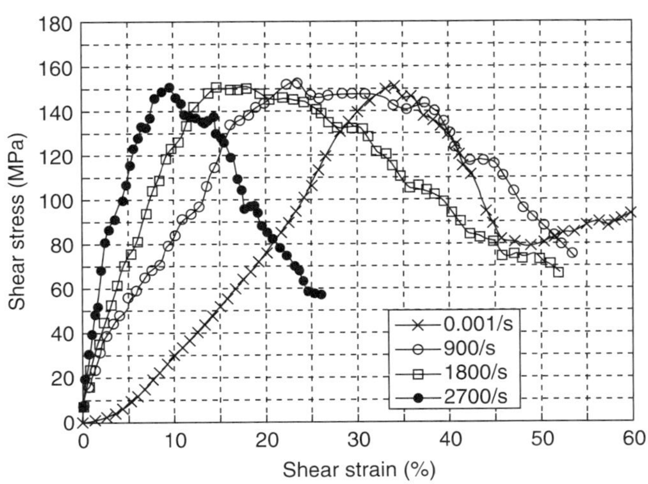

4.17 Stress–strain curves of 3-D orthogonal woven composites at various strain rates in the warp direction (adapted from Baozhong Sun and Bohong Gu, 2006).

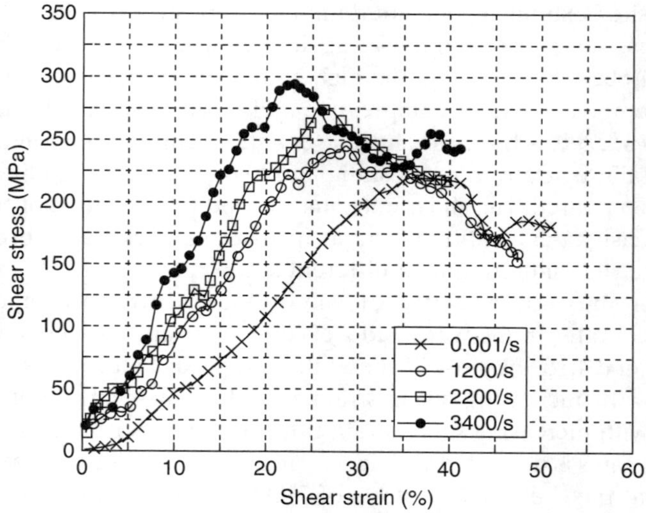

4.18 Stress–strain curves of 3-D orthogonal woven composites at various strain rates in the weft direction (adapted from Baozhong Sun and Bohong Gu, 2006).

Table 4.2 Mechanical properties of 3-D orthogonal woven composites at various strain rates in the warp direction

Shear strain (s^{-1})	Shear stiffness (GPa)	Shear failure (MPa)	Failure strain (%)
0.001	0.376	151.00	33.71
900	0.81	152.19	22.01
1800	1.27	152.00	13.64
2700	2.01	151.00	9.64

Source: adapted from Baozhong Sun and Bohong Gu, 2006.

Table 4.3 Mechanical properties of 3-D orthogonal woven composites at various strain rates in the weft-direction

Shear strain (s^{-1})	Shear stiffness (GPa)	Shear failure (MPa)	Failure strain (%)
0.001	0.630	220.02	36.01
1200	0.83	246.01	27.11
2200	1.12	270.98	23.96
3400	1.33	294.02	22.00

Source: adapted from Baozhong Sun and Bohong Gu, 2006.

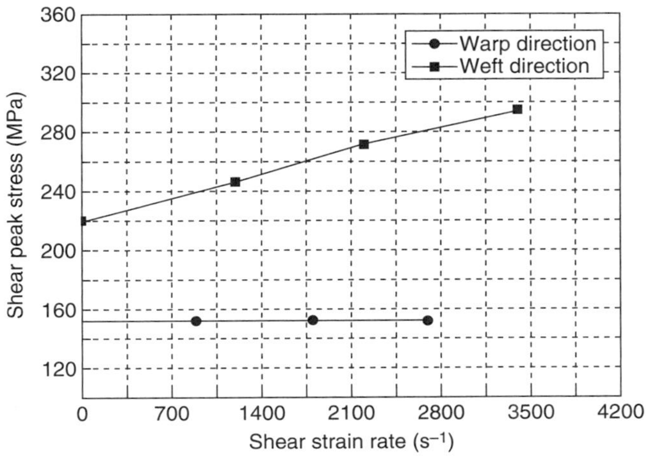

4.19 Shear failure stress of 3-D orthogonal woven composites in the warp and weft directions at various strain rates (adapted from Baozhong Sun and Bohong Gu, 2006).

From Fig. 4.19, the shear failure stress of 3-D orthogonal woven composites in the warp direction loading is nearly invariable with increasing shear strain rate. However, the shear failure stress in the weft direction loading increased with increasing shear strain rate.

The failure strain of the 3-D orthogonal woven composites in the warp and weft directions at various strain rates is shown in Fig. 4.20. The failure strain in the two directions decreases non-linearly with an increase in strain rate. Furthermore, the failure strain in the warp direction at higher strain rates decreases more rapidly compared with that in the weft direction. This shows that the failure strain in warp-direction loading is more rate-sensitive than that in weft-direction loading. From Fig. 4.20, the 3-D woven composites have larger shear failure strain in weft-direction loading than in warp-direction loading. This shows that the failure strains for loading in both the warp and weft directions are different at the same strain rate because different fibres are used for the warp and weft directions, respectively.

Figure 4.21 depicts the shear stiffness of 3-D orthogonal woven composites in two directions at various strain rates. The shear stiffness linearly increases with the strain rate in both the warp and weft directions. The shear stiffness of 3-D orthogonal woven composites in the weft direction is greater than that in the warp direction under quasi-static loading. This shows that the shear stiffness is more sensitive to shear strain rate in warp-direction loading than in weft-direction loading.

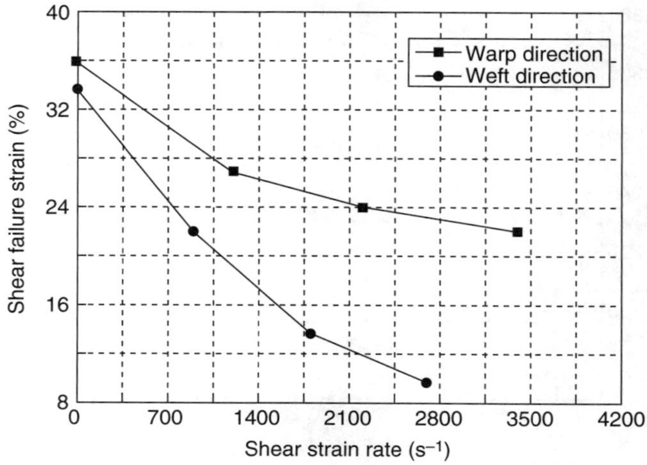

4.20 Shear failure strain of 3-D orthogonal woven composites in the warp and weft directions at various strain rates (adapted from Baozhong Sun and Bohong Gu, 2006).

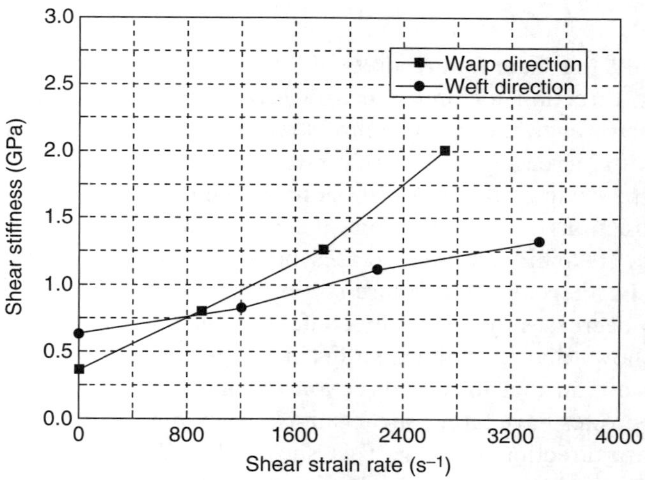

4.21 Shear stiffness of 3-D orthogonal woven composites in the warp and weft directions at various strain rates (adapted from Baozhong Sun and Bohong Gu, 2006).

Compressive behaviour

In order to understand the compressive behaviour of 3-D woven fabrics, researchers have attempted to study their effect on the properties of the final composites. In a study on the effect of fabric structure on the mechanical behaviour of 3-D woven composites, Youjiang Wang and Dongming

Table 4.4 Compression and flexural properties of 3-D woven composites

Type of fabric	Compression		Flexure	
	E (GPa) (CV%)	*σ* (MPa) (CV%)	*E* (GPa) (CV%)	*σ* (MPa) (CV%)
3-D woven	20.6 (3.1)	425 (5.4)	12.4 (7.2)	507 (5.6)
3-D woven (twisted yarn)	13.7 (4.8)	337 (7.7)	14.6 (3.4)	418 (4.2)

Zhao (2006) studied the compressive behaviour of 3-D woven fabrics and of 3-D woven fabrics with twisted yarns. Compressive properties such as modulus and strength were measured on the composites produced from these fabrics.

The measured compressive moduli for both composites were found to be lower than their respective tensile values (Table 4.4). This is attributed to the effect of fibre crimping, similar to the case when a bundle of wires are twisted together and then subjected to tensile and compressive loads; however, the compressive strengths for these composites are generally close to the corresponding tensile strengths. It was noticed that the specimens essentially failed in shear, with local yarn buckling and kink band formation. Such compressive failure modes are distinguished from those for composites reinforced with 2-D plain weave fabrics, in which failure associated with large-scale delamination and buckling is often observed. The bonding effect of the through-thickness yarns substantially decreased the possibility of delamination and limited the size of the failure zone.

4.5 Applications of multilayer woven fabrics

Multiple-layer woven textile fabrics are becoming one of the most important forms of reinforcement for composite materials because of their good resistance to delamination over the laminated reinforcements. Because of the 3-D integrated fibre assemblage, such structures are less prone to delamination and can offer high impact resistance.

- Multilayer woven fabrics are being used in a number of applications as composite reinforcements for defence, marine, automotive, aerospace and transportation industries. They also find applications in geotechnical engineering, the biomedical field and as protective clothing.
- In automotives, multilayer woven composites are used as drive shafts, side rails, doors, cross-members, oil pans, suspension arms, leaf springs,

wheels, quarter panels, trunk decks, hoods, hinges, transmission support, bumpers, seat frames and wheels.

- In the defence industry, 3-D woven fabric composites are ideal materials for lightweight mobile, easily transportable vehicles for tactical shelters, ballistic combat and logistic applications. They are ideal materials for aircraft and aerospace applications where high strength-to-weight ratio is required. For example, the strength of glass fibre composite is five times that of aluminium.
- In addition to structural body composites, composites are also used for interior parts such as overhead luggage compartments, sidewalls, ceilings, floors, galleys, partitions, cargo liners, etc.
- In aerospace applications, 40% of the composites are used for military applications. The applications include the wing box, forward fuselage, horizontal stabilizer, elevators, rudder, over-wing surfaces, etc.
- Three-dimensional multilayer woven fabric composites are used in space structures such as missiles, rockets and satellites. Space structures require low weight, high stiffness, a low coefficient of thermal expansion and dimensional stability, which are the main properties offered by the 3-D fabrics. The application areas of 3-D woven composites in missile systems include rocket motor cases, nozzles, skirts and interstage structures, control surfaces and guidance structural components. Structural components used in space include trusses, platforms, shells, pressure vessels and tanks.
- In the marine industry, 3-D multilayer woven fabrics are used for applications such as hull structures, fairwaters, sonar domes, antennas, floats, buoys, masts, spars, deckhouses, etc. They are also used in mine warfare vessels, tankers, trawlers, ferries, sonar domes, submarines, powerboats, racing yachts, pleasure boats, luxury yachts and laminated sailboats.
- In the medical field, multilayer woven fabrics are being used for manufacturing stents, prostheses, artificial joints and organs, implants, scaffolds, etc.
- The use of 3-D multilayer woven fabrics is also increasing in sporting goods. In particular, the graphite–boron and Kevlar–epoxy composites are used in golf carts, surf boards, hang-glider frames, javelins, hockey sticks, sail planes, sail shafts, fishing rods, snow and water skis, bows, arrows, tennis rackets, pole-vaulting poles, skateboards, bats, helmets, bicycle frames, canoes, catamarans, oars, paddles, etc.
- Other applications for multilayer woven fabrics include narrow-width webbing products where strength or abrasion resistance are desired: belting products for conveyors, dryers and paper machine clothing; filtration products; ballistic materials; ablatives; constant-thickness structural composites in which damage tolerance or through-thickness

> mechanical properties are essential; and biomedical applications that utilize the high compression strength afforded by the z-fibres. In all cases there are driving technical reasons that make the 3-D woven structure more desirable than a fabricated 2-D fabric alternative.

Recently, these materials have been finding increased usage in more commercial applications, particularly in marine structures and industrial components that are very cost sensitive. Due to the availability of heavyweight fabrics/reinforcements, and the subsequent reduction in lay-up labour, 3-D fabrics can reduce the cost of the finished composite structure.

4.6 Summary

This chapter summarizes the structure, manufacture and properties of multilayer woven fabrics. Three-dimensional multilayer woven (MLW) fabrics are becoming increasingly important owing to their excellent performance: permeability, compressibility, drapeability, ease of handling and ability to conform to complex shapes. Three-dimensional multilayer woven fabrics are textile structures having fibres oriented along the three directions of a unit cell. A 3-D fabric should have three or more yarns in the thickness direction in order to distinguish itself from a planar fibre assembly. Compared with 2-D woven fabrics, multilayer woven fabrics exhibit higher through-thickness and interlaminar properties because of their integrated structure in the presence of orthogonal and/or angle interlock constructions. They offer better processability and eliminate delamination of fabrics. These 3-D woven preforms with various architectures can be fabricated using different weaving methods. Multiwarp weaving methods are used for weaving orthogonal and/or angle interlocked multilayer woven fabrics.

MLW fabrics are finding wide application in composites, sports and the aerospace industry because of their superior properties compared to 2-D woven fabrics. Three-dimensional MLW fabric preforms offer many advantages over both 2-D fabrics and other categories of 3-D preforms. Other categories of 3-D preforms require extensive modifications to existing equipment or new equipment, whereas multilayer woven fabrics can be produced using existing textile manufacturing techniques on conventional equipment with few modifications.

Three-dimensional (3-D) orthogonal woven fabric composites have high in-plane tensile stiffness and strength in the warp and weft directions because there is no crimp in the warp and weft yarns in the in-plane direction. Furthermore, the composites have higher interlaminar shear strength than the laminates because the z-direction tows exist in the through-thickness direction.

4.7 References

Adanur S (1995), Textile structural composites, in *Wellington Sears Handbook of Industrial Textiles,* Technomic Publishing, Lancaster, PA, 231–271.

Baozhong Sun and Bohong Gu (2006), Shear behavior of 3D orthogonal woven fabric composites under high strain rates, *Journal of Reinforced Plastics and Composites*, **25**, 17, 1833–1845.

Chen X, Knox R T, McKenna D F and Mather R R (1992), Simulation of multiple layer woven textile structures for composite reinforcements, *Proc. First Int. Conf. on Intelligent Systems Engineering (Conf. Publ. No. 360)*, Edinburgh, UK, 104–110.

Goerner D (1989), Double, treble and four ply cloths, in *Woven Structure and Design, Part 2 Compound Structures,* BTTG Wira House, Leeds, UK, 16–58.

Gu H and Zhili Z (2002), Tensile behaviour of 3D woven composites by using different fabric structures, *Materials and Design*, **23**, 671–674.

Guo X, Li W, Gu B and Qiu Y (2007), Effect of strain rate on compression loading of 3D woven composites in three directions, *Pigment and Resin Technology*, **36**, 1, 39–43.

Hosur M V, Adya M, Vaidya U K, Mayer A and Jeelani S (2003), Effect of stitching and weave architecture on the high strain rate compression response of affordable woven carbon/epoxy composites, *Composite Structures*, **59**, 4, 507–523.

Khokar N and Peterson E (1998), 3D fabrics through the 'true' 3D-weaving process, *World Textile Congress on Industrial, Technical and High Performance Textiles,* University of Huddersfield, UK, 15–16 July, 175–181.

Ko F K (1989), Three-dimensional fabrics for composites, in *Textile Structural Composites* (ed. Tsu-Wei Chou and Frank K. Ko), Elsevier, New York, 129–171.

Mohamed M H (1990), Three-dimensional textiles, *American Scientist*, **78**, 6, 530–541.

Naik N K, Azad N M and Durga Prasad P (2002), Stress and failure analysis of 3D angle interlock woven composites, *Journal of Composite Materials*, **36**, 93–123.

Pastore C M and Cai Y J (1990), Applications of computer aided geometric modeling for textile structural composites, *Proc. 2nd Int. Conf. on Computer Aided Design in Composite Material Technology*, Brussels, Belgium, 25–27 April.

Potluri P, Porat I and Sharma S (2000), Three-dimensional weaving and moulding of textile composites, *Textiles Magazine*, Issue 4, 14–17.

Sun B Z, Gu B H and Ding X (2005a), Compressive behavior of 3D angle-interlock woven fabric composites at various strain rates, *Polymer Testing*, **24**, 4, 447–454.

Sun B Z, Yang L and Gu B H (2005b), Strain rate effect on four-step three-dimensional braided composite compressive behavior, *AIAA Journal*, **43**, 5, 994–999.

Sun B Z, Liu F and Gu B Z (2005c), Influence of strain rate on the uniaxial tensile behavior of 4-step 3D braided composites, *Composites, Part A*, **36**, 11, 1477–1485.

Watson W (1955), *Advanced Textile Design* (3rd edn), London: Longmans, Green.

Yi H L and Ding X (2004), Conventional approach on manufacturing 3D woven preforms used for composites, *Journal of Industrial Textiles*, **34**, 1, July, 39–50.

Youjiang Wang and Dongming Zhao (2006), Effect of fabric structures on the mechanical properties of 3-D textile composites, *Journal of Industrial Textiles*, **35**, 3, January, 239–256.

5

Tensile properties of multiaxial warp-knitted fabrics

Abstract: With the increasing application of multiaxial warp-knitted (MWK) reinforced composites in the automotive and other industries, the need to determine their mechanical properties, especially the tensile-extension behaviour, impact on the energy absorption capacity and damage tolerance properties, is becoming more and more important to ensure the stability and safety of the designed structures. The tensile properties of an MWK structure in a particular direction are governed by those of the yarns inserted in that direction. Many researchers have studied MWK fabrics, but very few have attempted to model their mechanical properties. In this chapter, a macroscopic modelling approach dealing with fabric structure under uniaxial tensile deformation is described to give an understanding of the tensile properties of MWK fabrics.

Key words: multiaxial warp-knitted (MWK) fabrics, tensile behaviour of MWK fabrics, modelling uniaxial tensile deformation, stress–strain relationship.

5.1 Introduction

Multiaxial warp-knitted (MWK) fabrics were formerly developed as substrates for flexible coated fabrics. They have an important role in the field of industrial applications due to their outstanding mechanical properties. At the beginning of the 1990s, MWK fabrics entered the field of rigid structural composites to produce aerospace-quality components, marine parts and automotive frames. As one of the 3-D composite preforms, MWK fabrics are attracting more and more interest due to their low production costs, high production efficiency, structural integrity, flexibility in design, high tear resistance and improved through-thickness strength. In addition to their good ease of handling and production economics, MWK structures also provide conformability to complex shapes, flexibility in the principal yarn directions and improved through-thickness strength.

Nowadays, MWK structures have a wide range of applications – from geotextiles, pneumatic materials and construction materials to automobiles, aerospace-quality components as well as vessel-body parts due to their desired mechanical properties, flexibity in design and low production cost (Kaufmann, 1991). In MWK structures, all the layers of insertion yarns are placed in perfect order and show the uniformity of the non-crimped parallel yarns. While the insertion yarns play a principal role in the plane reinforcement, the stitch yarns can provide the through-thickness reinforcement, and therefore produce a significant increase in damage tolerance, structural integrity and out-of-plane strength.

With the increasing application of MWK-reinforced composites in automotive and other industries, the need to determine their mechanical properties – especially their tensile-extension behaviour, impact on energy absorption capacity and damage tolerance properties – is becoming increasingly important in order to ensure the stability and safety of the designed structures.

MWK structures can be produced in a single knitting process. In addition to good ease of handling and production economics, MWK structures also conform to complex shapes and are flexible in the principal yarn directions. The mechanical properties of MWK structural composites, especially out-of-plane strength and impact behaviour, may be superior to those of conventional woven laminated composites due to the elimination of fibre crimp in the insertion yarns and to the presence of the through-thickness reinforcing stitch loops.

5.2 Tensile behaviour of multiaxial warp-knitted fabrics

Recent textile and composite literature reflects the growing interest in the strength properties of MWK-reinforced composite materials. The tensile performance of MWK fabrics has also been a much-discussed topic within the published literature. As MWK fabrics have more complex geometric architectures than woven and braided fabrics, most of the investigation has focused upon the general tensile properties (Dexter and Hasko, 1996). Only limited attention has been given to modelling the strength properties and knitting parameters of MWK fabrics, such as loop lengths, shapes and densities, that will allow a more detailed and quantitative analysis. However, a number of researchers have been working on this area and they have produced some meaningful results and conclusions.

Typical MWK fabrics have a complex structure in three dimensions to improve damage tolerance and reliability. All layers of the insertion yarns in a MWK structure are placed in an expected order and show the uniformity of the non-crimped parallel yarns. The insertion yarns play a principal role in the plane layer and with the increase of binding points, the symmetry of binding distribution and the increase of stitch yarn linear densities, the properties of MWK structures can be greatly improved. Therefore, MWK structure can provide a significant increase in damage tolerance, structural integrity and out-of-plane strength of the reinforced structure (Ko *et al.*, 1980, 1982, 1985, 1986).

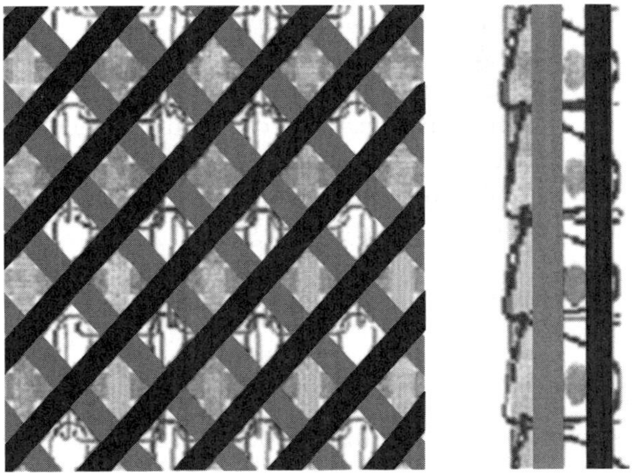

(a) Common MWK structure (b) Improved MWK structure

5.1 An improved MWK structure with double-loop pillar stitch.

Zhou Rongxing *et al.* (2004) studied the tensile properties of MWK structures produced with different basic stitches for composite reinforcement. In order to compare these structures, an experimental study of the tensile properties of MWK structures was carried out (as per Chinese National Standard GB7689.6-89 and GB/T7690-1989 respectively) using tricot stitch and double-loop stitch, as shown in Fig. 5.1. The results demonstrated that double-loop pillar stitches have better mechanical properties. The in-plane mechanical properties of an MWK structure in a test direction mainly depend on the properties of the insertion yarns in that direction. The tensile properties were better in the wale direction than in other directions, and the properties in the diagonal were the poorest. In addition, the authors deduced that the basic stitch has an obvious influence on the mechanical properties of the MWK structure.

Chen Nanliang (2002) has investigated the effect of the materials and structures of warp-knitted ground on the tensile strength of the composite reinforced with MWK fabrics. He reported that the material has a great influence on tensile strength: the tensile property of glass fibre used as warp-knitted ground is much better than that of polyester fibre in both the horizontal and vertical testing directions. In addition, different testing directions have different and varied effects on tensile properties. In the horizontal testing direction, the tensile property shows no obvious difference when using different warp-knitted binding stitches, i.e. plain tricot ground and cord stitch. In the vertical testing direction, however, tensile strength with a cord stitch ground is apparently better than with tricot.

Shen Wei (2002) tested and compared the tensile property of biaxial warp-knitted fabric with that of plain weave fabric in a study that is very meaningful for the understanding of the tensile characteristics of MWK fabrics. In this research, the relationship between the tensile strength of biaxial warp-knitted fabric and the density of the yarns was studied by means of regression analysis, and some characteristics and advantages of the biaxial warp-knitted fabric were also explored. It was observed that the tensile behaviour of biaxial warp-knitted fabric and plain woven fabric will show very different tensile strength curves. The potential of the yarn's tensile resistance can contribute more to the whole fabric's tensile performance, especially insertion yarns with high modulus. The tensile strength of biaxial warp-knitted fabric has a linear relationship with the insertion yarns along the tensile stress direction and the density of the insertion yarns with high tensile resistance. The tensile potential of yarns in biaxial warp-knitted fabric can be used more efficiently than in plain woven fabric, and insertion methods will cause different utilization of the yarn's strength.

The tensile behaviour of MWK fabrics has been demonstrated by Chen Nanliang and co-workers (2001, 2002). Their tests confirmed that the tensile stress, strain and elastic modulus of MWK fabric were superior to those of woven fabrics and that the tensile property depends on the structural parameters, such as the insertion methods of the binding yarns and the yarn's strength and density in the test direction. These conclusions are in agreement with the work of Shen Wei (2002). Although Shen Wei's conclusions were based on the testing of biaxial warp-knitted fabrics, the principle – especially the tensile mechanism – is very similar. In addition, Chen Nanliang suggested that existing test methods for tensile properties cannot at present meet the testing requirement for the diagonal yarn's exact contribution to the tensile properties of the whole fabric.

The interweave crimp of woven fabric is a factor influencing the tensile stiffness and strength of woven fabric, and its reinforced composites, because of the interlacement of warp and weft yarns. Contrary to the construction of woven fabrics, in a multiaxial, multilayer warp-knitted (MMWK) fabric the warp (0°), weft (90°) and bias ($\pm\theta$) yarns are held together by a chain or tricot stitch yarn through the thickness direction. There are no crimps of the warp, weft and bias yarns in MMWK fabrics. Hence, tensile stiffness and tensile strength will reach a maximum value compared with other kinds of fabrics. MMWK fabric-reinforced composites, like unidirectional composites, also have high values of tensile stiffness and strength along warp, weft and bias directions.

The deformation behaviour of MWK fabric has been studied under bending, tension, shear and friction, and the research revealed that biaxial and quadraxial fabrics demonstrated fairly isotropic behaviour in bending, tension and friction, but they displayed anisotropic deformation behaviour in shearing under low load (Lomov *et al.*, 2003). In a study of MWK fabrics, the tensile extension properties and deformation mechanisms of multiaxial non-crimp fabrics have been investigated (Kong *et al.*, 2004). This study evaluated the extension properties of a variety of biaxial and triaxial fabrics, consisting of continuous E-glass yarns stitched together with 0.14 mm diameter polyester thread. The results may be summarized as follows.

Figure 5.2 represents curves of load against gross longitudinal strain for the biaxial fabrics. The curves show a gradual rise in the load resistance until a strain of 0.15, above which the resistance to bias deformation increases rapidly for biaxial fabric type I. However, both types of biaxial fabric show a lower resistance to bias extension than plain woven fabric. Figure 5.3 shows the load–strain curves for the triaxial fabrics that were warp knitted with different line tensions on the stitch thread. It can be seen that the resistance to bias loading of the triaxial fabric increases with the line tension of the stitches, although in all cases the deformation load resistance was lower than for woven fabric.

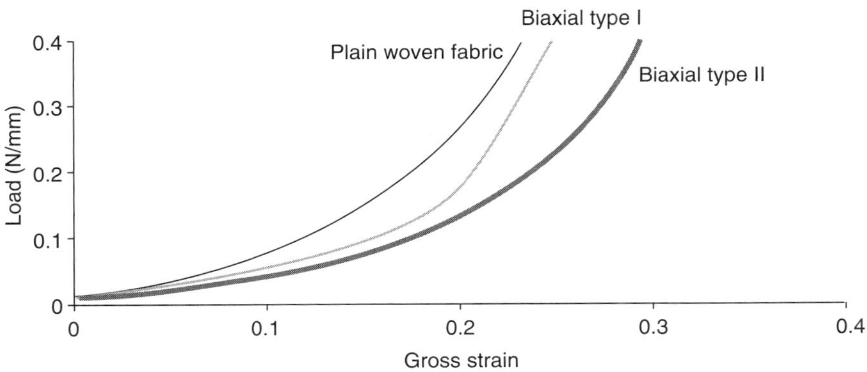

5.2 Gross bias extension curves for biaxial and plain woven fabrics.

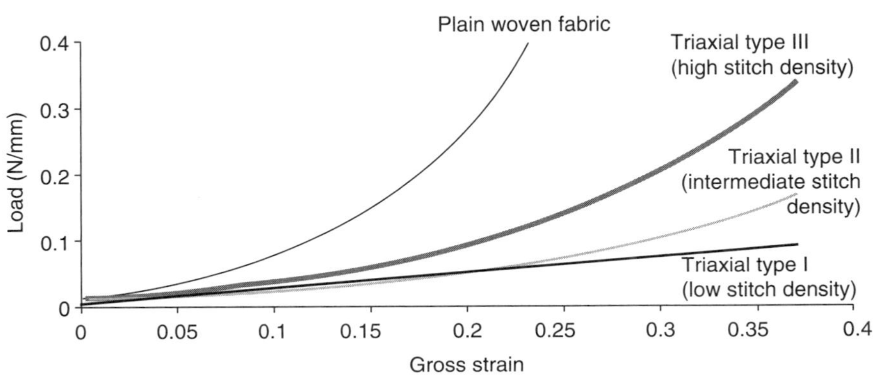

5.3 Bias extension curves for triaxial-type fabrics with different stitch-line tensions and for plain woven fabric.

In another interesting study (Pattyn *et al.*, 1999), the tensile extension properties of MWK fabrics and multilayered, multiaxial stitched preforms were compared. It was observed from the results that both types of fabrics exhibited a high in-plane performance. For stitched multiaxial fabrics, the elastic behaviour is more or less isotropic, while the strength behaviour is anisotropic (more pronounced for carbon). For glass-epoxy, the elastic performance is slightly better for the stitched multiaxial than for the regular MWK fabric, and there was no difference for carbon-epoxy. The results for strength and elongation properties are presented in Tables 5.1 and 5.2.

The mechanical properties of an improved MWK fabric structure (Fig. 5.4) for composite reinforcement were reported by Zhou Rongxing *et al.* (2004). The MWK structure with double-loop pillar stitch was used in this study and showed better mechanical properties than structures using other stitches. The tensile properties of different MWK structures in the weft and warp directions are presented in Fig. 5.4 and test result data are summarized in Table 5.3.

The above study showed that MWK structures with double-loop pillar stitch have better mechanical properties than the common MWK structure with tricot stitches. It can be seen that the breaking force for MWK structures with double-loop pillar stitch is about 5–7% higher than for MWK structures with tricot stitch, in both weft and warp directions. This result confirms that the MWK structure with double-loop pillar stitch exhibits better mechanical properties. However, the increased breaking force depends on the test direction. The increase in the weft direction was a little higher than that in the warp direction. This phenomenon was explained by the fact that even if the underlaps in the double-loop pillar stitches were more oriented in the weft and warp directions than in the tricot stitch, their contribution to the breaking forces in the weft direction was more important than in the warp direction. The results also showed that the stitch yarn linear density had an obvious influence on the breaking force. This is normal because the stitch yarn with higher linear density can undertake higher yarn-breaking force.

The effects of the basic stitches on the breaking elongation were not obvious, however. The differences in the breaking elongation of the improved MWK structure and the common MWK structure in all test directions were very small. This result indicates that the MWK structures are stable and their breaking elongations are independent of the basic loop structures.

Furthermore, the tensile properties of the MWK structures were visibly regular in different test directions. As illustrated in Fig. 5.4, the tensile breaking force of all the specimens in the warp yarn direction was a little higher than those in the weft yarn direction. There are two reasons for this.

Table 5.1 Elastic properties of multiaxial warp-knitted fabrics

Fabric	Angle	V_f	Young's modulus E (GPa)				Poisson's ratio ν			
			Exp.	Standard deviation	Theor.	Exp./theor.	Exp.	Standard deviation	Theor.	Exp./theor.
Glass-epoxy										
LIBA	0°	51.2%	16.2	0.6	17.63	92%	0.30	0.03	0.30	102%
	90°	51.2%	17.4	1.3	17.63	99%	0.33	0.01	0.30	110%
	67.5°	51.2%	17.6	0.5	17.42	101%	0.32	0.03	0.31	103%
Stitched multiaxial	0°	45.0%	16.4	0.3	15.63	105%	0.33	0.00	0.30	112%
	90°	45.0%	16.8	0.7	15.63	107%	0.35	0.02	0.30	116%
	67.5°	45.0%	15.2	0.1	15.44	99%	0.35	0.02	0.31	114%
Carbon-epoxy										
LIBA	0°	52.6%	41.7	1.3	42.34	99%	0.34	0.01	0.27	124%
	90°	52.6%	43.3	2.4	42.34	102%	0.32	0.05	0.27	117%
	67.5°	52.6%	43.9	1.1	40.99	107%	0.37	0.05	0.29	126%
Stitched multiaxial	0°	40.8%	34.6	1.7	33.25	104%	0.35	0.03	0.28	127%
	90°	40.8%	33.2	1.7	33.25	100%	0.34	0.02	0.28	123%
	67.5°	40.8%	34.4	1.1	32.22	107%	0.34	0.02	0.30	114%

Source: adapted from Pattyn *et al.*, 1999.

Table 5.2 Strength properties of multiaxial warp-knitted fabrics

Fabric	Angle	V_f	*S* (MPa)			
			Exp.	Standard deviation	Theor. (LPF)	Exp./theor.
Glass-epoxy						
LIBA	0°	51.2%	319	15	334	96%
	90°	51.2%	359	13	334	108%
	67.5°	51.2%	301	4	330	91%
Stitched multiaxial	0°	45.0%	320	5	256	125%
	90°	45.0%	361	15	256	141%
	67.5°	45.0%	329	15	175	188%
Carbon-epoxy						
LIBA	0°	52.6%	548	27	585	94%
	90°	52.6%	573	30	585	98%
	67.5°	52.6%	354	15	560	63%
Stitched multiaxial	0°	40.8%	547	25	476	115%
	90°	40.8%	538	16	476	113%
	67.5°	40.8%	394	20	468	84%

Source: adapted from Pattyn *et al.*, 1999.

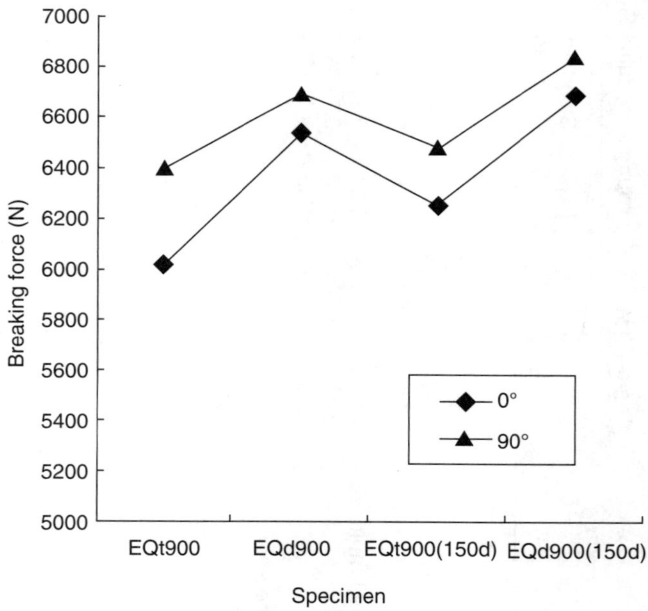

5.4 Tensile properties of different MWK structures in the weft and warp directions.

Table 5.3 Tensile data on MWK fabrics

	Specimen	Stitch yarn, dtex			
		75		150	
		Testing direction		Testing direction	
		0°	90°	0°	90°
Breaking force, N	EQt	6217.6	6402.7	6258.2	6486.5
	EQd	6538.1	6691.2	6693.1	6849.4
Increase of breaking force, %	–	5.15%	4.51%	6.95%	5.59%
Breaking elongation, mm	EQt	7.49	7.12	6.98	7.11
	EQd	7.58	7.71	6.20	6.14

Source: adapted from Zhou Rongxing *et al.*, 2004.

Firstly, the yarn densities of the insertion yarns in different directions are slightly different due to the setup requirements of the knitting machine; and secondly, the weft yarns are subjected to more damage than the warp yarns when the needles go through the fabric during the knitting process.

In addition, the tensile strength of the improved MWK structure depends on the direction. Improvement in the weft direction was better than that in the warp direction. The basic stitches had no influence on the breaking strength and elongation. The tensile properties of the MWK structure in a direction are governed by those of the yarns inserted in that direction.

5.3 Modelling tensile properties of multiaxial warp-knitted fabrics

Unlike traditional woven fabrics, MWK fabric does not possess an interlaced structure. Each inserting system has a series of uncrimped parallel yarns, which account for some intrinsic advantages of these fabrics, such as fast response to stress, which means a higher initial Young's modulus and a high percentage of the usable yarn potential (Chou and Ko, 1989).

Many researchers have studied MWK fabrics, but very few have attempted to model their mechanical properties. A meso-modelling approach for coupled drape and failure simulation of a biaxial non-crimp fabric with tricot stitch was carried by Creech and Pickett (2006).

There are probably three major reasons for the little use of modelling approaches. Firstly, since the inserting yarns are not interlaced with one another as in woven structures, slippage between yarns occurs very easily and is always accompanied by yarn rotation. This will obviously make any analysis complex and difficult. Particularly when the force method (Kawabata *et al.*, 1973) is used, the stress distribution on a single inserting yarn in the deformed state is nearly impossible to determine if both the frictional force and the interaction between the inserting yarns and stitching loops are considered. In addition, Kilby's trellis method (Kilby, 1963) cannot be applied here since the assumptions made in his paper are impossible for MWK fabrics.

Secondly, incorporating a stitching yarn system such as chain or tricot provides MWK fabrics with the properties of both woven and knitted fabrics to some degree. It is thus very difficult to measure the effect of the stitching yarn system. Thirdly, the non-uniform deformation of different geometric unit cells, meaning that different unit cells will deform differently, is an unavoidable problem in modelling the mechanical properties of MWK fabrics. This is illustrated in Fig. 5.5 (after deformation, the line p–p is changed into p*–p*). The deformations of cells 1, 2 and 3 are different, so there is extra complexity with the unit cell method (Du and Ko, 1996).

In this chapter, a macroscopic approach is developed, dealing with fabric structure under uniaxial tensile deformation, in order to tackle the above-mentioned difficulties. A model for uniaxial tensile deformation is obtained, which is justified by Instron 4466 tensile testing. In addition, a formula for calculating the tensile modulus of the fabric in any direction is presented.

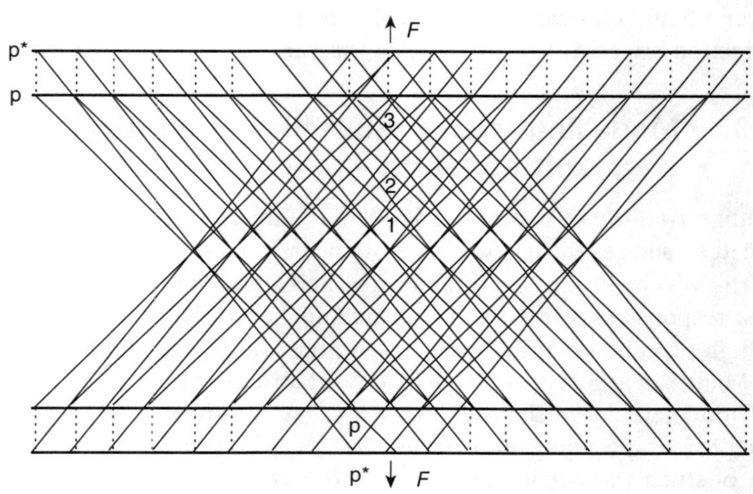

5.5 Different deformations of different unit cells in an MWK fabric under uniaxial stretch.

5.3.1 Theory

The modelling of the tensile stress–strain relationship of MWK fabrics is illustrated in this section. The following notations are valid throughout this chapter:

Subscript i ($i = 1$–4), where 1 = warp yarn, 2 = weft yarn, 3 = bias yarn ($+\theta$) and 4 = bias yarn ($-\theta$)
E = tensile modulus of the fabric, kg/cm
E_L = tensile modulus of a single inserting yarn, kg/cm
F = tensile force exerted on unit length of fabric, kg/cm
F_{is} = resultant of forces along the axis of a single inserting yarn ($i = 1$–4), kg
F_{isR} = tensile force exerted on a single inserting yarn in the stretching direction ($i = 1$–4), kg
F_L = tensile force exerted on unit length of tricot fabric, kg/cm
$f_i(\ )$ = function of the stress–strain relationship of a single inserting yarn ($i = 1$–4)
f_{is} = resultant frictional forces along the axis of a single inserting yarn ($i = 1$–4), kg
L_0 = sample length in the undeformed state, cm
L_0' = fabric length in the deformed state, cm
L_i = yarn length in the undeformed state, cm
L_i' = yarn length in the deformed state, cm
ΔL = elongation of the fabric after deformation, cm
n_i = number of yarns per unit length of fabric along the direction perpendicular to the yarn's axis, /cm
n_i' = number of yarns per unit length of fabric along the direction perpendicular to the stretching direction, /cm
W_0 = sample width, cm
ε = strain of the fabric after deformation, %
ε_i = strain of inserting yarns after deformation, %
θ = angle between the warp yarn axis and the tensile force direction along the clockwise direction, arc
$\pm\theta_0$ = angle between the bias yarn axis and the warp yarn axis, arc (the plus sign represents the angle from the warp yarn axis to the bias yarn axis along the clockwise direction, and minus is along the counter-clockwise direction)
$\Delta\theta_i$ = angle of an inserting yarn's rotation after deformation, arc

5.3.2 Basic assumptions and approximations

In order to proceed, it is necessary to make four assumptions to simplify the analysis. First, the fabric is in a small planar strain state; second, the yarns in each inserting system remain straight and parallel to one another, both before and after deformation; third, $\Delta\theta_i$ values are small enough to permit the neglect of the second-order and higher terms when expanding the related functions into Taylor's series; and fourth, one can neglect the possible waviness of inserting yarns.

5.3.3 Modelling of uniaxial tensile deformation

The model of a tensile stress–strain relationship is based on the simplified fabric structure in Fig. 5.6. A macroscopic approach, rather than a unit cell structure, is used for the analysis. For clarity, stitching loops in MWK fabrics are also omitted. In Fig. 5.6, the bold lines represent inserting yarns in the deformed state, while the fine lines represent those in the undeformed state.

Relationships between ε and ε_i

According to Fig. 5.6, it is easy to obtain the following equation:

$$\varepsilon = \frac{\frac{1}{2}\Delta L + \frac{1}{2}\Delta L}{L_0} = \Delta L / L_0 \tag{5.1}$$

In order to determine the relationship between ε and ε_i (i = 1–4), we take the warp inserting yarn as an example, as shown in Fig. 5.7. The relationship between ε and ε_i belongs to the scope of a plane-strain transformation (from

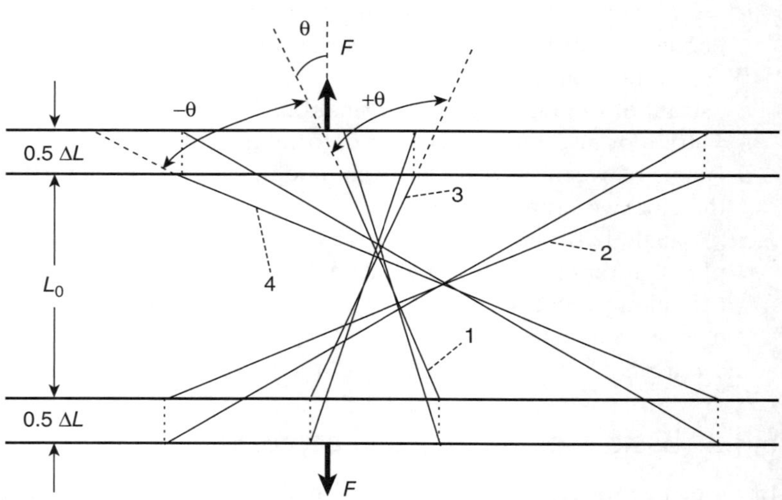

5.6 Uniaxial tensile deformation of an MWK fabric.

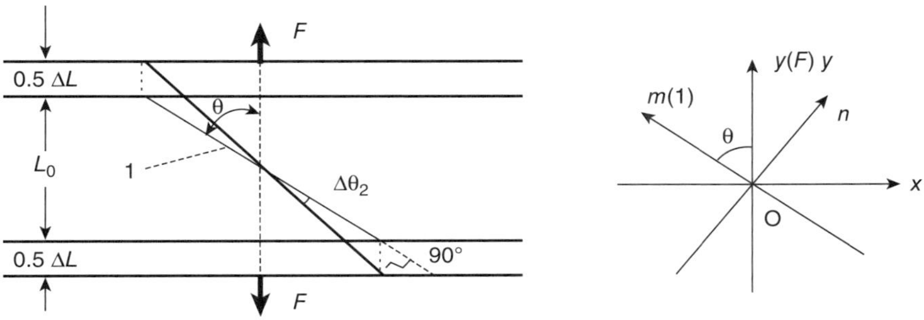

5.7 Tensile deformation of a single warp inserting yarn.

off-axis to on-axis) in elasticity. The formula for transformation is as follows (see the coordinates in Fig. 5.7):

$$\begin{bmatrix} \varepsilon_m \\ \varepsilon_n \\ \frac{1}{2}\gamma_{mn} \end{bmatrix} = [T] \begin{bmatrix} \varepsilon_x \\ \varepsilon_y \\ \frac{1}{2}\gamma_{xy} \end{bmatrix} \quad 5.2$$

where γ is a constant and

$$[T] = \begin{bmatrix} \cos^2\theta & \sin^2\theta & 2\sin\theta\cos\theta \\ \sin^2\theta & \cos^2\theta & -2\sin\theta\cos\theta \\ -\sin\theta\cos\theta & \sin\theta\cos\theta & \cos^2\theta - \sin^2\theta \end{bmatrix}$$

For uniaxial tensile deformation (low strain), Equation 5.2 can be simplified into Equation 5.3, in which the Poisson's ratio and shear strain are ignored both off-axis and on-axis, and only the tensile strain along the yarn's axis is particularly emphasized:

$$\begin{bmatrix} \varepsilon_m \\ 0 \\ 0 \end{bmatrix} = [T] \begin{bmatrix} \varepsilon_x \\ 0 \\ 0 \end{bmatrix} \quad 5.3$$

According to Fig. 5.8 and Equation 5.3, we can obtain Equation 5.4:

$$\varepsilon_1 = \varepsilon\cos^2\theta \quad 5.4$$

Similarly, we can work out the following equations:

$$\varepsilon_2 = \varepsilon\cos^2(\theta - \pi/2) \quad 5.5$$

$$\varepsilon_3 = \varepsilon\cos^2(\theta_0 - \theta) \quad 5.6$$

$$\varepsilon_4 = \varepsilon\cos^2(\theta_0 + \theta) \quad 5.7$$

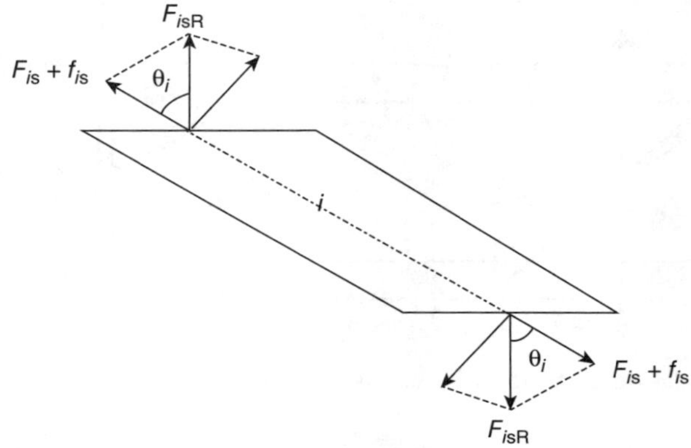

5.8 Forces exerted on a single inserting yarn.

Then Equations 5.4–5.7 can be written into one equation:

$$\varepsilon_i = \varepsilon \cos^2 \theta_i \quad (i = 1\text{–}4) \qquad 5.8$$

The stress–strain relationship of a multiaxial warp-knitted fabric under uniaxial stretch

When an MWK fabric is subjected to a uniaxial stretch, the forces exerted on a single inserting yarn are illustrated in Fig. 5.8. Here, we limit our consideration to the resultant force along the neutral line of a single yarn. From Fig. 5.9, we can obtain Equation 5.9:

$$F_{is} = F_{isR} \cos \ {}_i - f_{is}$$
$$F_{isR} = (F_{is} + f_{is})/\cos \ {}_i \quad (i = 1\text{–}4) \qquad 5.9$$

Through a simple geometrical relationship, it is easy to obtain Equation 5.10:

$$n_i' = n_i \cos \theta_i \ (i = 1\text{–}4) \qquad 5.10$$

The load–elongation relationship of a single inserting yarn can be represented by Equation 5.11:

$$F_{is} = f_i(\varepsilon_i) \ (i = 1\text{–}4) \qquad 5.11$$

From Figs 5.7 and 5.9, we can obtain Equation 5.12:

$$F = \frac{1}{W_0} \sum_{i=1}^{4} n_i' F_{isR}(W_0 - L_0 \tan \theta_i) + F_L(\varepsilon) \qquad 5.12$$

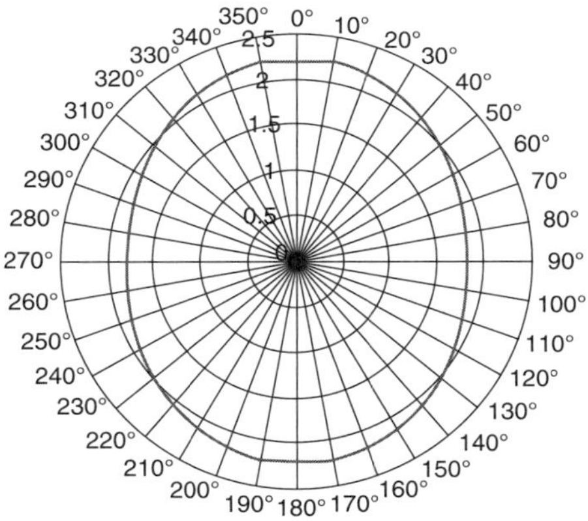

(a) $\theta_0 = 38° * 3.1415927/180°$

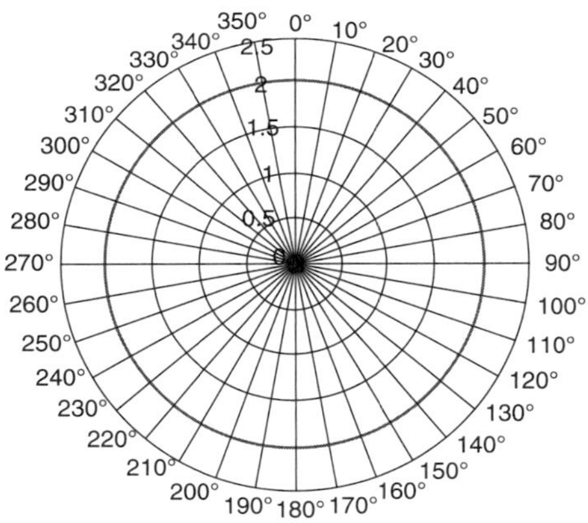

(b) $\theta_0 = 45° * 3.1415927/180°$

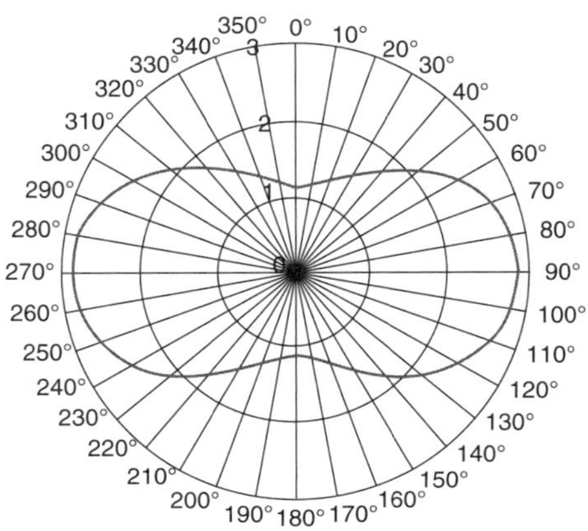

(c) $\theta_0 = 75° * 3.1415927/180°$

5.9 Tensile moduli in different directions of an MWK fabric.

We then apply Equations 5.9, 5.10 and 5.11 to Equation 5.12, in order to obtain Equation 5.13:

$$F = \frac{1}{W_0} \sum_{i=1}^{4} n_i [f_i(\varepsilon_i) + f_{is}(\varepsilon_i)](W_0 - L_0 \tan \theta_i) + F_L(\varepsilon) \quad 5.13$$

In Equation 5.13, n_i and ε_i stand for the structural parameters of an MWK fabric; $f_i(\ )$ and $F_L(\ )$ stand for the inserting yarns' mechanical properties and the behaviour of the stitching system respectively; and $f_{is}(\ )$ stands for the interactions between inserting yarn systems. On the right-hand side of Equation 5.13, all the terms can be determined from simple tests, except for $f_{is}(\ )$, which is very difficult to evaluate, even though the frictional coefficient between glass fibre yarns can be obtained.

Formulation of the tensile modulus

Equation 5.8 can be rewritten as Equation 5.14 when $W_0 >> L_0$, as follows:

$$F = \sum_{i=1}^{4} n_i [f_i(\varepsilon_i) + f_{is}(\varepsilon_i)] + F_L(\varepsilon) \quad 5.14$$

Generally speaking, when the fabric is only subjected to a small planar strain, the terms $f_{is}(\ )$ and $F_L(\ )$ may be neglected. If the four inserting yarn systems have exactly the same density (n) and tensile modulus (E_L), and the terms $f_{is}(\ )$ and $F_L(\ )$ are neglected, then Equation 5.14 can be rewritten as Equation 5.15:

$$F = nE_L \sum_{i=1}^{4} \varepsilon_i \quad 5.15$$

We can then substitute Equation 5.8 in 5.15 to obtain Equation 5.16:

$$F = nE_L \varepsilon \sum_{i=1}^{4} \cos^2 \theta_i \quad 5.16$$

From Equation 5.16, we can obtain Equation 5.17:

$$E = nE_L \sum_{i=1}^{4} \cos^2 \theta_i \quad 5.17$$

Substituting Equations 5.4–5.7 into Equation 5.17, and letting $dE/d\theta = 0$, we can get θ values for E_{max} (maximum) and E_{min} (minimum):

$$\sin 2\theta \cos 2\theta_0 = 0 \quad 5.18$$

$$E = \begin{cases} E_{max}, \text{ when } \quad \theta = \begin{cases} 0, & \pi/6 < \theta_0 < \pi/4 \\ \pi/2, & \pi/4 < \theta_0 < \pi/2 \end{cases} \\ E_{min}, \text{ when } \quad \theta = \begin{cases} 0, & \pi/4 < \theta_0 < \pi/2 \\ \pi/2, & \pi/6 < \theta_0 < \pi/4 \end{cases} \end{cases} \quad 5.19$$

$$E = 2nE_L \ (\theta_0 = \pi/4) \qquad 5.20$$

According to Equation 5.20, we can conclude that the fabric is nearly isotropic when $\theta_0 = \pi/4$. Three polar diagrams for Equation 5.17 are given in Fig. 5.9, which will be of great value in the design and application of MWK fabrics. Note that the real value for the tensile modulus is equal to the product of nE_L and the radius coordinate value.

5.4 Experimental methods and validation

The specifications of the samples used in the present research are listed in Table 5.4. The inserting yarns are all glass filaments and the tricot yarns are polyester filaments. The sample size is 3 cm (L_0) × 6 cm (W_0). Uniaxial tensile testing of samples is done along four directions (warp, weft and bias ±45°) on an Instron 4466. Single yarn load–elongation curves are obtained as shown in Fig. 5.10.

Table 5.4 Specifications of samples ($\theta_0 = \pi/4$)

Direction	Tex	n_i
Warp – 1	900	4.6
Weft – 2	70	2.5
Bias (+45°) – 3	300	6.8
Bias (–45°) – 4	300	6.8
Tricot loops – 5	9	5

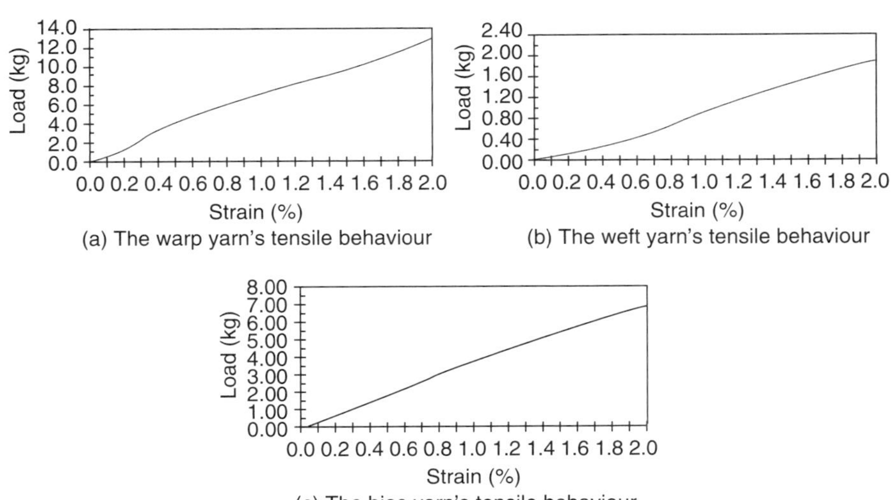

5.10 Tensile behaviour of an inserting yarn.

5.4.1 Validation of the model

In order to verify the accuracy of the model established above, comparison was made between the theoretical and tested results. In practical calculation, the effective width of sample for every inserting system should be $W_0 - L_0\tan\theta_i$, since the aspect ratio of the sample is not infinite. In Tables 5.5–5.7, $f_{is}(\varepsilon_i)$ and $F_L(\varepsilon)$ are neglected, which proves to be acceptable. The theoretical calculations are listed in Tables 5.5–5.7 where the input data include ε and f_{is} and output results of the calculation are ε_i and F^*.

Three sets of stress–strain curves are presented in Fig. 5.11, according to the theoretical calculations in Tables 5.5–5.7. The experimental curves are also shown for the sake of comparison. For simplicity, the curves for $\theta = -\pi/4$ are not given since they are basically the same as those for $\theta = +\pi/4$. Obviously, the two sets of results are in good agreement. According to Fig. 5.11, it can also be noted that the experimental curves are always below the theoretical ones. This is different from the relationship indicated by

Table 5.5 Theoretical calculations for warp directional stretch ($\theta = 0$)

ε	1		2		3		4		F^* (see note)
	ε_1	n_1f_{1s}	ε_2	n_2f_{2s}	ε_3	n_3f_{3s}	ε_4	n_4f_{4s}	
0	0	0.00	0	0	0	0.00	0	0.00	0.00
0.001	0.001	1.84	0	0	0.0005	0.82	0.0005	0.82	2.66
0.002	0.002	5.52	0	0	0.0010	1.36	0.0010	1.36	6.88
0.003	0.003	9.66	0	0	0.0015	2.45	0.0015	2.45	12.11
0.004	0.004	14.26	0	0	0.0020	3.54	0.0020	3.54	17.80
0.005	0.005	17.94	0	0	0.0025	4.90	0.0025	4.90	22.84
0.006	0.006	21.62	0	0	0.0030	6.53	0.0030	6.53	28.15
0.007	0.007	24.84	0	0	0.0035	8.02	0.0035	8.02	32.86
0.008	0.008	28.06	0	0	0.0040	9.32	0.0040	9.32	37.38
0.009	0.009	30.36	0	0	0.0045	11.02	0.0045	11.02	41.38
0.010	0.010	32.66	0	0	0.0050	12.24	0.0050	12.24	44.90
0.011	0.011	35.42	0	0	0.0055	13.53	0.0055	13.53	48.95
0.012	0.012	36.80	0	0	0.0060	15.03	0.0060	15.03	51.83
0.013	0.013	39.10	0	0	0.0065	15.98	0.0065	15.98	55.08
0.014	0.014	41.86	0	0	0.0070	17.61	0.0070	17.61	59.47
0.015	0.015	44.16	0	0	0.0075	18.56	0.0075	18.56	62.72
0.016	0.016	46.92	0	0	0.0080	19.86	0.0080	19.86	66.78
0.017	0.017	50.14	0	0	0.0085	21.42	0.0085	21.42	71.56
0.018	0.018	52.90	0	0	0.0090	22.85	0.0090	22.85	75.75
0.019	0.019	56.12	0	0	0.0095	24.21	0.0095	24.21	80.33
0.020	0.020	58.88	0	0	0.0100	25.30	0.0100	25.30	84.18

Note: $F^* = \dfrac{1}{W_0}\sum_{i=1}^{4} n_i f_i(\varepsilon_i)(W_0 - L_0 \tan\theta_i)$

Table 5.6 Theoretical calculations for weft directional stretch ($\theta = \pi/2$)

ε	1		2		3		4		F^* (see note)
	ε_1	$n_1 f_{1s}$	ε_2	$n_2 f_{2s}$	ε_3	$n_3 f_{3s}$	ε_4	$n_4 f_{4s}$	
0	0	0	0	0.00	0	0.00	0	0.00	0.00
0.001	0	0	0.001	0.15	0.0005	0.82	0.0005	0.82	0.97
0.002	0	0	0.002	0.25	0.0010	1.36	0.0010	1.36	1.61
0.003	0	0	0.003	0.35	0.0015	2.45	0.0015	2.45	2.80
0.004	0	0	0.004	0.55	0.0020	3.54	0.0020	3.54	4.09
0.005	0	0	0.005	0.78	0.0025	4.90	0.0025	4.90	5.67
0.006	0	0	0.006	1.05	0.0030	6.53	0.0030	6.53	7.58
0.007	0	0	0.007	1.33	0.0035	8.02	0.0035	8.02	9.35
0.008	0	0	0.008	1.65	0.0040	9.32	0.0040	9.32	10.97
0.009	0	0	0.009	1.90	0.0045	11.02	0.0045	11.02	12.92
0.010	0	0	0.010	2.20	0.0050	12.24	0.0050	12.24	14.44
0.011	0	0	0.011	2.53	0.0055	13.53	0.0055	13.53	16.06
0.012	0	0	0.012	2.80	0.0060	15.03	0.0060	15.03	17.83
0.013	0	0	0.013	3.08	0.0065	15.98	0.0065	15.98	19.06
0.014	0	0	0.014	3.35	0.0070	17.61	0.0070	17.61	20.96
0.015	0	0	0.015	3.60	0.0075	18.56	0.0075	18.56	22.16
0.016	0	0	0.016	3.88	0.0080	19.86	0.0080	19.86	23.73
0.017	0	0	0.017	4.13	0.0085	21.42	0.0085	21.42	25.55
0.018	0	0	0.018	4.40	0.0090	22.85	0.0090	22.85	27.25
0.019	0	0	0.019	4.68	0.0095	24.21	0.0095	24.21	28.88
0.020	0	0	0.020	4.88	0.0100	25.30	0.0100	25.30	30.17

Note: $F^* = \frac{1}{W_0} \sum_{i=1}^{4} n_i f_i(\varepsilon_i)(W_0 - L_0 \tan \theta_i)$

Equation 5.13, where the solid (experimental) curves should be higher than the broken (theoretical) ones when $f_{is}(\ )$ and $F_L(\ )$ are ignored. There are three possible reasons for this discrepancy: firstly, when the sample is clamped for testing, it does not stay in a perfect straight state; secondly, the inserting yarns do not deform uniformly even in the same system; and lastly, slippage may occur during stretch.

5.5 Conclusions

There are many difficulties in modelling the mechanical properties of MWK fabrics. This chapter presents a simplified method of dealing with deformation using the macroscopic structure of MWK fabrics, in which a uniaxial tensile model is established and a formula worked out for calculating tensile moduli in any direction. Comparison of the model and experimental results demonstrates that the model is justified, and the results obtained from the theoretical calculation and experiments show good agreement with each other.

Table 5.7 Theoretical calculations for bias directional stretch ($\theta = +\pi/4$)

ε	1		2		3		4		F^* (see note)
	ε_1	n_1f_{1s}	ε_2	n_2f_{2s}	ε_3	n_3f_{3s}	ε_4	n_4f_{4s}	
0	0	0.00	0	0.00	0	0.00	0	0	0.00
0.001	0.0005	1.15	0.0005	0.08	0.001	1.36	0	0	1.97
0.002	0.0010	1.84	0.0010	0.15	0.002	3.54	0	0	4.53
0.003	0.0015	3.91	0.0015	0.23	0.003	6.53	0	0	8.60
0.004	0.0020	6.30	0.0020	0.25	0.004	9.32	0	0	12.59
0.005	0.0025	8.05	0.0025	0.28	0.005	12.24	0	0	16.40
0.006	0.0030	9.66	0.0030	0.35	0.006	15.03	0	0	20.03
0.007	0.0035	12.19	0.0035	0.43	0.007	17.61	0	0	23.92
0.008	0.0040	14.26	0.0040	0.55	0.008	19.86	0	0	27.26
0.009	0.0045	16.01	0.0045	0.70	0.009	22.85	0	0	31.20
0.010	0.0050	17.94	0.0050	0.83	0.010	25.30	0	0	34.68
0.011	0.0055	19.55	0.0055	0.95	0.011	27.88	0	0	38.13
0.012	0.0060	21.62	0.0060	1.05	0.012	30.26	0	0	41.60
0.013	0.0065	23.46	0.0065	1.20	0.013	32.64	0	0	44.97
0.014	0.0070	24.84	0.0070	1.33	0.014	35.70	0	0	48.78
0.015	0.0075	26.68	0.0075	1.48	0.015	38.28	0	0	52.36
0.016	0.0080	28.06	0.0080	1.65	0.016	40.53	0	0	55.38
0.017	0.0085	29.50	0.0085	1.78	0.017	42.36	0	0	58.00
0.018	0.0090	30.36	0.0090	1.90	0.018	43.86	0	0	59.99
0.019	0.0095	31.51	0.0095	2.05	0.019	45.36	0	0	62.14
0.020	0.0100	32.66	0.0100	2.20	0.020	46.78	0	0	64.21

Note: $F^* = \frac{1}{W_0}\sum_{i=1}^{4} n_i f_i(\varepsilon_i)(W_0 - L_0 \tan\theta_i)$

On the basis of the foregoing analysis, it may be concluded that the deformation of different unit cells is quite different, as is shown in Fig. 5.6. As a result, the unit cell approach (Du and Ko, 1996) is difficult to apply when modelling the tensile properties of MWK fabrics. The basic assumption of Kilby's trellis method (Kilby, 1963) is that the woven yarns pivot together at the intersections. This cannot be applied to MWK fabrics, where the inserting yarns tend to slip very easily and this is always accompanied by the rotation of the yarn.

The force method often begins with a chosen unit cell, as in Kawabata's study (Kawabata *et al.*, 1973). However, it is very difficult – and indeed nearly impossible – to determine the forces exerted on a single inserting yarn when an MWK fabric is under tensile force, since this will involve many unexpected force moments (e.g. the frictional force moment), as well as many additional forces, such as frictional and shear forces. Thus, the equilibrium equation is too complex to solve easily.

According to the results obtained, the macroscopic approach that deals with fabric structure under tensile deformation is workable. It is reasonable

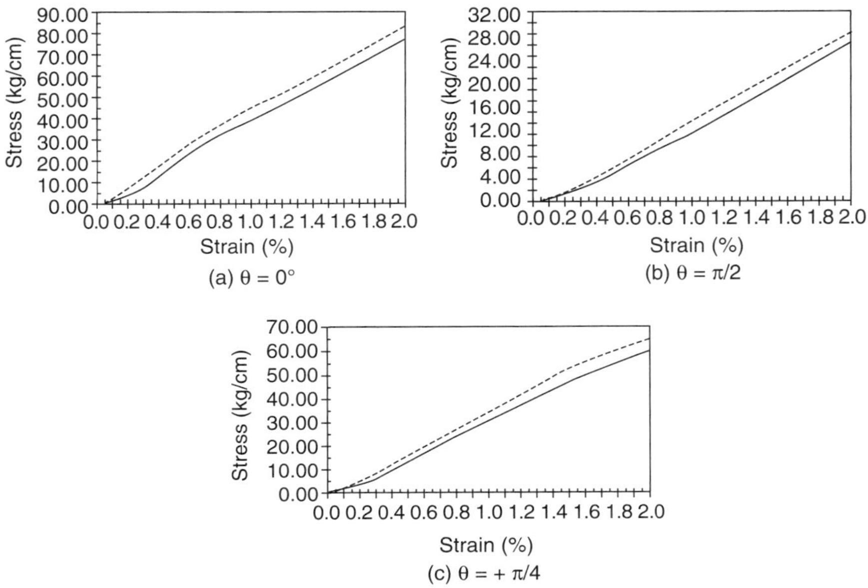

5.11 The stress–strain curves of MWK fabrics (the broken line is from the model, the solid line from experimental).

to neglect the effect of frictional force and the stitching system in predicting the uniaxial in-plane tensile properties of MWK fabrics. The model for the uniaxial stress–strain relationship and the formula for tensile modulus are of great value in designing and engineering applications for MWK fabrics.

5.6 References

Chen Nanliang (2001), Research on the tensile property of the composite reinforced with multiaxial warp-knitted fabrics (in Chinese), *Journal of Donghua University*, **27**, 2, April, 99–101.

Chen Nanliang (2002), Effect of the materials and structures of the warp-knitted ground on the tensile strength of the composite reinforced with multiaxial warp-knitted fabrics (in Chinese), *Journal of Donghua University*, **28**, 1, February, 105–106.

Chou T-W and Ko F K (eds) (1989), *Textile Structural Composites* 3, Elsevier, New York, 29–169.

Creech G and Pickett A K (2006), Meso-modelling of non-crimp fabric composites for coupled drape and failure analysis, *Journal of Material Science*, **41**, 6725–6736.

Dexter H B and Hasko G H (1996), Mechanical properties and damage tolerance of multiaxial warp-knit composites, *Composites Science and Technology*, **56**, 3, 367–380.

Du G-W and Ko F (1996), Analysis of multiaxial warp-knit preforms for composite reinforcement, *Composites Science and Technology*, **56**, 3, 253–260.

Kaufmann J R (1991), Industrial applications of multiaxial warp knit composites, Chapter 5 in *High-Tech Fibrous Materials* (ed. Tyrone L Vigo and Albin F Turbak), American Chemical Society, Washington, DC, 81–89.

Kawabata S, Niwa M and Kawai H (1973), The finite-deformation theory of plain weave fabrics, Part I: The biaxial-deformation theory; Part II: The uniaxial-deformation theory; Part III: The shear-deformation theory, *Journal of Textile Institute*, **64**, 21, 47, 62.

Kilby W F (1963), Planar stress–strain relationships in woven fabrics, *Journal of the Textile Institute*, **54**, T9–27.

Ko F K, Bruner J, Pastore A and Scardino F (1980), Development of multi-bar weft insertion warp-knit fabric for industrial applications, *ASME Paper No. 90-TEXT-7*, October.

Ko F K, Krauland K and Scardino F (1982), Weft insertion warp-knit for hybrid composites, in *Progress in Science and Engineering of Composites, ICCM-V, Fourth International Conference on Composites*, 982–987.

Ko F K, Fang P and Pastore C (1985), Multilayer multidirectional warp-knit fabrics for industrial applications, *Journal of Industrial Fabrics*, **4**, 2, 4–12.

Ko F K, Pastore C, Yang J M and Chou T W (1986), Structure and properties of multilayer multidirectional warp-knit fabric reinforced composites, in *Proc. 3rd US–Japan Conf. on Composites*, Tokyo, 21–28.

Kong H, Mouritz A P and Paton R (2004), Tensile extension properties and deformation mechanisms of multiaxial non-crimp fabrics, *Composite Structures*, **66**, 249–259.

Lomov S V, Verpoest I, Barburski M and Laperre J (2003), Carbon composites based on multiaxial multiply stitched preforms, Part 2. KES-F Characterisation of the deformability of the preforms at low loads, *Composites, Part A*, **34**, 359–370.

Pattyn H, Verpoest I, Ivens J and Villalon E (1999), Comparison of warp-knitted and stitched non-crimp fabrics, *Duracosys 99, Proc. 4th Int. Conf. on Durability Analysis of Composite Systems*, Brussels, Belgium, 11–14 July, 435–440.

Shen Wei (2002), Research on tensile property of bi-axial warp-knitted structure (in Chinese), *Journal of Donghua University*, **28**, 6, December, 105–110.

Zhou Rongxing, Hu Hong, Chen Nanliang and Feng Xunwei (2004), A study on the tensile properties of the MWK structures for composite reinforcement, *Journal of Donghua University (Eng. Ed.)*, **21**, 6, 121–123.

6
Bending properties of multiaxial warp-knitted fabrics

Abstract: The bending properties of fabrics govern many aspects of fabric performance, such as hand and drape, and they are an essential part of the complex fabric deformation analysis. To understand the behaviour of multiaxial warp-knitted (MWK) fabrics in general under bending is very important, since their functional properties are closely related to their mechanical properties, such as bending, shear and tensile. The main concern in this chapter is to establish a predictive model for assessing the bending behaviour of MWK fabrics. In addition, an elaborate description and interpretation of the bending properties of MWK fabrics are presented, based on many bending hysteresis curves obtained from KES-FB-2.

Key words: multiaxial warp-knitted (MWK) fabrics, bending behaviour of MWK fabrics, modelling bending behaviour of MWK fabrics, KES-FB-2, bending hysteresis.

6.1 Introduction

The bending properties of fabrics govern many aspects of fabric performance, such as hand and drape, and they are an essential part of the complex fabric deformation analysis. Like other mechanical properties, the bending behaviour of a woven fabric directly influences its performance. It is well known that fabric bending rigidity, along with its shear rigidity, determines the drapeability of a fabric. Research results show that fabric bending rigidity is also an important contributor to the fabric's formability, handle, buckling behaviour, wrinkle resistance and crease resistance. In the case of industrial fabrics, such as air-supported structures or fabric-reinforced flexible-composite conveyor belts, the bending behaviour of the fabrics is critically important.

Thus the bending of woven fabrics has received considerable attention in the literature. A substantial amount of literature is available on woven fabrics, especially on plain weaves. Bending of fabrics, however, is generally non-linear. Woven fabrics are made of large numbers of fibres that may have considerable freedom of motion, relative to each other, within the fabric structure. As a result, the fibre strains which develop during bending are considerably lower than those which develop in bending of corresponding solid sheet materials. With this mobility, the potential flexibility of the fibres can be realized and the fabric structure will, in turn, have a low bending rigidity. The inter-fibre friction associated with the fibre movement is believed to be the major cause of fabric non-linear bending behaviour. Computational models for solving large-deflection elastic problems from theoretical models have been applied to specific fabric engineering and apparel industry problems, for example the prediction of the robotic path for controlling the laying of fabric onto a work surface (Brown *et al.*, 1990; Clapp and Peng, 1991). The non-linear bending behaviour of woven fabrics can be separated into two components: a non-linear component due to friction and a linear component. The bending resistance is made up of three components: (i) the bending resistance of the threads lying in the direction of bending, (ii) the interaction between the threads, and (iii) a frictional restraint. A typical bending hysteresis curve of plain weaves is given in Fig. 6.1, which is basically symmetrical about the origin O and regular (like a leaf) for both warp-wise and weft-wise bending.

To understand the behaviour under bending of knitted fabrics in general is very important, since their functional properties are closely related to their mechanical properties, such as bending, shear and tensile. For example, the drape properties of fabrics are affected by both bending and shear properties. An increase in bending and shear parameters, such as bending and shear rigidity, and hysteresis of bending, results in a decrease in the drape structure of the fabric, something undesirable in most cases (Gaucher and King, 1983). Another example of the importance of the bending of knitted fabrics involves their handle properties. Handle is the sum total of sensations of the physical and mechanical properties of fabric when it is handled by touching, flexing by the fingers, smoothing, etc. In most cases, lower bending and shear parameters and lower roughness for knitted fabrics

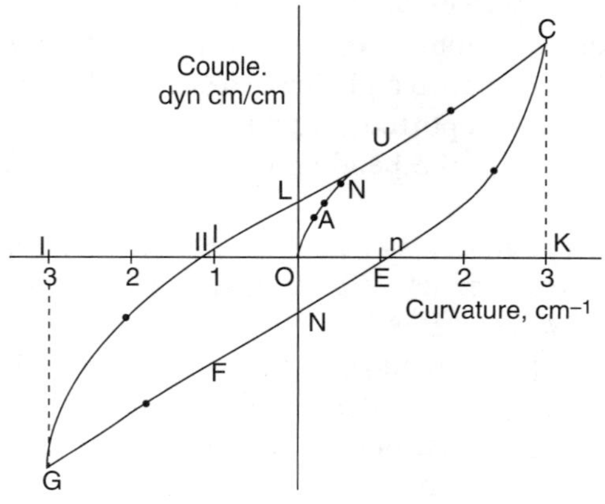

6.1 Typical bending-hysteresis curve for woven fabrics.

are necessary for the best handle (Chen *et al.*, 1992). Many examples exist showing the relationship between a fabric's functional properties and its bending properties.

In this chapter, an elaborate description and interpretation of the bending properties of MWK fabrics are presented, based on many bending hysteresis curves obtained on KES-FB-2. Further, a predictive bending model to assess the MWK fabrics based on KES-F experiments (Kawabata, 1980) in different bending directions is described. In the modelling process, an important finding is that the bending paths of non-orthogonally bent yarn systems follow different cylindrical helices. In addition, the theoretical model is derived directly from the structural parameters of the fabric, and the bending properties of its constituent inserting yarns can be used to calculate all the bending hysteresis curves of an MWK fabric in different bending directions. Furthermore, the bending rigidity at an arbitrary curvature can also be calculated with this model.

6.2 Bending properties of multiaxial warp-knitted fabrics

The bending behaviour of MWK fabrics is quite different from that of plain weaves and other apparel materials. From the experimental analysis, it may be understood that the bending properties of MWK fabrics are dependent not only on bending directions (such as warp-wise, weft-wise and bias-wise) but also on bending sequences, which means that which of the two sides of an MWK fabric is bent inwards or outwards first will lead to different bending results even if the bending direction (say, in the warp-wise direction) remains unchanged. According to the bending hysteresis curves obtained from KES-FB-2 (Kawabata's evaluation system), MWK fabric bending involves not only the spreading of filaments in inserting yarns, but also the buckling of inserting yarns, which accounts for the more remarkable irregularity and non-symmetry of MWK fabric bending hysteresis curves compared with those of plain weaves. A simple procedure for measuring the bending properties of fabrics is explained here. In order to measure the bending properties of the fabric the sample is bent between the curvatures –2.5 and 2.5 cm^{-1}, the radius of the bend being 1/curvature as shown in Fig. 6.2. The bending moment required to give this curvature is continuously monitored to give the curve shown in Fig. 6.3. The following quantities are measured from this curve:

Bending rigidity B = slope of the bending moment–curvature curve
Moment of hysteresis $2HB$ = hysteresis width of the curve

Furthermore, the inserting yarns not completely parallel to the applied bending moment direction will undergo both bending and torsional

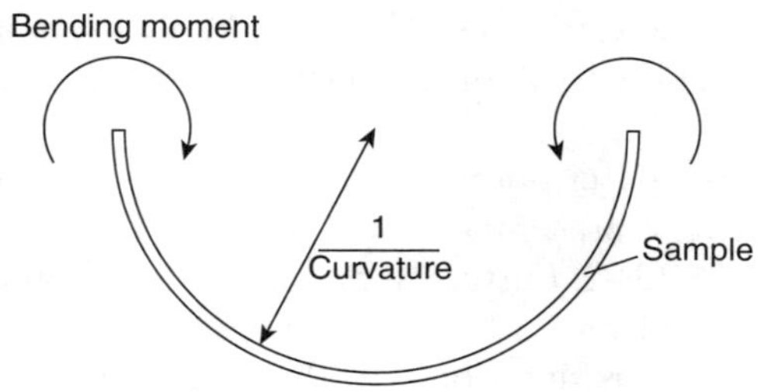

6.2 Forces involved in fabric bending.

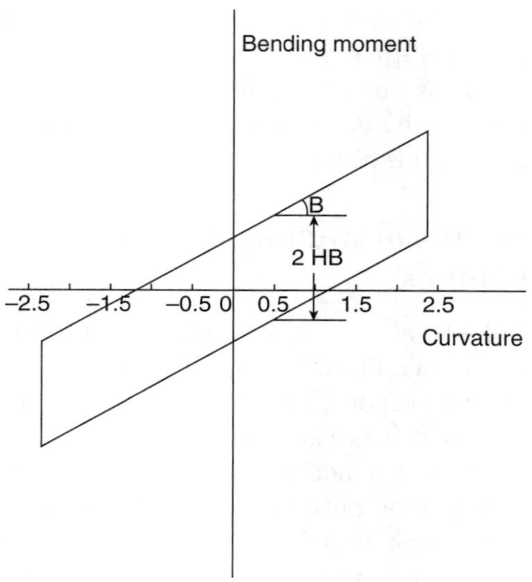

6.3 Plot of bending moment against curvature.

deformation during a bending cycle. Experimental observation and theoretical study have shown that the bending paths of those non-orthogonally bent yarn systems follow different cylindrical helices.

There is no elaborate theory and modelling method currently available as far as bending properties of MWK fabrics are concerned, but a thorough and systematic research approach in this direction is needed in product design, especially composite design (in the moulding process, the conformability of the fabric will depend to a great degree on the bending properties). Davies and Owen, 1971; Delaney, 1981a,b; Gibson and Postle, 1978; Ko *et al.*, 1986; Vigo and Turbak, 1991 and other authors (Cooper, 1960; Livesey and Owen, 1964; Grosberg, 1966; Grosberg and Swani, 1966; Owen, 1968) made a considerable contribution to the study of bending properties of plain weaves; unfortunately, these works are less helpful when studying MWK fabrics, which are more complex than plain weaves.

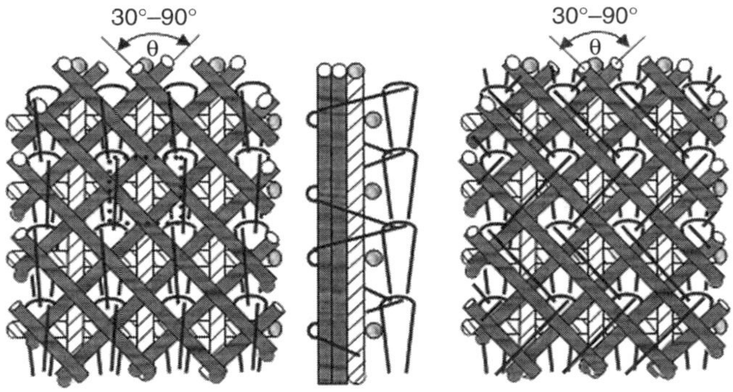

(a) Chain structure (b) Tricot structure

6.4 Typical structure of an MWK fabric.

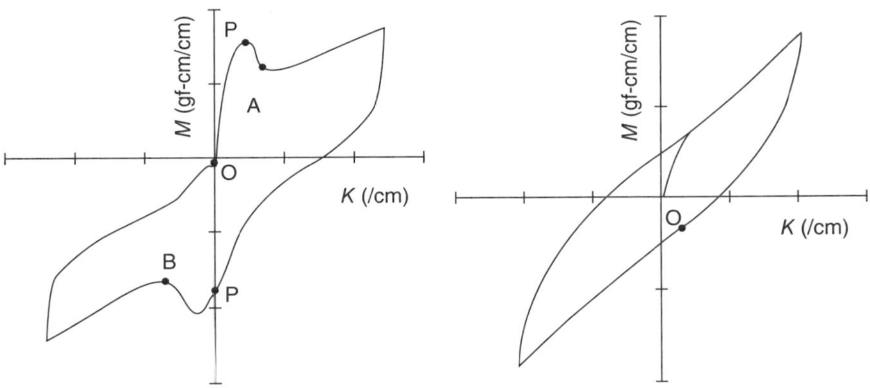

(a) MWK fabric (b) Plain weave

6.5 Comparison of typical bending hysteresis curves of an MWK fabric and a plain weave.

Figure 6.4 shows typical structures of MWK fabrics (Du and Ko, 1996) with two categories of stitching systems – tricot structure and chain loops. Structure (b) using glass filament yarns as the inserting system is chosen for modelling, the results of which can be applied to other MWK fabric structures.

The bending properties of MWK fabrics are characterized by KES-FB-2, and are quite different from those of plain weaves. Figure 6.5 shows typical bending hysteresis curves of an MWK and a plain weave. It can be observed that the bending hysteresis curve of an MWK fabric looks more irregular and non-symmetrical (about the origin O) than that of a plain weave. Two points, A and B, are particularly worth noting because they indicate the special bending characteristics of MWK fabrics with glass filament yarns as inserting systems.

With the aid of KES-FB-2, experiments on bending properties were carried out using three kinds of MWK fabrics. The inserting yarns are all glass filament yarns, and the tricot loops are polyester yarns. The details of the sample fabrics are as follows:

- **Sample 1.** Biaxial warp-knitted fabric (Karl Mayer type)
  There are two inserting yarn systems, weft and warp respectively, which are held together by tricot loops. Density: 5.0 threads/cm for weft, the same for warp. The yarn count is 618 Tex for weft inserting yarn, 596 Tex for warp inserting yarn, and 9 Tex for stitching yarns (tricot loops).
- **Sample 2.** Triaxial warp-knitted fabric (LIBA type)
  There are three inserting yarn systems, one for warp and the other two for bias (30° to the warp direction), held together by tricot loops. Density: 4.1 threads/cm for warp, four yarns, 1100 Tex for warp and 9 Tex for stitching yarns.
- **Sample 3.** Multiaxial warp-knitted fabric (as shown in Fig. 6.4)
  There are four inserting yarn systems, one for warp, one for weft and the other two for bias (45° to the warp direction), also held together by tricot loops. Density: 4.6 threads/cm for warp, 2.5 threads/cm for weft and 6.8 threads/cm for bias. The yarn count is 300 Tex for bias yarns, 900 Tex for warp, 70 Tex for weft and 9 Tex for stitching yarns. The sample size is 10 cm (length) × 3 cm (width).

6.2.1 Results of experiments

In interpreting the bending properties of MWK fabrics, one thing must be pointed out first. The inserting yarns employed in MWK fabrics in most cases are glass fibre yarns since they are generally used as industrial and engineering materials. Polyester yarns, especially with high strength and high modulus, can also be used as inserting yarns. For the purpose of a typical study, however, MWK fabrics with glass filament bundles as inserting yarns were chosen for research.

Glass filament yarns are quite different from those made of polymer filaments in mechanical behaviour. They possess no viscosity but perfect elasticity within a tensile strain of about 23%. As a result, for bending hysteresis curves of this kind of fabric, only the frictional effect between filaments, between inserting yarn systems, and between inserting yarns and tricot loops at the contact regions, will contribute to the hysteresis, and the viscosity has no effect.

In the following analysis, the effect of the stitching system has not been considered, since the count of stitching yarns is far smaller than that of inserting yarns. The tricot loops, however, determine the maximum slipping distance of inserting yarns and relate to motional friction within fabrics.

6.3 Bending hysteresis curves of multiaxial warp-knitted fabrics

In Fig. 6.6, typical bending hysteresis curves of MWK fabrics are given, of which (a) is the most typical according to the experiments. It is quite different from that of a plain weave. Compared with the curve for plain weave in Fig. 6.1, it can be seen that the bending hysteresis curve of an MWK fabric presents even more remarkable irregularity and non-symmetry (about the origin O). For convenience, here we divide the curve in Fig. 6.6(a) into six segments, marked OA, AB, BC, CD, DE and EO.

Figure 6.6 shows that the shape of segments OA and AB is basically the same for all the curves (a) to (d). For segment OA, no slippage occurs and the high initial Young's modulus is due to the combined contribution of the high bending rigidity of glass yarns (bundle of filaments, being bent as an unseparated integrity) and the static friction between filaments and between yarn systems. The bending curvature at point A is very small, nearly equal to zero. This segment possesses good linearity.

After point A, small relative slippage at regions with low frictional restraint begins to occur. With the bending process under way, the slippage phenomenon becomes more and more visible till point B. It can be seen from Fig. 6.6(a) that segment AB also presents a substantial linearity, although no better than that of segment OA. Here, we call the motional friction occurring in segment AB the 'stable motional friction', which is quite different from that in segment BC.

The experiment shows that a very special phenomenon occurs in segment BC (at point P), that the filaments in the inserting yarns are spread laterally at the bending point, and that the cross-section of these yarns is changed from the shape of 'race track' to 'a thin piece' (a series of filaments nearly lined up). This is illustrated in Fig. 6.7, which may account for the sharp decrease of bending moment at point P.

It can be understood from Fig. 6.6(a) and (c) that the motional friction in segment BC is very unstable and segment BC shows a remarkable non-linearity, which may be due to the spread of filaments to a great degree. However, this conclusion is not always true. Referring to Fig. 6.6(b), it can be found that the corresponding segment is basically linear, which is caused by the higher density and higher count of the inserting yarns being bent. In this situation, the effect of the spread of filaments plays only a secondary role.

The shape of segments CD and EO is just analogous to those of plain weaves (Fig. 6.1). However, it can also vary a lot according to different

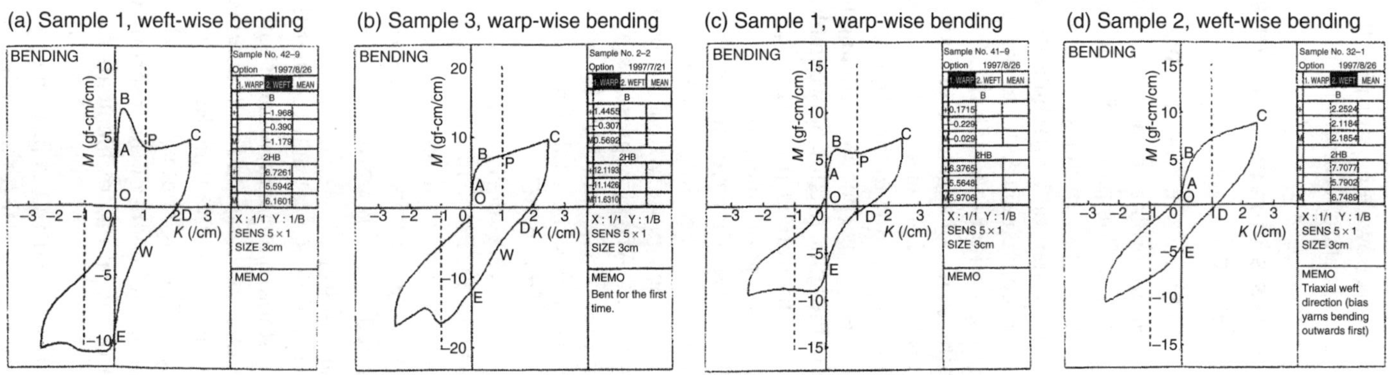

6.6 Typical bending hysteresis curves for MWK fabrics.

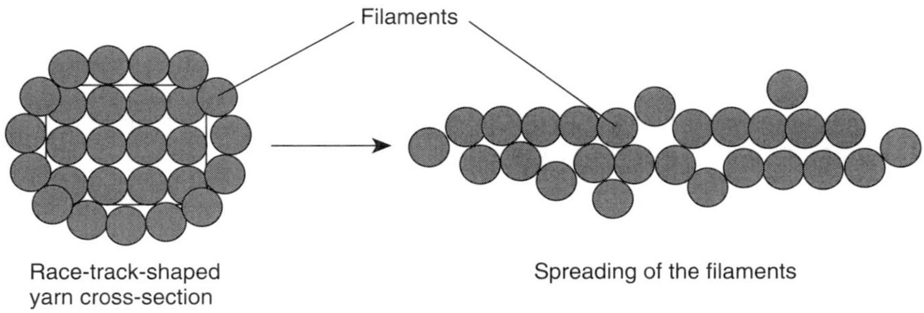

6.7 Spread of filaments in a bent yarn at the bending point.

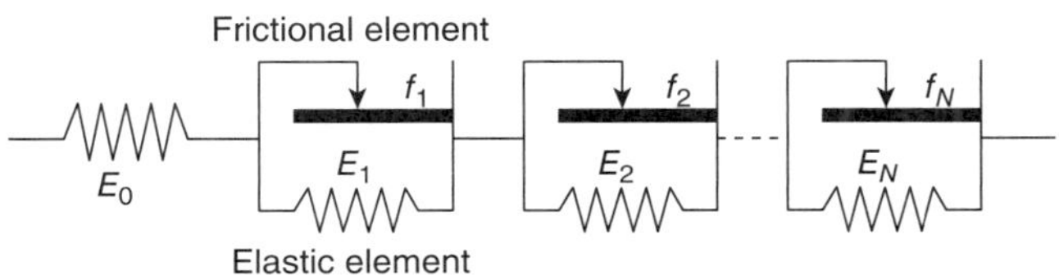

6.8 A modified Olofsson's model.

fabric parameters and different bending conditions. Here, observations show that point S is most interesting, where the inserting yarns being bent are still in a bent state even though the bending curvature here is zero. This will lead to the succeeding buckling phenomenon in segment DE (Fig. 6.6(a) and (b), point W), which will be analyzed in detail later.

As far as the bending process of segments OA and AB is concerned, a modified Olofsson's model of assembly of frictional and elastic elements may be applied here to give a clearer description. As shown in Fig. 6.8, an extra elastic element (E_0) is connected in series with the original model. The elements in Fig. 6.6 are arranged in the sequence below:

$$\sigma(E_0) < \sigma(f_1) < \sigma(f_2) < \ldots < \sigma(f_N), \quad E_0 > E_N > E_{N-1} > \ldots > E_2 > E_1$$

For segment OA, it is an elastic and static frictional bending, in which only E_0 takes effect and all frictional elements do not move. With the increase of bending curvature (from point A to point B), the frictional element f_1 first begins to move, then f_2, f_3, etc., till f_N. In this segment, however, the shape of the cross-section of inserting yarns hardly changes, which accounts for the good linearity of segment AB.

In Fig. 6.6(d), the curve is visibly different from the others, being more like the bending hysteresis curve of a plain weave (Fig. 6.1). The sample for Fig. 6.6(d) is triaxial warp-knitted fabric, LIBA type, in which the inserting yarns (except warp yarns) are penetrated randomly by the tricot loops. Compared with Karl Mayer type fabrics, the LIBA type fabric is a little looser in structure. As a result, the frictional effect is less remarkable, which leads to the lower initial Young's modulus and comparatively regular shape of the bending hysteresis curve.

In summary, it can be seen from Fig. 6.6 that the shape of the bending hysteresis curves of an MWK fabric is quite like a 'dog's bone'.

6.4 Buckling of the bent-inserting yarns

The outstanding characteristic of the bending properties of MWK fabrics is the buckling of bent-inserting yarns, which can account for the special shape of segment DE in Fig. 6.6(a). This phenomenon is very complex and dependent on many factors, such as the yarn count, the interaction between the filaments (which is related to the surface finishing of the filaments) and the frictional effect between yarn systems, as well as the pressure exerted by the two clamps of the KES device.

For the curve in Fig. 6.6(a), the warp yarns do not become straight at point S, though the bending curvature here is zero. In Fig. 6.9, a schematic diagram of this state is illustrated. The buckles B_1, B_2 and B_3 are kept in their shapes by the frictional restraint resulting from the two clamps (C_1 and C_2) and the interaction between yarn systems. When the bending process reaches point D (where the bending moment becomes large enough to overcome the frictional restraint), the buckles will disappear suddenly, resulting in the sharp decrease in bending moment at point W. This may also account for the non-symmetry of the bending hysteresis curve to some degree.

6.5 Effect of bending sequence on bending hysteresis curves

According to experiments, it was found that when the bending direction keeps unchanged, the bending hysteresis curves present differences in terms of different bending sequences. This phenomenon is even more visible in biaxial warp-knitted fabrics. As shown in Fig. 6.10, two curves are obtained from sample 1, both being bent warp-wise but in inverse bending sequence.

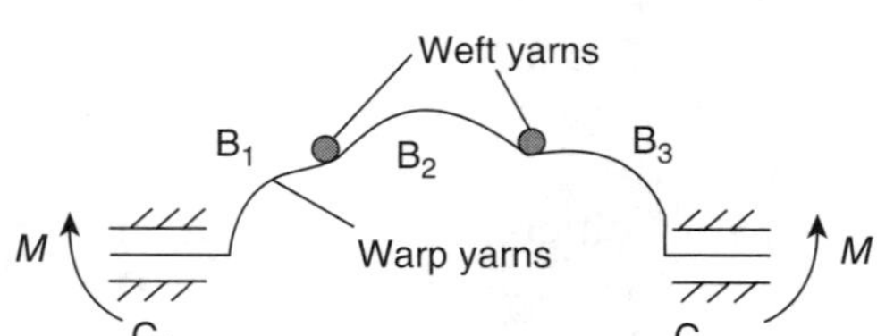

6.9 A schematic diagram of buckling (exaggerated).

(a) Sample 1, warp-wise bending
(Warp yarns are bent outwards first)

(b) Sample 1, warp-wise bending
(Warp yarns are bent inwards first)

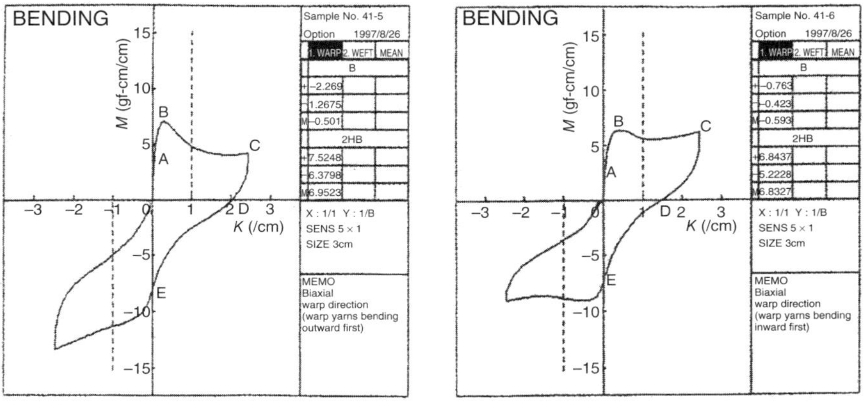

6.10 The effect of bending sequence on the bending hysteresis curves.

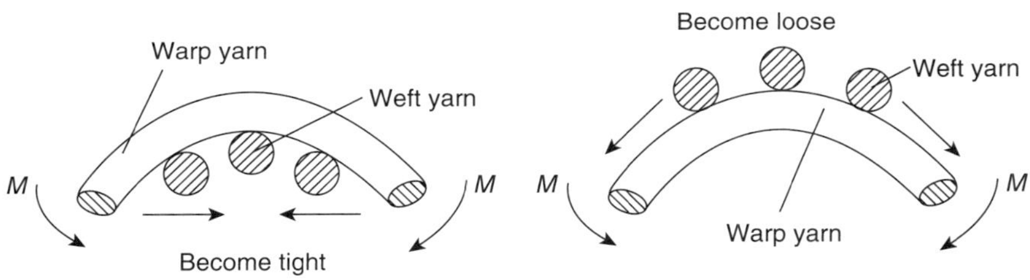

(a) (b)

6.11 A schematic diagram of different bending sequences.

It can be noted that the two curves are different in shape, especially for segments BC and DE. This phenomenon may be explained by Fig. 6.11 as follows. It is the different frictional effect in Fig. 6.11(a) and (b), resulting from the different slippage of non-bent yarns, which leads to the different bending hysteresis curves. This must be considered in product design, especially in moulding. It should be pointed out that the slippage of inserting yarns depends on the size and density of tricot loops to a great degree.

6.6 Cyclic bending

Although the bending hysteresis curves of MWK fabrics look irregular and non-symmetrical, they tend to become regular and symmetrical after cyclic bending, as shown in Fig. 6.12. It can be deduced from these curves

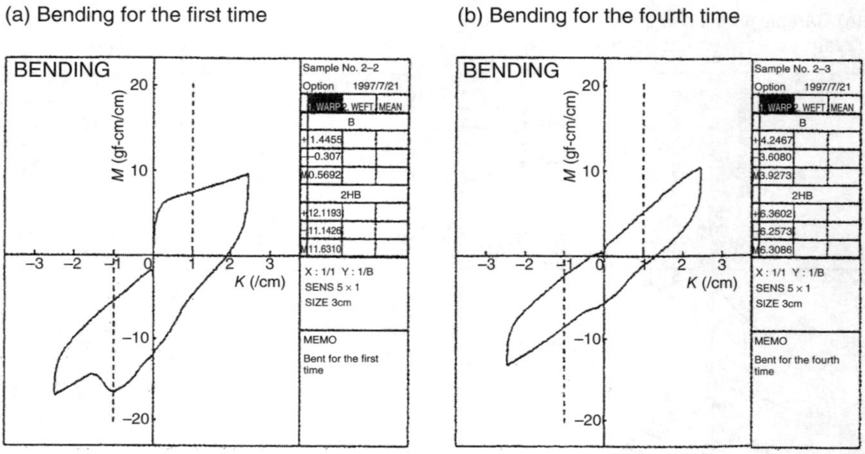

6.12 Sample 3, warp-wise bending.

that the final shape of the bending hysteresis curve of an MWK fabric is very close to a parallelogram, as shown in Fig. 6.12(b) (the broken lines). In fact, the static friction within an MWK fabric will become smaller and smaller, and the fabric structure will become looser and looser after cyclic bending.

As a result, the non-linear component will decrease and the buckling phenomenon will disappear. The final 'parallelogram' may be of more practical value, since it represents a relatively stable state quite different from the situation in the first bending which is easily influenced by contingent factors. From this point of view, the bending with unstable bending results can be regarded as 'mechanical conditioning'. How many cycles an MWK fabric will need to be mechanically conditioned obviously relates to the different structural parameters of the fabric, different yarn finishes and different yarn materials.

6.7 Modelling bending properties of multiaxial warp-knitted fabrics

From experimental observations, it is found that point A in Fig. 6.6(a) is produced by the spreading of glass filaments in inserting yarns at the bending point, as illustrated in Figure 6.7. However, to what degree the moment value at point A will go down with respect to the peak point P_o depends on different structures of MWK fabrics, although this kind of spreading occurs in all situations. Generally, the thicker the bent inserting yarns are, the smaller the degree to which this spreading will influence the shape of the hysteresis curve, since the larger bending moment of the thicker yarns will be dominant in determining the magnitude of the final bending moment.

The other point (B) comes from the common contribution of both the aforesaid spreading of glass filaments and the buckling of bent inserting yarns, and the buckling phenomenon dominates the moment magnitude of point B to a great degree. This kind of buckling is illustrated in Fig. 6.9. Actually, when the bending process reaches point P, the fabric does not really become completely straight, despite the vanishing of bending curvature at this point. The buckles B_1, B_2, and B_3 keep their shapes from the frictional restraint in the two clamps (C_1 and C_2) and the interaction of the yarn systems. After point P, when the bending moment becomes large enough to overcome this frictional restraint, the buckles disappear suddenly at point B, which may account for the sharp decrease of bending moment at this point.

The main concern in this chapter is to establish a predictive model for assessing the bending behaviour of MWK fabrics. As the analysis above shows, it is very difficult to consider both the spreading and buckling phenomena in the model. Fortunately, however, the bending behaviour of a single inserting yarn is quite analogous to that of an MWK fabric, which also shows the two special characteristics – the spreading of glass filaments and the buckling phenomenon – as shown in Fig. 6.13.

Based on the previous discussion, according to the experimental results from the bending of single inserting yarns and the structural parameters of the fabric, we have developed a predictive model for calculating the entire bending hysteresis curve of an MWK fabric being bent in an arbitrary direction, using the following notation:

1 = warp yarn
2 = weft yarn
3 = bias yarn ($+\theta_0$)
4 = bias yarn ($-\theta_0$)
K = bending curvature of the fabric, /cm
K_i = bending curvature of an inserting yarn system (i = 1–4), /cm
L_0 = sample length, cm
M = bending moment exerted on the fabric, gf·cm/cm
M_i = bending moment exerted on a single inserting yarn (i = 1–4), gf·cm
$M_L(K)$ = bending moment of the stitching system, gf (gram force)·cm/cm
$M_S(K)$ = frictional moment between yarn systems, gf·cm/cm
n_i = number of yarns per unit length of fabric along direction perpendicular to corresponding yarn's axis (i = 1–4), /cm

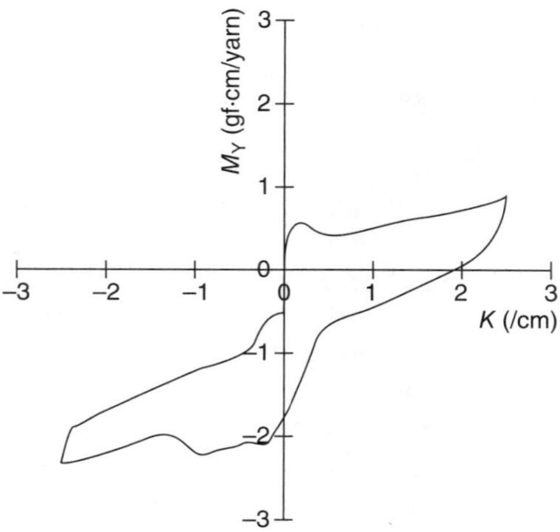

6.13 Typical bending hysteresis curve of a single glass filament yarn (900 Tex).

n_i' = number of yarns per unit length of fabric along direction perpendicular to bending moment direction (i = 1–4), /cm
W_0 = sample width, cm
θ = clockwise angle from bending moment direction to warp yarn axis
$\pm\theta_0$ = angle between bias yarn axis and warp yarn axis (the plus sign represents the angle from the warp yarn axis to the bias yarn axis along the clockwise direction, and the minus indicates the counterclockwise direction)

As far as the bending properties of fabrics are concerned, perhaps the biggest difficulty in modelling is the introduction of frictional restraint, which should account for most of the deviation. To model the bending behaviour of plain weaves, two methods are available to determine the frictional restraint: one is the cantilever method (Grosberg and Swani, 1966), which requires many experiments and is therefore tedious, and the other is the parallel plates method (Grosberg, 1966), which requires the frictional coefficient μ and the internal pressure force V. From the viewpoint of practical application, neither method can be readily applied. For MWK fabrics with glass filament yarns as inserting systems, the frictional restraint can be separated into two components: the frictional interaction between glass filaments within single inserting yarns, and the frictional interaction between yarn systems. The first component has been contained in our model. For the second, it is reasonable to ignore the acceptable deviation as far as our model is concerned. The modelling is described in the foregoing section.

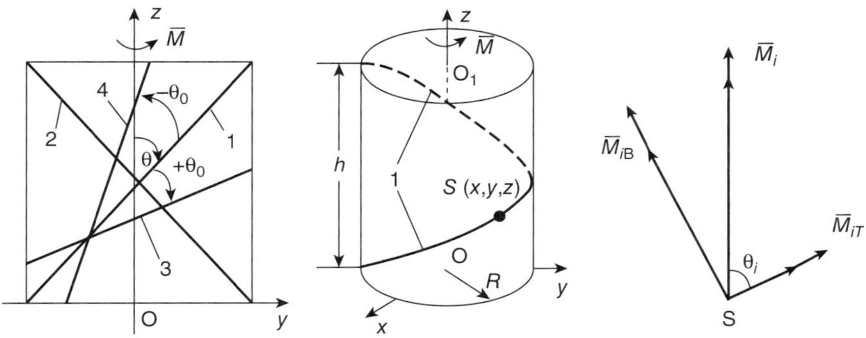

6.14 Bending of an MWK fabric along an arbitrary direction.

From Fig. 6.6, it is obvious that when an MWK fabric is bent along an arbitrary direction, the inserting yarns non-orthogonal to the bending moment direction will undergo both bending and torsional moment. According to our experimental study and mathematical analysis, the bending path of these yarns follows a cylindrical helix, as illustrated in Fig. 6.14. The parametric equation of the cylindrical helix can be written as:

$$S(x, y, z): \begin{cases} x = R\cos t \\ y = R\sin t \\ z = Ct \end{cases} \tag{6.1}$$

where C is a constant and t is the parameter. Using Equation 6.1, we can obtain two other equations:

$$\tan(\pi/2 - \theta) = h/(2\pi R) \tag{6.2}$$

and

$$C = h/2\pi \tag{6.3}$$

From Equations 6.2 and 6.3 we have:

$$C = R\tan^{-1}\theta \tag{6.4}$$

According to Equations 6.1 and 6.4, the curvature of an arbitrary point S on the helix can be obtained as follows:

$$K_S = \sqrt{\frac{(\dot{x}^2 + \dot{y}^2 + \dot{z}^2)(\ddot{x}^2 + \ddot{y}^2 + \ddot{z}^2) - (\dot{x}\ddot{x} + \dot{y}\ddot{y} + \dot{z}\ddot{z})}{(\dot{x}^2 + \dot{y}^2 + \dot{z}^2)^3}} = \frac{\sin^2\theta}{R} \tag{6.5}$$

Based on the mathematical derivation above, we have

$$K_i = K\sin^2\theta_i \quad (i = 1, 2, 3, 4) \tag{6.6}$$

where

$$K=\frac{1}{R},\quad \theta_1=\theta,\quad \theta_2=\theta+\frac{\pi}{2},\qquad\qquad 6.7$$
$$\theta_3=\theta+\theta_0,\quad \theta_4=\theta-\theta_0.$$

In addition, the bending moment exerted on a single yarn can be decomposed into two components – the bending moment $M_{i\mathrm{R}}$ and the torsional moment $M_{i\mathrm{T}}$ (Fig. 6.14):

$$\begin{aligned} M_{i\mathrm{R}} &= M_i \sin\theta_i \\ M_{i\mathrm{T}} &= M_i \cos\theta_i \end{aligned} \quad (i=1\text{–}4) \qquad\qquad 6.8$$

From Equation 6.8, the magnitude of the moment exerted on a single inserting yarn can be calculated as

$$M_i=\frac{M_{i\mathrm{B}}}{\sin\theta_i} \qquad\qquad 6.9$$

Figure 6.15 illustrates how a sample fabric is clamped in KES testing. For convenience, only warp yarns are drawn here.

It is very easy to derive the relationship between n_i and n_i' from Fig. 6.15:

$$n_i'=n_i\sin\theta_i \qquad\qquad 6.10$$

Also, the effective bending length should be the length of PQ, which relates to the corresponding inserting yarn system. Here 'effective' means that both ends of the inserting yarns being bent are held by the clamps. PQ can be obtained from Equation 6.11 as follows:

$$|\mathrm{PQ}|_i=L_0-W_0\left|\tan^{-1}\theta_i\right| \qquad\qquad 6.11$$

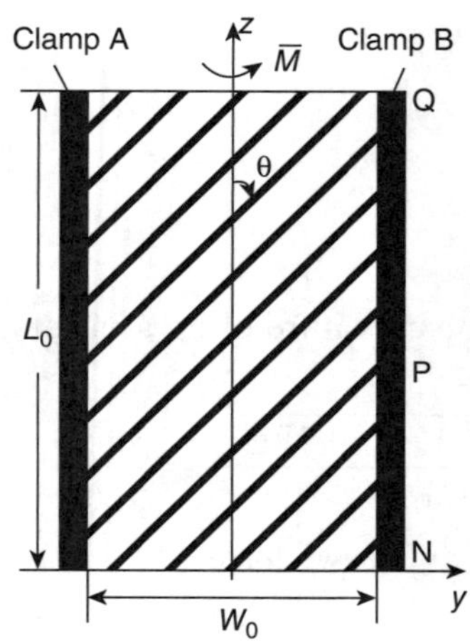

6.15 Illustration of a clamped sample fabric.

Table 6.1 Specifications of samples ($\theta_0 = 45°$)

Direction	Tex	n_i
Warp – 1	900	4.6
Weft – 2	70	2.5
Bias (+45°) – 3	300	6.8
Bias (–45°) – 4	300	6.8
Tricot loops – 5	9	5

According to Equations 6.9, 6.10, and 6.11, the bending moment exerted on an MWK fabric can be calculated as:

$$M = \sum_{i=1}^{4} n_i' M_i \frac{|PQ|_i}{L_0} + M_S(K) + M_L(K) \tag{6.12}$$

Equation 6.12 can be rewritten as Equation 6.13:

$$M = \sum_{i=1}^{4} n_i M_{iR}\left(1 - \lambda\left|\tan^{-1}\theta_i\right|\right) + M_S(K) + M_L(K), \quad \lambda = \frac{W_0}{L_0} \tag{6.13}$$

We tested the sample fabrics (Fig. 6.4(b)) along four bending directions (warp, weft and bias $\pm\theta_0$) on the KES-FB-2. The specifications of samples used in our research are listed in Table 6.1, and the sample size is 3 cm (L_0) × 1 cm (W_0). We also tested the bending behaviour of single inserting yarns at the same time. Fig. 6.16 presents three sets of bending hysteresis curves obtained on KES. Here, only the bias +45° bending is given, since the other bias bending (–45°) is analogous.

6.8 Model validation

As far as the model given by Equation 6.13 is concerned, the most difficult term to determine is $M_S(K)$, which denotes the frictional restraint between yarn systems. As analyzed earlier, this kind of restraint is very complex and relates to many parameters, such as the number of inserting systems, the density and pattern of the stitching system, the inserting yarn count, the finishing of glass filaments, and the bending direction. Actually, it is very difficult to represent $M_S(K)$ directly by an analytical function with an acceptable deviation. The term $M_S(K)$, however, can be neglected to a great degree as far as a relatively open MWK structure is concerned, and the open condition is basically always satisfied by MWK fabrics in view of their industrial applications. Furthermore, it is clear from Fig. 6.17 that the model without $M_S(K)$ is in good agreement with the experiments.

The other term to consider is $M_L(K)$, which denotes the bending of the stitching system. Since the stitching yarns are much finer than the inserting yarns, $M_L(K)$ can also be neglected in order to simplify the predictive

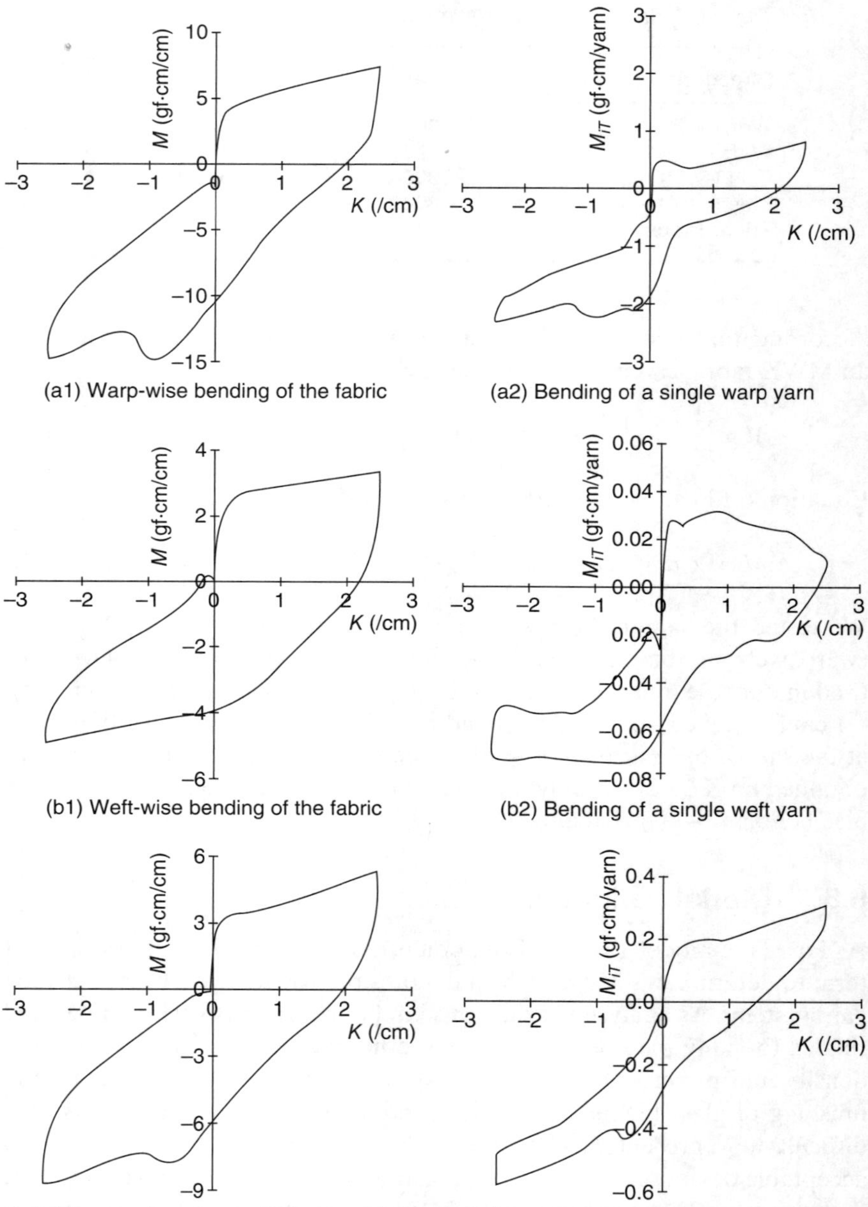

6.16 Experimental bending hysteresis curves of an MWK fabric and its single inserting yarns.

model. Of course, this term can be introduced very easily only if the bending properties of the stitching system are tested.

According to the analysis above, Equation 6.13 can be simplified into the form of Equation 6.14 as:

$$M = \sum_{i=1}^{4} n_i M_{iB}(1 - \lambda|\tan^{-1}\theta_i|), \quad \lambda = \frac{W_0}{L_0} \tag{6.14}$$

Based on Equation 6.14, calculations are made along three bending directions (warp-wise, weft-wise and bias-wise ($+\theta_0$)). The comparison of the calculated results and experimental data is given in Fig. 6.17, which shows that the model set forth earlier (Equation 6.13) is in good agreement with the experiments.

6.9 Conclusions

This chapter presents the bending properties of MWK fabrics and their modelling to describe the bending behaviour. The bending properties of MWK fabrics are quite different from those of traditional wovens and apparel materials, and the bending hysteresis curves are not only irregular but non-symmetrical. This presents more complexity and variability. The bending process involves both the slippage of non-bent-inserting yarns and the buckling of bent-inserting yarns, and different bending sequences will lead to different bending hysteresis curves. The shape of the bending hysteresis curve is liable to be affected by many factors, which leads to difficulty in modelling. Generally, the bending hysteresis curve of an MWK fabric obtained in the first bending cycle is quite like a ‘dog’s bone’ and close to a parallelogram after cyclic bending. As far as the frictional restraint between inserting yarn systems is concerned, it is more difficult to determine even if the frictional coefficient (which should be some function of the frictional direction) between the glass filament yarns can be obtained. Actually, this kind of frictional restraint may not be represented directly by an analytical function, but it can be ignored to a great degree as far as a basic estimation is concerned, which seems reasonable from Fig. 6.17.

The following conclusions can be drawn from this chapter. The bending process of MWK fabrics involves two special deformation phenomena, i.e. spreading of the glass filaments in inserting yarns and buckling of inserting yarns being bent. The spreading and buckling of single inserting yarns can reflect those of an MWK fabric consisting of these yarns. The bending path of a non-orthogonally bent inserting yarn follows a cylindrical helix. The frictional restraint between yarn systems can be ignored to a great degree for a basic estimation of the bending behaviour of MWK fabrics. The whole bending hysteresis curve of an MWK fabric being bent in an arbitrary

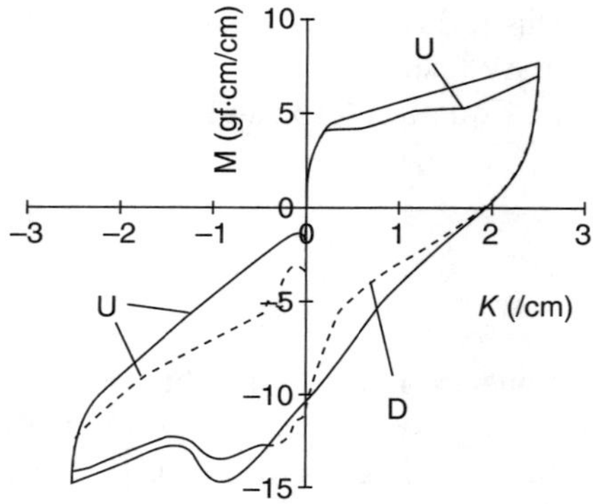

(a) Warp-wise bending

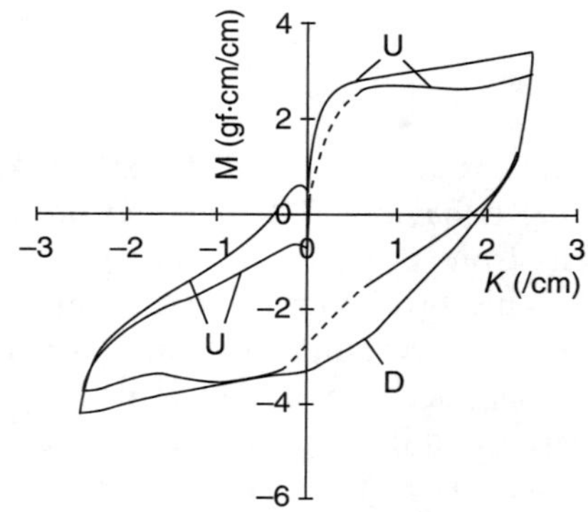

(b) Weft-wise bending

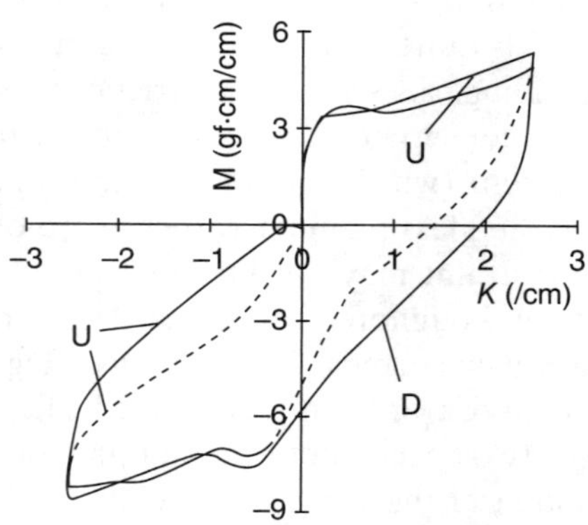

(c) Bias-wise bending (+θ)

6.17 Comparison of calculated results and experimental data.

direction can be calculated through the model established in this chapter. The prediction seems consistently off in the unloading portion of the bending hysteresis curves, which could be improved by including the frictional component.

6.10 References

Brown III P R, Buchanan D R and Clapp T G (1990), Large deflection bending of woven fabric for automated material handling, *Journal of the Textile Institute*, **81**, 1.

Chen P L, Barker R L, Smith G W and Scruggs B (1992), Handle of weft knit fabrics, *Textile Research Journal*, **83**, 200–210.

Clapp T G and Peng H (1991), A comparison of linear and non-linear bending methods for predicting fabric deformation in automated handling, *Journal of the Textile Institute,* **82**, 341.

Cooper D N E (1960), The stiffness of woven textiles, *Journal of the Textile Institute*, **51**, T317.

Davies I and Owen J D (1971), The bending behaviour of warp-knitted fabrics, *Journal of the Textile Institute*, **6**, 42, 181.

Delaney P (1981a), The bending properties of some plain and twill weave wool fabrics, *SAWTRI Tech. Rep. no. 480*, September.

Delaney P (1981b), The bending properties of some punto-di-roma wool fabrics, *SAWTRI Tech. Rep. no. 489*, November.

Du G-W and Ko F (1996), Analysis of multiaxial warp-knit preforms for composite reinforcement, *Composites Science and Technology*, **56**, 3, 253–260.

Gaucher M L and King M W (1983), Predicting the drape coefficient of knitted fabrics, *Textile Research Journal*, **53**, 297–303.

Gibson V L and Postle R (1978), An analysis of the bending and shear properties of woven, double-knitted, and warp-knitted outerwear fabrics, *Textile Research Journal*, **48**, 14.

Grosberg P (1966), The mechanical properties of woven fabrics, Part II: The bending of woven fabrics, *Textile Research Journal*, **36**, 205.

Grosberg P and Swani N M (1966), The mechanical properties of woven fabrics, Part III: The buckling of woven fabrics; Part IV: The determination of the bending rigidity and frictional restraint in woven fabrics, *Textile Research Journal*, **36**, 332, 338.

Kawabata S (1980), *The Standardization and Analysis of Hand Evaluation* (2nd edn), Textile Machinery Society of Japan, Osaka.

Ko F K, Pastore C, Yang J M and Chou T W (1986), Structure and properties of multilayer multidirectional warp knit fabric reinforced composites, in *Proc. 3rd US–Japan Conf. on Composites*, Tokyo, 21–28.

Livesey R G and Owen J D (1964), Cloth stiffness and hysteresis in bending, *Journal of the Textile Institute,* **55**, T516.

Owen J D (1968), The bending behaviour of plain-weave fabrics woven from spun yarns, *Journal of the Textile Institute*, **5**, 3, 391.

Vigo T L and Turbak A F (1991), *High-tech Fibrous Materials: Composites, Biomedical Materials, Protective Clothing, and Geotextiles*, American Chemical Society, Washington, DC, 81–89.

7
Formability of multiaxial warp-knitted fabrics

Abstract: The understanding of the unique behaviour of multiaxial warp-knitted (MWK) fabrics during forming and the practical viability of various forming processes is crucial in assessing their utilization and applicability. In this chapter, a model for the prediction of the formability of an MWK fabric to a 3-D surface is described. For this purpose, a detailed characterization of the forming behaviour of MWK fabrics containing two bias inserting yarns (TBMWK fabrics) is made. A mathematical model is established for predicting the deformations of TBMWK fabrics during the hemisphere-forming process. The shape of flat TBMWK fabric that can yield the corresponding hemisphere during the forming process as well as local deformations can be calculated through this model.

Key words: multiaxial warp-knitted (MWK) fabrics, formability of MWK fabrics, deformation mechanisms, two-bias MWK (TBMWK) fabrics, modelling formability of MWK fabrics, hemisphere.

7.1 Introduction

Over the past decade, there has been a significant amount of research interest in the field of composites, especially in 3-D fabric-reinforced sheet materials. The understanding of the unique behaviour of these 3-D fabrics during forming and the practical viability of various forming processes is crucial in assessing their utilization and applicability. A relatively recent innovation in this field is the use of MWK fabrics as the reinforcement. Knitted fabrics offer a number of potential advantages over woven fabrics: they can be stretched in both directions during forming, thus increasing the potential for forming complex and deeply curved components (Mayer *et al.*, 1998; Mouritz *et al.*, 1999; Savci *et al.*, 2000), they have excellent impact, fracture toughness and energy absorption properties and (in the case of 3-D knits) they exhibit improved out-of-plane mechanical properties.

So far, the bulk of the literature has concentrated on characterizing the mechanical properties of both thermosetting and thermoplastic-based knitted composite materials. Modelling of the mechanical properties (mainly stiffness and tensile) has also been popular. However, limited literature exists on the forming property characteristics of knitted fabric-reinforced thermoplastics (Lim *et al.*, 1998; Takano *et al.*, 2001) and their processing properties are still poorly understood. In fact, most of the literature on forming properties deals with unidirectional, mostly woven and, to a lesser extent, braided reinforcements, rather than knitted reinforcements.

Although the actual mechanism of sheet forming can be quite complex, a general understanding of the process and material characteristics can be gained through a series of simple experiments including hot-tension, dome-forming, cup-drawing, V-bending and picture-frame experiments. They are designed to subject the material to selective modes of deformation, allowing a more systematic approach to the analysis. One way of obtaining a physical measure of the forming behaviour is to develop kinematic models based purely on geometrical changes. These changes can be measured using grid strain analysis (GSA) where grid points before and after the forming process are used to calculate forming strains. The experiments not only provide a general understanding, but also serve as a check for any predictive work (Duhovic and Bhattacharyya, 2005).

When sufficient data regarding the material and other processing parameters are available, numerical simulations can provide time-saving predictions of the forming behaviour. For materials with complex reinforcing structures, such as knitted fabrics, the material model must be able to accommodate the most important parameters, so that the material can be simulated accurately. Since the forming behaviour of textile composites is governed by their reinforcing structures, the examination of the reinforcement can identify these important parameters. It has been shown that the best results can be achieved using a low-viscosity matrix, i.e., when the lubricating effects are significant. Therefore, the understanding of the deformation behaviour of the reinforcing structure alone is of critical importance. To investigate these parameters physically would be very difficult, because of the scale and level of detail required; however, using advanced numerical simulations, a large quantity of information may be obtainable.

Textile composite materials offer an attractive alternative to metals in the automotive and aerospace industries. However, fibre and yarn movement during fabric forming can cause adverse effects such as wrinkling and thinning, which will lead to a decrease of the mechanical properties of the finished composite (Kaufman, 1991). In addition, the high level of waste generated by subsequent trimming operations is unacceptable. Hence there exists a need to establish a model to predict the deformation and possible wrinkles of MWK fabrics during the forming process in order to enable waste-free design and defect predictability. Before doing this, an understanding of the detailed characterization of the forming behaviour of multiaxial warp-knitted (MWK) fabrics containing two bias inserting yarns (TBMWK fabric) is very important.

In this chapter, a model for the prediction of the formability of a multiaxial warp-knitted (MWK) fabric to a 3-D surface has been described. For this purpose, a detailed characterization of the forming behaviour of MWK fabrics containing two bias inserting yarns (TBMWK fabric) was made. Through experimental observation, it was found that the two bias inserting yarns always tend to gather along the weft direction. The angle between the two bias yarns has a linear relationship with the perpendicular distance from the measured points to the longitudinal axis of the hemisphere during the forming process. The slope of this linear relationship is also linear with the magnitude of the radius of the pressing hemisphere, provided that the radius is larger than 7 cm.

Based on the above finding, a mathematical model is established for predicting the deformations of TBMWK fabrics during the hemisphere-forming process. The shape of flat TBMWK fabric that can yield the corresponding hemisphere during the forming process as well as local deformations can be calculated through this model. The hemisphere-forming experiments show that the present model is workable and accurate. The results from both the model and experiments suggest that the shape of flat TBMWK fabric that can yield the corresponding hemisphere is close to rectangular, not square as presented by woven fabric.

The method developed in this chapter has laid a foundation for further modelling of the forming behaviour of MWK fabrics onto other 3-D surfaces. More importantly, it is of great value to find that the two bias inserting yarns always tend to gather along the weft direction of the fabric, which is a starting point for modelling of the forming behaviour of MWK fabrics.

7.2 Textile composite deformation mechanisms

In thermoforming textile composite materials, it is useful to understand the deformation mechanisms that take place inside them, so that the forming process can be optimized to produce parts of the best quality. The experimental observations and measurements are helpful for a general understanding, but do not provide enough detail to reveal the actual mechanisms. The following sections attempt to investigate these mechanisms in more detail.

The hierarchy of deformation modes for this family of composite materials can be divided into three categories: prepreg flow mechanisms, macro-level fabric deformation modes, and micro-level fabric deformation modes, each of which contains a number of different mechanisms.

7.2.1 Prepreg flow mechanisms

When the textile fabric reinforcement is combined with the matrix to form the composite prepregs, a set of deformation modes are introduced, which may be referred to as top-level deformation modes since they involve the movements of the reinforcement (macro- and micro-level fabric deforma-

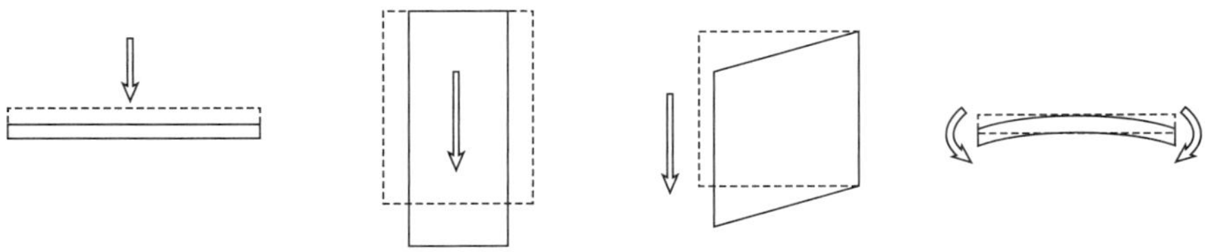

(a) Transverse compression (b) In-plane tension (c) In-plane shear (d) Out-of-plane bending

7.1 Macro-level fabric deformation modes.

tion mechanisms) and the matrix. They are in fact the conformation modes of the composite prepreg sheet or group of sheets, as is usually the case, during the forming process. The hierarchy of the top-level deformation modes consist of transverse flow, resin percolation, interply shear and intraply shear, as summarized by Cogswell (1992) and revisited by Martin *et al.* (1997).

7.2.2 Macro-level fabric deformation modes

The four types of macro-level fabric deformation modes, as shown in Fig. 7.1, describe the deformations observed when looking at the fabric as a whole. However, the way in which each fabric complies to these modes is different and can be attributed to the deformations occurring within the textile structure itself. These sub-structure or micro-level deformation modes are the real mechanisms behind textile deformations and need to be identified in order to understand the material's behaviour.

7.2.3 Micro-level fabric deformation modes

Micro-level fabric deformation modes exist through the interaction of structured yarns within the fabric. Figure 7.2 shows what are generally believed to be the eight micro-level deformation modes for textile fabrics. Inter-yarn slip, as shown in Fig. 7.2(a), occurs when the yarns that construct the fabric move over one another. It is one of the modes of deformation belonging almost exclusively to knitted fabrics. In this mode of deformation, the friction between the yarns becomes important since it determines where the onset of buckling will take place, as well as the magnitude of the forming forces required. Fortunately, the matrix and fibre chemical sizing (coatings) usually lubricate the yarn to help this mode of deformation.

Inter-yarn shear is a common mode of deformation in many woven fabrics. This is where the yarns rotate about their crossover points to accommodate the required deformation, as shown in Fig. 7.2(b). In fact, this type of mechanism has been reported to occur in multilayered continuous fibre-reinforced composites also, as outlined by Krebs and Bhattacharyya (1998),

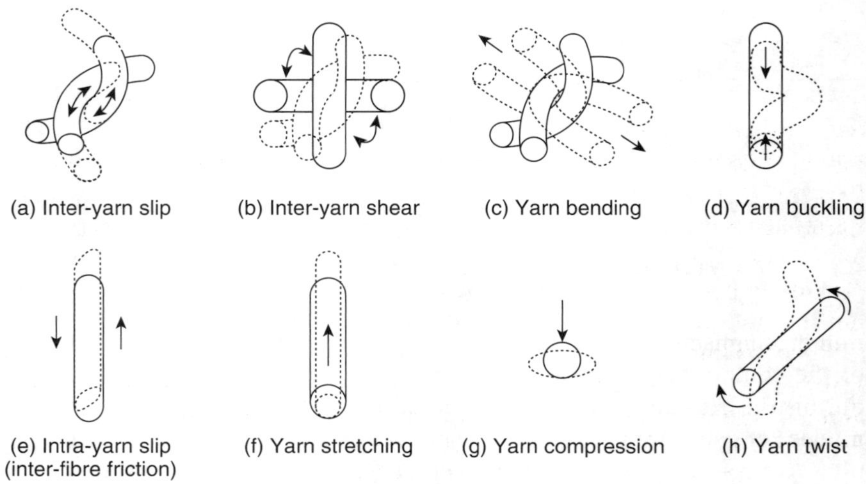

7.2 Micro-level fabric deformation modes.

Martin *et al.* (1997) and Christie (1997), and is commonly referred to by many researchers as the 'trellising effect'. In knitted fabrics, depending on the orientation of the reinforcement, large in-plane tension and in-plane shear can be accommodated, but not in the same direction.

The yarn bending or 'straightening' shown in Fig. 7.2 is in many cases the most significant deformation mode in many textiles. It is the most influential mode in knitted fabrics because of the knit loop geometry. Straightening also occurs to a lesser extent in woven and braided fabrics depending on the amount of crimp or yarn undulation present in the fabric structure. Out of all the different deformation modes, fibre buckling is the only unfavourable one since the material movement through this mode creates what is considered as defects, although it is quite difficult to observe with complex structures, such as knits and braids. Out-of-plane buckling usually occurs when the inplane modes cannot accommodate the required deformation. In-plane buckling can also occur, but is less likely due to in-plane geometric constraints – see Fig. 7.2(d).

The intra-yarn slip shown in Fig. 7.2(e) coupled with yarn bending, Fig. 7.2(c), are the biggest contributors to a textile fabric force–displacement curve. Intra-yarn slip is where the continuous fibres within the yarn slide past one another along the length of the fibre because of changes in fibre curvature during bending and unbending. The yarn stretching, Fig. 7.2(f), while not so prominent in the early stages of fabric deformation, is certainly present and contributes significantly to the deformation at larger strains.

Another fabric deformation mechanism to consider is yarn compression, Fig. 7.2(g), where forces at yarn crossover points compress the filaments in the yarn and cause them to flatten out and conform to the curvature of perpendicular yarns. Like fibre stretching, this can also be considered relatively insignificant; it only starts to contribute to the load–extension curve once the aforementioned mechanisms have been exhausted.

Finally, the yarn twist, Fig. 7.2(h), which has been observed in knitted fabrics and not so much in woven fabrics, imparts further resistance to the fabric deformation. This is where the yarn is subjected to one full turn during the manufacture of the fabric in order to create the looping structure of the knit. The twist creates a resistance to the increase in yarn curvature during fabric deformation.

7.3 Structure of multiaxial warp-knitted fabrics

The basic structure of MWK fabrics has been discussed in previous chapters (e.g. Chapter 3). The typical structure of an MWK fabric is illustrated in Fig. 7.3. A variation of this structure contains two bias inserting yarn systems, hereafter called TBMWK fabric, which is the main concern of this chapter and will be introduced in later sections.

MWK fabrics may still be categorized as 2-D reinforcements even though there is a stitching system (chain or tricot loops) which partly orients along the thickness direction and accordingly improves the through-thickness strength as well as the interlaminar shear resistance. This kind of fabric thus can be applied for conforming to a 3-D surface, which is called the forming process. There have been many papers (Mack and Taylor, 1956; Robertson *et al.*, 1981, 1984; Heisey and Haller, 1988; Amirbayat and Hearle, 1989; Bassett and Postle, 1990; Bergsma, 1993; Aono *et al.*, 1994, 1996; Laroche and Vu-Khanh, 1994; Long, 1994; Hu *et al.*, 1998) published on the forming behaviour of woven fabrics, most of which are based on the main and

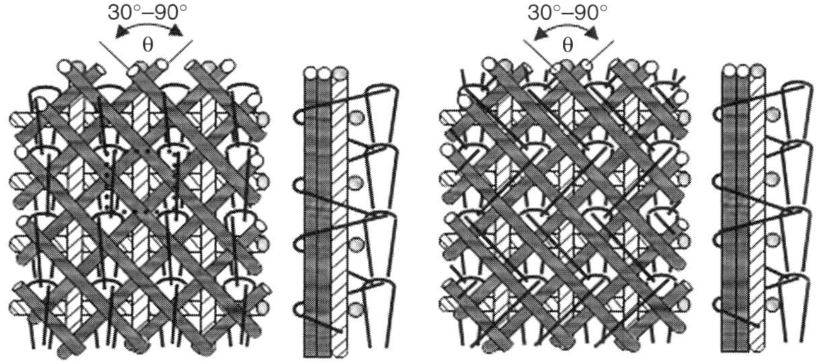

(a) Chain structure (b) Tricot structure

7.3 Typical structure of an MWK fabric.

important assumption that the two interlaced yarn systems in a plain weave are pin-jointed together at crossovers.

In the 2-D fabric forming process, the main deformation modes usually include in-plane tension, transverse compression, in-plane shear and out-of-plane bending, of which the in-plane shear is the most important. It can be regarded as the dominant factor which dictates whether the desired 3-D surface can be formed or not.

7.4 Deformation characteristics of woven fabrics during the forming process

For woven fabrics, the in-plane shear deformation directly determines the magnitude of the locking angle (Bergsma, 1993), which determines the jamming state during forming. It can be seen according to experiments that woven fabrics are formed into a 3-D surface just through the changes of the original right angle between the warps and wefts, as illustrated in Fig. 7.4 (a woven fabric with yarns of glass filament bundles). This has also been shown by many other authors (Mack and Taylor, 1956; Robertson *et al.*, 1981, 1984). This means that, under the assumption of inextensibility of the constituent yarns, the in-plane shear deformation (trellis effect) plays the leading role in the forming process of woven fabrics. If the fabric can be sheared to a great degree, which means the locking angle is sufficiently small, better forming will result and a more complex 3-D surface may be formed. Once the locking angle is exceeded, wrinkles or buckling will occur.

7.5 Deformation characteristics of multiaxial warp-knitted fabrics during the forming process

For MWK fabrics the situation is quite different due to the different geometrical structures. Our earlier study (Hu *et al.*, 1998) suggests that the typical structure of an MWK fabric (Fig. 7.3) presents an isotropy in the in-plane tensile properties, which is the unique mechanical advantage of this kind of fabric. However, this is at the same time a disadvantage owing to the possibility of deformation during the forming process. The dotted

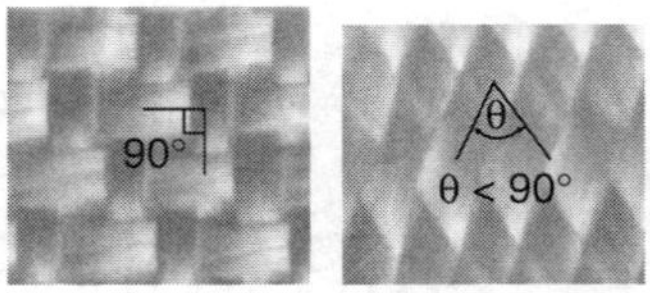

(a) Undeformed state (b) Deformed state

7.4 Trellis effect of the woven fabric.

square marked in Fig. 7.3(a) represents a unit cell which can never be sheared easily no matter how the in-plane shear force is exerted, since in all situations there will be one or more systems of inserting yarns stretched by the shear force. This means that the typical structure of MWK fabric is more difficult to deform, i.e. to conform to a 3-D surface. As illustrated in Fig. 7.5(a), the hemisphere pressing experiment gives a proof to this analysis in which wrinkles occur. However, a variant structure of MWK fabric (as shown in Fig. 7.6) shows quite good conformability, as suggested by

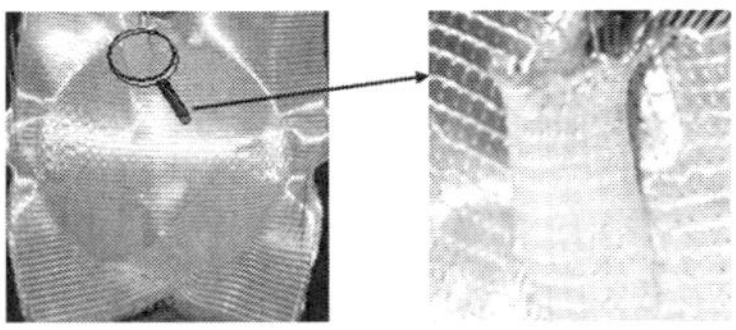

(a) Typical structure (as shown in Fig. 7.3, inserting yarns are glass fibre bundles)

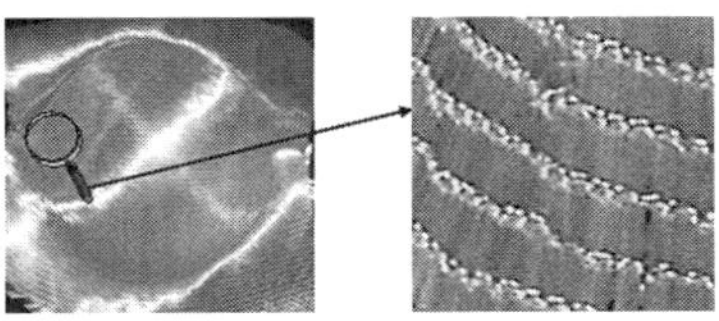

(b) MWK fabric with only two bias inserting systems (as shown in Fig. 7.4)

7.5 Hemisphere-pressing experiments on MWK fabrics.

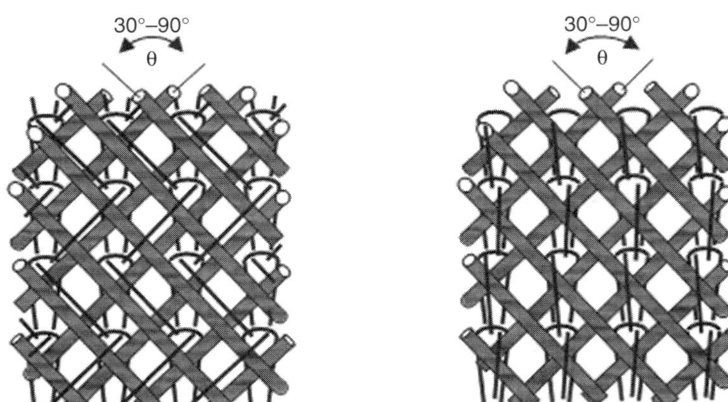

(a) Tricot structure (b) Chain structure

7.6 A variation of MWK fabric (TBMWK) containing only two bias inserting yarn systems.

Fig. 7.5(b), in which the inserting yarns come gathered together along the weft direction.

It can be inferred from Fig. 7.5 that the typical structure of MWK fabric is more suitable for plate or low-curvature-shell composite material, and the structure shown in Fig. 7.6 (hereafter in this chapter called two-bias multiaxial warp-knitted (TBMWK) fabric) has great potential in forming 3-D complex preforms. In the subsequent sections of this chapter, discussions will be concentrated on deformations of TBMWK fabrics.

7.6 Deformation behaviour of two-bias multiaxial warp-knitted fabrics

7.6.1 Relative movement of two inserting yarn systems

As shown in Fig. 7.6, the TBMWK fabric contains only two bias inserting yarn systems. Since the inserting yarn systems are not interlaced as the woven fabric but overlapped together, there are obviously fewer frictional constraints at the contact area. This means that the relative movement between the two inserting systems will be easier. Actually, the angle between the two systems can nearly approach 0° after deformation, as illustrated in Fig. 7.7(a), in which both inserting yarn systems become nearly parallel to the weft direction. However, the analogous phenomenon, that both inserting yarn systems become nearly parallel to the warp direction when the exerted forces change direction, does not occur as might be expected, as denoted in Fig. 7.7(b).

Figure 7.7 gives the deformations of the TBMWK fabric with tricot loops (Fig. 7.6(a)). The other structure, TBMWK fabric with chain loops (Fig. 7.6(b)), presents a similar deformation, as shown in Fig. 7.8.

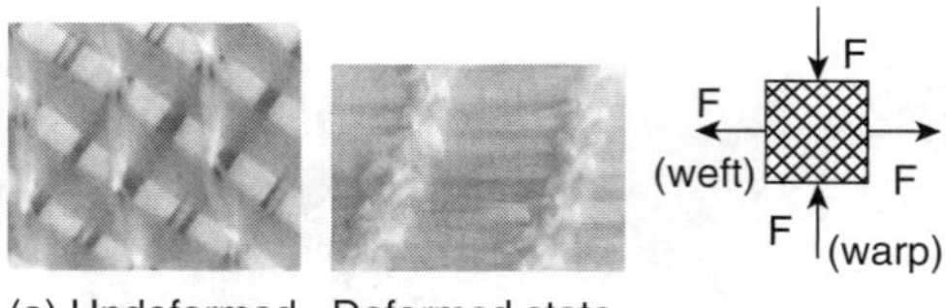

(a) Undeformed state Deformed state

(b) Undeformed state Deformed state

7.7 Deformations of a TBMWK fabric with tricot loops.

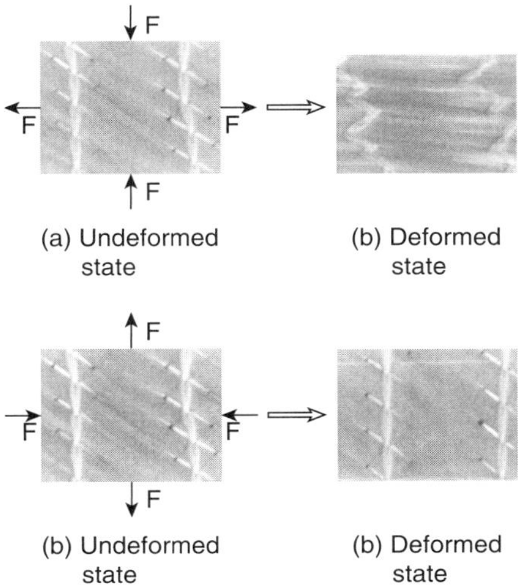

7.8 Deformations of a TBMWK fabric with chain loops.

7.6.2 Roles of stitching loops on the deformation of two-bias multiaxial warp-knitted fabrics

The inserting yarns always being gathered along the weft direction is the unique deformation characteristic of TBMWK fabric, which is quite different from woven fabrics. The deformation of woven fabric should be similar under the exerted force conditions of Fig. 7.7(a) and (b) or Fig. 7.8(a) and (b), i.e. the yarns will tend to orient along the ±45° directions (measured from the warp or weft direction). This has been proved from the experiment of Robertson *et al.* (1981) of shaping cotton cheesecloth around a bowling ball. Apparently, this special deformation behaviour of TBMWK fabric must result from the stitching system – tricot loops or chains. Microphotographs of deformations of these two kinds of stitches under stretching force can clearly explain this deformation characteristic, as shown in Fig. 7.9.

Although the tricot fabric can extend in both the weft and warp directions, the fabric elongation along the warp direction is much smaller than that along the weft direction (Fig. 7.9(a)). The extensibility of the chain loops (Fig. 7.9(b)) is also much smaller (3% or so for polyester filament yarn of 15 Tex). As shown in Figs 7.7(b) and 7.8(b), when the inserting yarns in the TBMWK fabric tend to orient along the warp direction, the stitching system (just being stretched in the same direction) constrains their movement. From this point of view, when the TBMWK fabric conforms to a 3D surface, the quick and easy response of the inserting yarns is to orient along the weft direction, as justified by Fig. 7.5(b). This deformation behaviour of

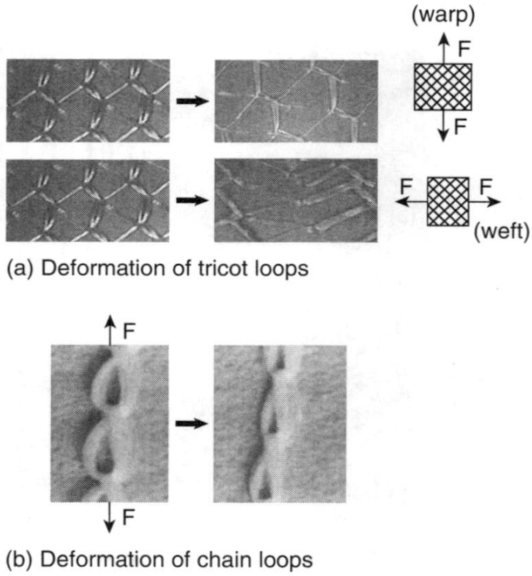

7.9 Deformations of tricot loops and chains.

TBMWK fabric is of great importance as far as the design of the forming process is concerned.

7.6.3 Hemisphere-pressing experiments

Hemisphere-pressing experiments were conducted on three sample fabrics in order to study the forming behaviour of TNMWK fabric. Specifications of the samples are listed below:

- Sample 1: TBMWK fabric, consisting of two bias (±45°) inserting yarn systems (glass filament bundles, count of 300 Tex, density of 6.8 yarns per cm, yarn width of 0.98 mm), held by tricot loops (polyester filaments, count of 15 Tex).
- Sample 2: TBMWK fabric, consisting of two bias (±45°) inserting yarn systems (glass filament bundles, count of 250 Tex, density of 10.3 yarns per cm, yarn width of 0.96 mm), held by chain loops (polyester filaments, count of 15 Tex).
- Sample 3: plain woven fabric, consisting of two interlaced yarn systems (glass filament bundles, count of 300 Tex, density of 11 yarns per cm, yarn width of 0.86 mm).

A series of hemispheres of different radii ranging from 5 to 15 cm are chosen in order to study the effect of the magnitude of the radius on the forming results. The angle changes between yarns are measured along the weft direction for TBMWK fabrics or along the ±45° directions for

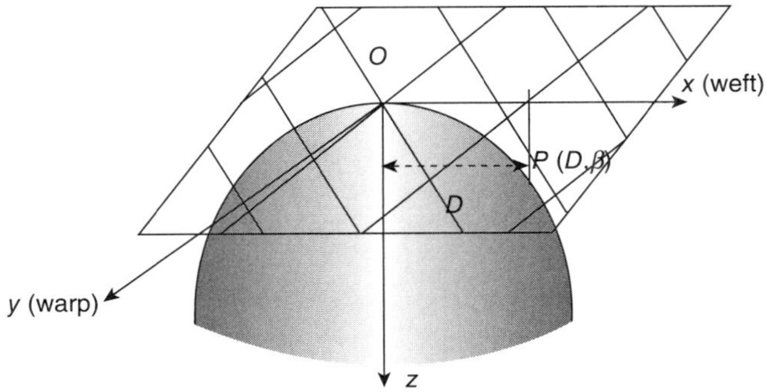

7.10 Illustration of measurement.

the plain-woven fabric. In addition, comparisons are also made between the patterns of flat TBMWK fabric and plain-woven fabric that can yield the corresponding hemisphere surface.

7.6.4 Theoretical analysis

As shown in Fig. 7.10, we assume that the centre point O (assumed to be some crossover point) of the original flat sample fabric happens to touch the pole of the pressing hemisphere after forming. Measurements are made of angles (denoted β in degrees) as well as the corresponding distances (denoted D in cm). Angles β are between the two bias inserting yarn systems along the weft direction for Samples 1 and 2, and between warps and wefts along the ±45° directions for Sample 3. D is the distance from the measured point (such as P) to the longitudinal axis of the hemisphere. In Fig. 7.11, the relation between D and β is presented for hemispheres with radii of 7.5 and 11 cm.

In order to consider comprehensively the effect of the magnitude of the radius of the pressing hemisphere on the relation between D and β, more pressing experiments (the radius of the pressing hemisphere ranging from 5 to 15 cm) were carried out on Sample 1. In Fig. 7.12, the relation between the slope of the trendline for the plot of D versus β and the radius of the pressing hemisphere, as well as the relation between the correlation coefficient of the trendline for the plot of D versus β and the radius of the pressing hemisphere, are presented. It can be inferred from Figs 7.11 and 7.12 that:

- The plot between D and β basically possesses good linearity, and the larger the radius of the pressing hemisphere, the higher the linearity.
- The slope of the trendline for the plot of D versus β is linear with the radius of the pressing hemisphere, and the larger the radius, the smaller the absolute value of the slope.

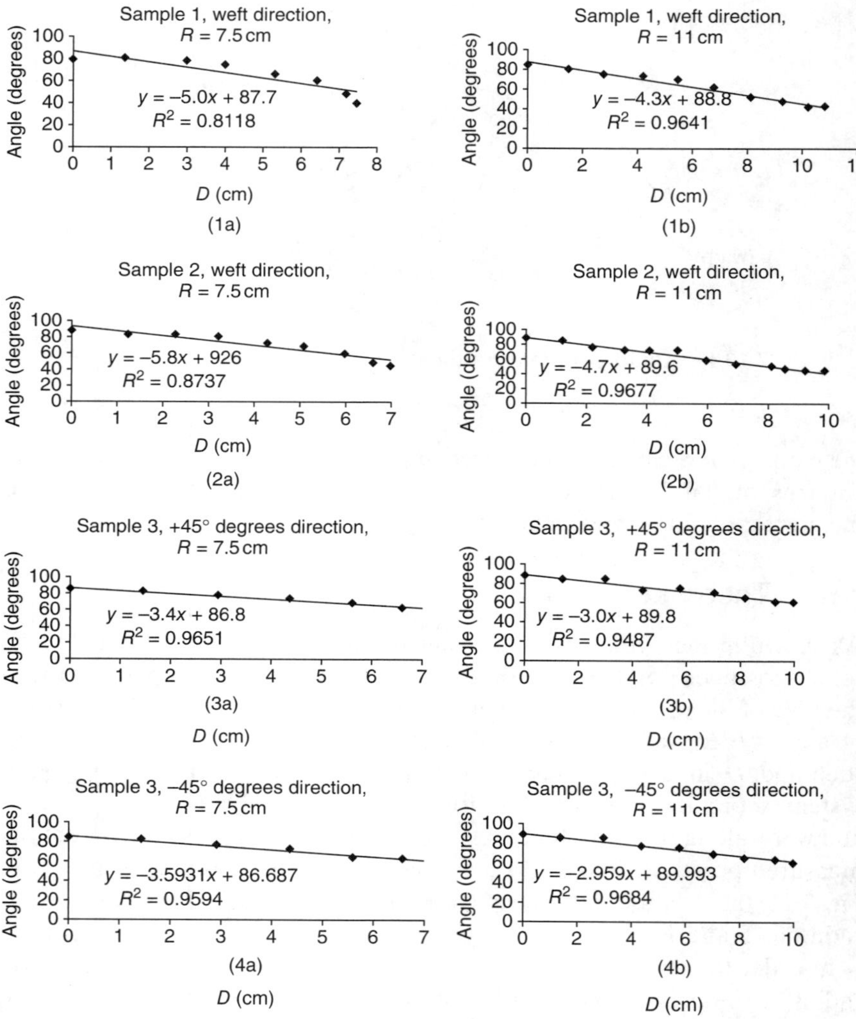

7.11 The relation between D and β for hemispheres with radii of 7.5 and 11 cm.

- The correlation coefficient of the trendline for the plot of D versus β is also linear with the radius of the pressing hemisphere, and the larger the radius, the larger the correlation coefficient (close to unity).
- The inserting yarns of TBMWK fabrics tend to become gathered always along the weft direction since the slope for the plot D versus β is always negative.
- The plots of D versus β for Sample 3 along the –45 and +45° directions are just analogous.

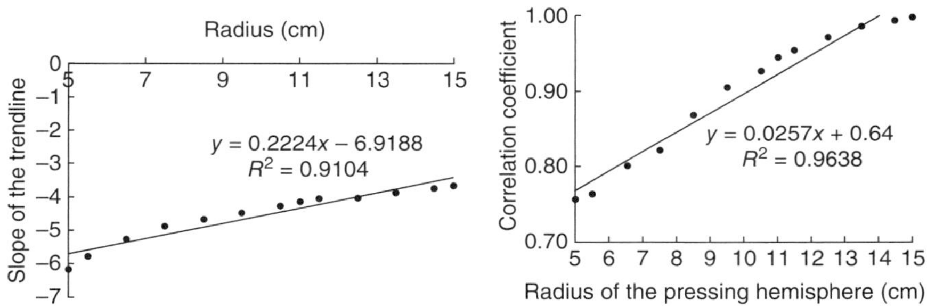

7.12 Effect of radius magnitude of a pressing hemisphere on the relation between D and β.

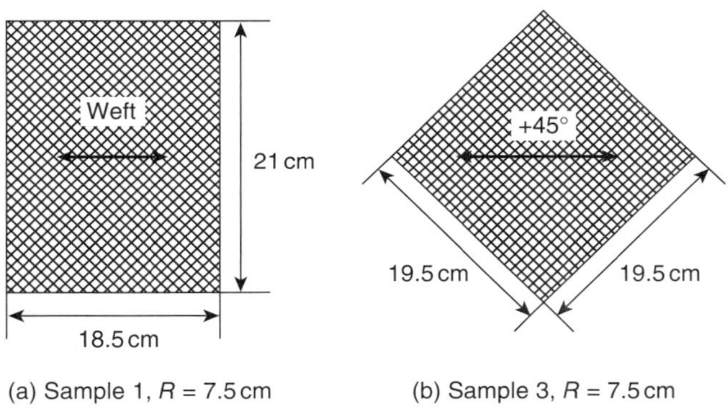

7.13 Schematic diagram of the flat patterns of both TBMWK fabric and woven fabric that can yield a hemisphere of diameter 7.5 cm.

Other hemisphere-pressing experiments show that the plot of D versus β is non-linear and the correlation coefficient of the trendline decreases below 0.8 when the radius of the pressing hemisphere is less than 7 cm. It is also impossible to represent the relationship between D and β by a simple equation under this condition.

Figures 7.11 and 7.12 just provide proofs to the foregoing analysis of the deformation behaviour of TBMWK fabrics as well as the woven fabrics. In addition, it can also be seen from the schematic diagram in Fig. 7.13 that the two inserting yarn systems of a TBMWK fabric tend to become gathered along the weft direction during the hemisphere-pressing process. Figure 7.13 gives the schematic flat patterns of both TBMWK fabric and woven fabric that can yield a hemisphere of diameter 7.5 cm, in which pattern (a) is close to rectangular while pattern (b) is close to square.

7.7 Modelling the formability of two-bias multiaxial warp-knitted fabrics

As suggested by Fig. 7.11, the relation between D and β presents very good linearity and the slope of the trendline is not sensitive to the diameter of the pressing hemisphere provided that the radius of the pressing hemisphere is more than 7 cm. Accordingly, it is reasonable to use one linear equation to represent the relation between D and β for a specified radius (more than 7 cm) of the pressing hemisphere. Based on this point of view, a model for predicting the deformation of TBMWK fabrics during the hemisphere-forming process is established. The model makes two main assumptions: (1) the distance between any two adjacent crossovers along any single inserting yarn remains unchanged; and (2) the relative slippage between the two inserting yarn systems is ignored in consideration of the constraints of the stitching system.

7.7.1 Modelling the forming process of two-bias multiaxial warp-knitted fabrics

An illustration of the vertical view of the sample fabric after deformation is given in Fig. 7.14, in which the xOy coordinate system simply corresponds to that in Fig. 7.10.

If we use the equation $\alpha = Kx + \alpha_0$ to represent the relation between D and β (Fig. 7.10), the coordinates of point A (1,1) in the 3-D Cartesian coordinate system (Fig. 7.14) are

$$\begin{cases} A_x(1,1) = L_0 * \sqrt{(\cos^2(\alpha_0/2) - L_0^2/4R^2)} \\ A_y(1,1) = -L_0 * \sin(\alpha_0/2) \\ A_z(1,1) = L_0^2/2R \end{cases}$$

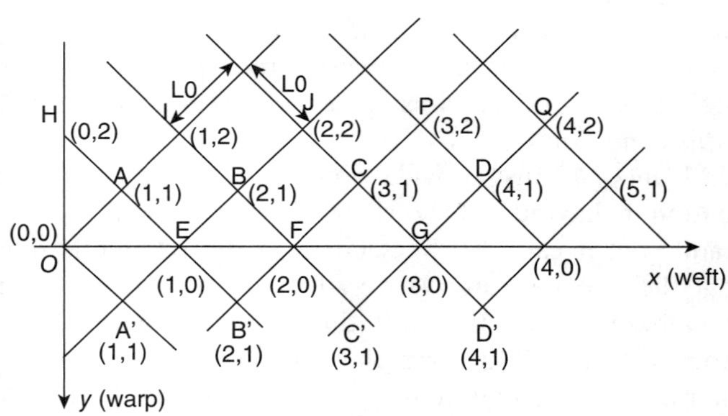

7.14 Illustration of the vertical view of the sample fabric after deformation.

in which α_0 corresponds to angle AOA'. Then the coordinates of the symmetrical point A' of A about plane xOz can be obtained. To locate the node point E (1,0), we consider the points of intersection of three spheres: one centred at A (1,1) of radius L_0, another centred at A' (1,1) of radius L_0, and the third the hemisphere of radius R. These three spheres will intersect at two points: O (0,0) and E (1,0). Here and in each subsequent step we will choose the point furthest from the pole. According to the angle BEB' (E_x (1,0)$K + \alpha_0$) and the coordinates of point E, the locations of B and B' can be determined. Following these steps, the locations of all the node points $O, E, F, G, \ldots$ and $A, B, C, D, \ldots$ can be calculated. The coordinates of all the other points can also be obtained through the same method, i.e. solving the group of equations of three spheres. For example, to locate the node point H (0,2), we consider the points of intersection of three spheres: one centred at A'' (1,1) of radius L_0 (A'' is the symmetrical point of A about the plane yOz), another centred at A (1,1) and the third the hemisphere of radius R. To locate the node point P (3,2), we consider the points of intersection of three spheres: one centred at C (3,1) of radius L_0, another centred at D (4,1) of radius L_0, and the third the hemisphere of radius R.

Since the angles between the two inserting yarn systems along the weft direction can nearly become 0° (as shown in Figs 7.5(a) and 7.6(a)), we have reason to use the distances (d_C) between the pairs of symmetrical points (AA', BB', CC', DD', . . .) as criteria to determine whether wrinkles occur or not. If d_C becomes less than the width of a single inserting yarn during theoretical calculation, we assume that wrinkles occur.

7.7.2 Verification of the model

In Fig. 7.15, the calculated vertical view (xOy plane) and side view (yOz plane) of Sample 1 after deformation in the hemisphere-pressing experi-

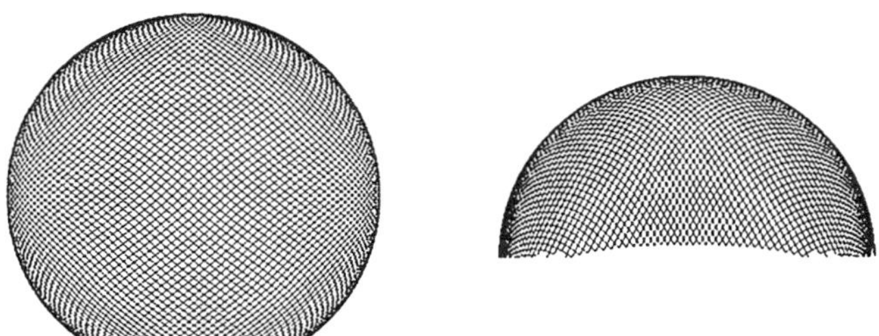

(a) Vertical view (b) Side view

7.15 Calculated vertical view and side view of Sample 1 after deformation in the hemisphere-pressing experiment (Sample 1, $R = 7.5$ cm).

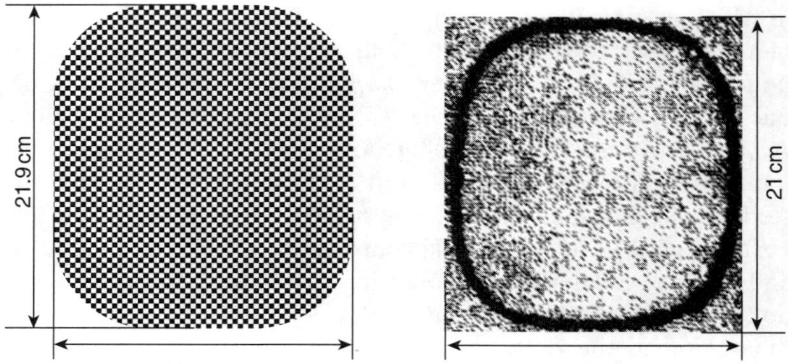

18.1 cm
(a) Calculated results

18.5 cm
(b) Experimental results

7.16 Comparison between the calculated flat shape and the experimental one.

ment (R = 7.5 cm) are given. As far as Sample 1 TBMWK fabric is concerned, no wrinkles occur even at the equator area of the hemisphere. In Fig. 7.16(a), the calculated flat shape of the fabric (Sample 1) which can yield the hemisphere (R = 7.5 cm) is shown. It fits the experimental result (Fig. 7.16(b)) very well since the differences between them are only about 2% and 4% along the length and width directions respectively. The results for Sample 2 are similar to those of Sample 1.

According to the model established before, all crossover points on the 2-D fabric can be transformed to 3-D coordinates. Therefore, the local deformations, i.e. the local angle changes between the two bias inserting yarn systems, can also be predicted. In Fig. 7.17, comparisons between the calculated and measured local deformations are illustrated, in which the radius of the pressing hemisphere is 11 cm. As shown in Fig. 7.10, if the z-axis minus is described as the north direction, then Fig. 7.17(a) represents latitude 60°N, and Fig. 7.17(b) represents latitude 30°N. All calculated or measured angles are from the +45° inserting yarn system to the –45° inserting yarn system. It can be noted that the calculated local deformations fit the experimental ones very well.

7.8 Summary

Based on the characterization of the deformation behaviour of TBMWK fabrics during the hemisphere-pressing process, a model for predicting possible wrinkles as well as the flat shape of fabric that can yield the hemisphere was established. Comparison between the theory and the experimental results suggested that the model fits the experimental results very well.

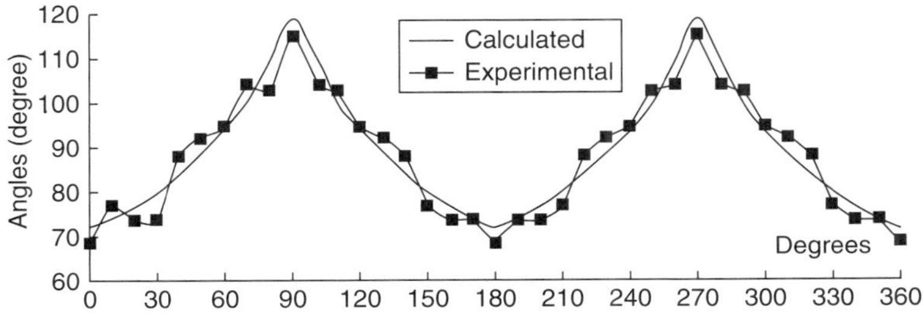

(a) Angle changes on latitude 60°N
(0° on the horizontal axis represents weft direction)

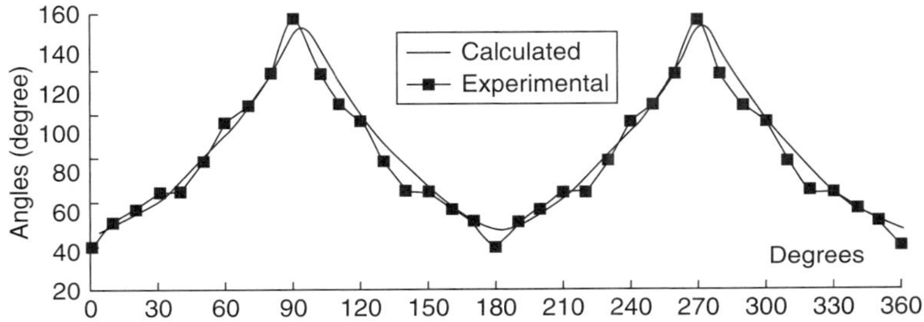

(b) Angle changes on latitude 30°N
(0° on the horizontal axis represents weft direction)

7.17 Comparisons between the calculated and measured local deformations.

It was found that the typical structure of MWK fabric does not possess satisfactory conformability, which is suitable for plate or low-curvature shell materials. TBMWK fabrics, however, possess much better conformability. It was noticed that the two bias inserting yarn systems tend to become gathered together always along the weft direction during the hemisphere-forming process, which is quite different from that of woven fabrics in which the yarns tend to orient along the ±45° directions. In addition, the angle between the two bias inserting yarn systems can reach nearly 0° during deformation; the woven fabrics, however, possess some locking angle, which is usually much larger than 0°.

An important phenomenon was found that the angles between the two bias inserting yarn systems in a TBMWK fabric along the weft direction during the hemisphere-forming process are basically linearly related to the perpendicular distances from the measured points to the longitudinal axis of the hemisphere, provided that the radius of the pressing hemisphere is larger than 7 cm. The larger the radius, the larger is the correlation coefficient (close to unity). The slope of the trendline is also linearly related to the magnitude of the radii of the pressing hemispheres, and the larger the radius, the smaller the absolute value of the slope. In addition, the shape of flat TBMWK fabric that can yield the corresponding hemisphere is close to rectangular, not square as presented by the woven fabric. In addition, the predicted local deformations of a TBMWK fabric during the hemisphere-pressing process also fit the experimental results very well.

7.9 References

Amirbayat J and Hearle J W S (1989), The anatomy of buckling of textile fabrics: drape and conformability, *Journal of the Textile Institute*, **80**, 1, 51–70.

Aono M, Breen D E and Wozny M J (1994), Fitting a woven-cloth model to a curved surface: mapping algorithms, *Computer-Aided Design*, **26**, 4, 278–292.

Aono M, Denti P, Breen D E and Wozny M J (1996), Fitting a woven cloth model to a curved surface: dart insertion, *IEEE Computer Graphics and Applications (Computer Graphics in Textiles and Apparel),* 60–69.

Bassett R J and Postle R (1990), Fabric mechanical and physical properties, Part 4: The fitting of woven fabrics to a three-dimensional surface, *International Journal of Clothing Science and Technology,* **2**, 1, 26–31.

Bergsma O K (1993), Computer simulation of 3D forming processes of fabric reinforced Plastics, *Proc. 9th Int. Conf. on Composite Materials (ICCM-9)*, 560–567.

Christie G R (1997), Numerical modeling of fibre-reinforced thermoplastic sheet forming, PhD Thesis, Department of Mechanical Engineering, University of Auckland.

Cogswell F N (1992), *Thermoplastic Aromatic Polymer Composites*, Butterworth-Heinemann, Oxford, UK and Boston, MA.

Duhovic M and Bhattacharyya D (2005), Deformation mechanisms in knitted fabric composites, in *Polymer Composites from Nano-to-Macro Scale* (ed. K Friedrich, S Fakirov and Z Zhang), Springer, New York, 265–288.

Heisey F L and Haller K D (1988), Fitting woven fabrics to surfaces in three dimensions, *Journal of the Textile Institute*, **79**, 2, 250–263.

Hu J, Jiang Y M and Ko F (1998), Modeling of bending properties of multiaxial warp knitted fabrics, *Textile Research Journal*, **68**, 11, 828–834.

Kaufmann J R (1991), Industrial applications of multiaxial warp knit composites, Chapter 5 in *High-Tech Fibrous Materials* (ed. Tyrone L Vigo and Albin F Turbak), American Chemical Society, Washington, DC, 81–89.

Krebs J E and Bhattacharyya D (1998), A direct comparison of matched-die versus diaphragm forming, *Composites, Part A*, **29**, 183.

Laroche D and Vu-Khanh T (1994), Forming of woven fabric composites, *Journal of Composite Materials,* **28**, 18, 1825–1839.

Lim T C, Ramakrishna S and Shang H M (1998), Improvement of knitted fabric composite sheet formability by simultaneous deep drawing and stretch forming, *First Asian–Australasian Conf. on Composite Materials (ACCM-1)*, **1**, 409.

Long A C (1994), A simulation of reinforcement deformation during the production of preforms for liquid molding processes, *I. Mech. E., J. Eng. Manuf.*, **208**, 269–278.

Mack C and Taylor H M (1956), The fitting of woven cloth to surfaces, *Journal of the Textile Institute,* **47**, 8, T477–T488.

Martin T A, Christie G R and Bhattacharyya D (1997), Grid strain analysis and its application in composite sheet forming, in *Composite Sheet Forming* (ed. D Bhattacharyya), Elsevier, New York, 217.

Mayer J, Haan J D, Reber R and Wintermantel E (1998), Knitted carbon fibre reinforced thermoplastics: An overview, *First Asian–Australasian Conf. on Composite Materials (ACCM-1)*, **1**, 401.

Mouritz A P, Bannister M K, Falzon P J and Leong K H (1999), Review of applications for advanced three-dimensional fibre textile composites, *Composites*, Part A, **30**, 1445–1461.

Robertson R E, Hsiue E S, Sickafus E N and Yeh G S Y (1981), Fiber rearrangements during the molding of continuous fiber composites. I. Flat cloth to a hemisphere, *Polymer Composites*, **2**, 3, 126–131.

Robertson R E, Hsiue E S and Yeh G S Y (1984), Fiber rearrangements during the molding of fiber composites. II. Flat cloth to a rounded cone, *Polymer Composites*, **5**, 3, 191–197.

Savci S, Curiskis J I and Pailthorpe M T (2000), A study of the deformation of weft knit preforms for advanced composite structures. Part 2: The resultant composite, *Composites Science and Technology*, **60**, 1943.

Takano N, Ohnishi Y, Zako M and Nishiyabu K (2001), Microstructure-based deep drawing simulation of knitted fabric reinforced thermoplastics by homogenization theory, *International Journal of Solids Structures*, **38**, 6333.

8

Permeability of multilayer woven fabrics

Abstract: In this chapter, a comprehensive study on the permeability modelling of multilayer woven (MLW) fabrics is presented. A framework for flow permeability measurement in resin transfer moulding is described. The Darcy law is used to model the flow through the reinforcement fibres where permeability is a measure of the resistance of the fibres to the flow. Two types of woven fabrics, one fabricated using monofilaments so as to eliminate the effects of other factors such as fibre bundle on the permeability, and the other by using multifilament yarns, are used for modelling. Two models are developed to explain the permeability of these two different varieties of multilayer woven fabrics.

Key words: multilayer woven (MLW) fabrics, permeability of MLW fabrics, Darcy's law, modelling permeability of MLW fabrics, fractal permeability model.

8.1 Introduction

Permeability is a geometric characteristic related to the structural features of the textile at several length scales. It is an important property that affects the comfortability of fabric. In the manufacture of composites with textile reinforcement, the permeability plays a key role (Guangbiao Xu and Fumei Wang, 2005). It is the most important parameter influencing the flow through the porous medium and relates the velocity of the flow to the pressure gradient within the medium. In other words, permeability is a measure of the ability of the resin to flow through the preform. In the Liquid Composite Moulding (LCM) process, such as Resin Transfer Moulding (RTM) and Reaction Injection Moulding (RIM), for fibre composites, the flow characteristics, such as the permeability and flow pattern, of the fibre reinforcements play a crucial role in the fill process and finished properties of the composites. Thus, if the permeability of the reinforcements can be acquired, the fluid progress and fill time during the injection process may be accurately predicted to obtain an artistically finished product with minimum void content.

The prediction of textile permeability is important due to the often encountered problems of non-uniform impregnation, which may even involve void and dry spot formation. Permeability can be highly directional (anisotropic), a serious concern for mould design in terms of the placement of inlets, and vents, and is directly affected by the amount of compaction of the preform in the mould as well as the preform architecture.

One of the key issues in the manufacture of polymer composites is the resin infiltration of the fibrous preform. The resin transfer moulding (RTM) method is generally used to manufacture composite parts. In this method, the reinforcing fibres are placed in a mould cavity and the resin is injected to fill up the empty spaces. After the resin cures, the mould is opened and the part ejected. The resin must infiltrate the whole preform and displace the air between the fibres within the time limits dictated by its processing parameters and pressures limited by the equipment and mould design. This is not necessarily easy to accomplish. To predict the necessary pressures and filling times and the proper locations for the inlet ports for resin injection and of the vents for air ejection, it is necessary to model the resin infiltration process. A key to this modelling is permeability which characterizes the resistance of fibres to the flow of infiltrating resin.

It is apparent that the permeability of a textile preform (transport of resin) is a key aspect of fibre-reinforced plastic composite fabrication processes such as RTM. As one of the elemental performances of fibre materials, the importance of the permeability lies in the fact that, firstly, it can determine the filling and packing time in injection, and mould design, locate the injection port and exhaust port, thus becoming a critical parameter for numerical simulation; secondly, it can determine fluid flow attribute, and degree of impregnation, thus becoming a key factor in the manufacture of high-performance products.

Problems related to the permeability include the formation of dry spots or voids within the preforms. These dry spots and voids will have an adverse effect on the performance of products, so it is important to eliminate them. But this problem is not easy to solve, particularly when the preform is composed of multiple layers of reinforcement. Only after analyzing the relationship between fabric permeability and fabric construction is it possible to eliminate dry spots and voids to improve the performance of the products.

Studies of the permeability of textile preforms have been reviewed by Lee (1997) and Rudd *et al.* (1997). The preform permeability can be obtained by experiments, empirical and analytical equations, and numerical simulations (Skartsis *et al.*, 1992; Berdichevsky and Cai, 1993; Wang *et al.*, 1994).

For resin flow through the interstices and between layers of fibre beds, the Newtonian flows in macroscopically homogeneous domains were described originally by Darcy (1856), who developed the constitutive description by observing water flow through beds of sand. Application of Darcy's model to composite processes, from laminates to moulded materials, is widespread. The percolation flow is generally applied to thermosetting resins in autoclave (laminate) or liquid moulding processes.

The familiar Darcy model may be written $\bar{v} = \frac{-K}{\mu} \nabla P$, where v is the average fluid velocity, μ is the fluid viscosity, K is the permeability of the porous medium, and ∇P is the pressure gradient and can be written in tensorial form for anisotropic preforms, where up to three scalar permeabilities are required in the plane case. Such implementations thus require determination of the permeability or components of a permeability tensor, by experimental or theoretical means. In application to actual processing, various factors are commonly used to account for flow 'tortuosity', or circuitousness of the path that must be traversed to penetrate the material, the shape of the particles, material anisotropy, and the average volume fraction. Several modifications to Darcy's law have thus been proposed in the polymer processing arena and in other areas of fluid–structure interaction; for example, Kozeny (1927) treated a permeated porous medium as a bundle of capillary tubes and obtained a relationship to adapt Darcy's law to include capillarity effects with an empirical relation; Blake (1922) derived a similar expression. Carman (1937) modified Kozeny's work by defining S, the specific surface with respect to a unit volume of solid, instead of a unit volume of porous medium. The 'Kozeny–Carman relationship' arose through these sequential contributions. Carman experimentally determined a range of 'Kozeny constants' for a variety of packing schemes and geometries of reinforcements.

Such relations have commonly been used to model polymeric flow in composite materials in the last 20 years (Coulter and Guceri 1988). Williams *et al.* (1974) considered the flow of several fluids through aligned reinforcements, both dry and presaturated with liquid. They obtained higher permeabilities for saturated than for unsaturated reinforcement, as did Martin and Son (1986). Many authors since the late 1980s have specifically studied the permeability of fibrous preforms for liquid moulding. More recent work has focused on complex geometries. Rudd *et al.* (1996) established a 'permeability map' for complex geometries. Smith *et al.* (1997) related permeabilities of sheared fabrics to ply angle. Lai and Young (1997) related similar experimental data to a geometry-based flow model.

Other closed-form modifications to Darcy's law have been developed to relate volume fraction and geometric or empirical constants such as the maximum packing fraction to the permeability of a periodic medium comprised of parallel cylinders. Gebart (1992) derived an equivalent permeability based on the assumption of hexagonally arranged fibres. Cai and Berdichevsky (1993) extended a classic self-consistent approach, wherein a heterogeneous element was assumed to be embedded in an equivalent homogeneous medium. The homogeneous medium was constructed such that the total flow and dissipation energy remained the same. Use of a no-slip boundary condition at the fibre surface and zero velocity gradient in the radial direction at the domain boundaries resulted in an equivalent permeability value. The improved self-consistent method took into account an additional parameter V_A, the maximum packing capacity of fibres, to increase the accuracy of the model.

Bruschke and Advani (1993) developed a closed-form solution for permeability by matching the lubrication solution for low porosities and an analytical cell model solution for high porosities. They found agreement between their hybrid, arrangement-specific closed-form solution and a numerical solution of the Navier–Stokes equations for flow around both hexagonal and square arrangements of cylinders (using a simulation package, POLYFLOW). Van der Westhuizen and Du Plessis (1996) used phase-averaged Navier–Stokes equations to calculate the permeability of representative unit cells, and reported agreement with experimental in-plane permeabilities. Their model did not assume any particular arrangement of fibres for longitudinal permeability, but used the maximum packing capacity for different arrangements of fibres to create an effective volume fraction for transverse permeability. Wang (1996) developed a similar relation for an array of rectangular packed fibres.

Computational fluid dynamics software can be used to solve the full Navier–Stokes equations to determine flow progression, although in practice this can be done only for small domains. Several workers have developed intermediate special-purpose codes to model fluid flow in fabrics that account for effects such as capillarity, race tracking, saturation, etc. For example, Advani and co-workers developed 'LIMS' (Pillai and Advani, 1998), which is able to simulate edge effects by implementation of a mass sink to the continuity equation to account for saturation. Chang and Hourng (1998) developed a two-dimensional model for tow impregnation, taking into account micro/macroscopic flow and void formation. Ambrosi and Preziosi (1998) developed a model for flow dynamics in an elastically deforming environment.

Flow in tows versus gaps or voids has been specifically studied by a number of workers. Shih and Lee (1998), for example, used six different types of glass fibre reinforcements to determine the effect of fibre architecture on apparent permeability. The reinforcements included four-harness woven, plain weave, random fibre and stitched fibre mats. They argued that the gap size between the tows and the connectivity of the gaps control permeability. Kolodziej *et al.* (1998) proposed a theoretical model which accounted specifically for gaps as caverns or fissures inside the bundle of fibres. Both diameter of fibres and also diameter of gaps inside the fibre bundles were incorporated in the model. They reported that tow heterogeneity can decrease or increase tow permeability, depending on the critical dimensionless radius of these gaps. Heterogeneity was identified as the probable cause of deviation of permeabilities from permeability models, which assume uniformity inside the fibre bundles.

The numerical simulation is usually made by the finite element analysis (Berdichevsky and Cai, 1993), finite difference (Gebart, 1992) and control volume methods (Lee, 1997) for idealized and/or simple preform structures. Regarding woven fabric, Rebenfeld and his students investigated the effective permeability in terms of the pore structure, weave type, fabric layering, fibre orientation and compressibility. When considering the heterogeneity, the combination of Navier–Stokes and Darcy equations with different permeability for flow in different portions of a preform extended to complicated fabrics (Lee, 1997).

The flow through multilayer assemblies is complex. Adams and Rebenfeld (1991) found that the permeability of multilayer assemblies differed from those of their constituent layers. In the homogeneous assemblies, the interlayer pores can increase the effective in-plane permeability, while in heterogeneous assemblies the permeability and anisotropy are governed by high-permeability layers or directions. They suggested a transverse flow mechanism to fill the low-permeability layers and keep the fluid front macroscopically uniform. Mogavero and Advani (1997) investigated the effect of varying the order of lay-up of a fixed number of plies and the impact of varying the thickness of individual layers. Batch and Cumiskey (1990) investigated the interface between layers in relation to compressibility and developed a model for calculating average volume fraction for multilayer assemblies, where it was found that increasing the number of layers decreased the permeability. They attributed this to the blocking at the interface of adjacent layers, which created tortuous flow paths. Loos *et al.* (1991) studied the permeability of carbon multiaxial warp knit preforms. It was found that the introduction of through-thickness fibres significantly increased the permeability of the preforms, especially for the preforms with high volume fraction. The Kozeny–Carman equation was found to be adequate to provide a quantitative relationship between permeability and the preform porosity. But the through-thickness fibres here are still different from those in 3-D multilayer woven fabrics.

Lekakou *et al.* (1996b) investigated the relationship between the compressibility and flow permeability of woven fabric assemblies. Multilayer assembly here means separate single-layer woven stacks. The interface is related to the sloughing or nesting (Mogavero and Advani, 1997), which is different from the stitching in integrally woven multilayer assemblies. Ko and Du (1997) reported that stitching in multiaxial warp-knitted fabrics can increase the fabric permeability, but the stitching here is still different from that in 3-D multilayer assemblies.

A large body of literature exists about the effects of fabric structures on permeability. For 3-D multilayer woven fabrics, the key feature of stitched fabric is its interbundle stitches, but very limited information is available on the relationship between permeability and the effect of stitches. Shih and Lee (1998) proposed a parallel permeability model for flow through bidirectional stitched fabrics. The flow between the bundles is set by the Kozeny–Carman equation, and the effect of stitching was not specially mentioned. Cairns *et al.* (1999) developed a model that incorporates Darcy's law in fibrous bundle regions and the channel flow equations between bundles. This paper reported that two experimental systems, the stitched ±45 system and the stitched ±45/0 system, cannot be predicted well by using the presented model. Cairns *et al.* suggested that the permeability parameters needed to be re-evaluated with some account for the extensive stitching that was presented in these preforms. Loos *et al.* (1991) studied the permeability of carbon multiaxial warp-knitted preforms. It was found that the introduction of through-thickness fibres significantly increased the permeability of the preforms, especially for preforms with high volume fraction. The Kozeny–Carman equation was found to be adequate to provide a quantitative relationship between permeability and preform porosity.

Lundström (2000) also proposed a model for non-crimp stitched fabrics through theoretical analysis. It has been found that the flow in the inter-bundle channels was the most important and that the bundle flow could be neglected in computations of the overall flow rate and consequently also with regard to the permeability. The results showed that the permeability of the fabrics varied considerably as a function of the direction of infiltration, although the geometrical variations were small. In the production direction of the fabrics the permeability was generally two to three times higher than it was in the perpendicular direction. On the other hand, the spread of the results between several samples was always measured to be higher in the high flow direction. The experiments also revealed that a choice of larger bundles in a fabric did not necessarily result in a higher permeability. The proposed simple model works well for certain cases, while it overrates the permeability in other cases. This suggests that if the real fabric should have the same geometry as the model, its permeability would be higher. Therefore, it is really significant and urgent to carry out a quantitative study on the question of how the inter-layer stitching affects the effective permeability. Such a study will be very helpful for the rational design of the stitched fabrics, the exact pre-simulation of the moulding process, and the global optimization of the processing parameters.

In this chapter, a comprehensive study on the permeability modelling of multilayer woven fabrics is presented. A framework for flow permeability measurement in resin transfer moulding will be described. The Darcy law is used to model the flow through the reinforcement fibres where the permeability is a measure of the resistance of the fibres to the flow. Two types of woven fabrics, one fabricated from monofilaments so as to eliminate the effects of other factors such as fibre bundle on the permeability, and the other made by using multifilament yarns, have been used for modelling. Two models were developed to explain the permeability of these two different varieties of multilayer woven fabrics.

8.2 Fabric compressibility

Compressibility of preforms is of importance in the RTM process, wherein mould closing force and attainable fibre volume fraction are directly dependent on the fabric compressibility (Padaki *et al.*, 2006). Fabric compressibility is very important in all RTM processes, and affects both the material and processing properties of the part. As the fabric is compressed by fluid pressure or the mould surface the fibres get compacted and the fibre volume fraction increases. This decreases the thickness of the part, decreases the permeability, and decreases the porosity. Compressibility is possibly a more important concept in one-sided moulding processes than in closed-mould processes. In a closed-mould process the permeability and fabric thickness are fixed at a certain value which is determined by the mould gap. Throughout the process the permeability is constant and independent of the injection pressure. In one-sided moulding processes the compaction of the fabric can lead to several important phenomena. In processes where the flow is in the plane of the fabric such as VARTM and SCRIMP™, a part with non-uniform thickness can be created since the net compaction pressure varies throughout the mould.

In processes where the resin is forced though the thickness, the pressure applied to the fluid is also the pressure compacting the fabric. Therefore, the permeability and fabric thickness can change throughout a process and can depend on the pressure at which the process is taking place. This can create an interesting competing mechanism in these types of processes. According to Darcy's law, an increase in pressure will increase the velocity of the fluid though the fabric. However, increasing the pressure of the fluid will increase the compaction pressure and lower the permeability. It could be possible in certain cases for an increase in pressure to increase injection time, although this is not common. For most fabrics the decrease in thickness tends to compensate for the decreased permeability in through-thickness flow. The effect of compaction on permeability is very dependent on the fabric architecture, which means some fabrics are more affected than others.

Fabric compaction also affects the porosity of the fabric, which will affect the saturation time for unsaturated flow. This fact adds yet another complication to the problem. Although permeability decreases with compaction, the decrease in porosity can increase the velocity of the fluid through a preform. Decreasing the porosity also increases the capillary pressure. However, in most cases these effects are minor.

As a fabric is compressed there are three distinct regimes that have been identified. The first is where the spacing in the fabric caused by the stitching and weaving is compressed. This occurs at very low pressures, and results in fibre-on-fibre contact. This region is also very linear in nature. In the second regime, both the solid and the voids are compressed. This is the most complex region, and is the most studied. Very complex models have been generated to predict the behaviour of the fabric in this region. Although the fibres are touching, they are still moving due to fibre bending, slippage and nesting. The third region is where the fabric has been fully compressed. Most fabrics are fully compressed with 1–2 MPa pressure. In the third regime, all the fibres have been manipulated into a stable position and cannot be moved any further. The only compression occurring in this regime is due to the solid material compressing. Many fabrics have compressed to half their original thickness by this point.

8.3 Permeability testing

The permeability characterizes the porous material in terms of resistance to the fluid flow for a given injection pressure. The greater the value of permeability, the faster the resin permeates through the fibrous reinforcement. In the case of an isotropic fabric, the permeability tensor reduces to a single scalar. The most general situation of orthotropic reinforcement requires three values, k_1, k_2 and k_3, for the permeability in the principal directions of the material. However, most RTM moulded parts are thin shells and the preform does not consist of a multilayer reinforcement. So it is not always necessary to determine the through-thickness permeability k_3. Hence it is essential to measure the first two principal permeabilities, k_1 and k_2, the so-called in-plane permeabilities (Ferland *et al.*, 1996).

For flow through the fabric, it is extremely difficult to calculate the permeability constant (K) by knowing only the geometric information. Micro-models exist for estimating the permeability of a fabric given fibre diameters, fibre spacing and other relevant information (Simacek and Advani, 1996; Cairns *et al.*, 1999). However, these models are very complex and have varying accuracy. In addition, tests must still be performed in order to determine some of the parameters needed as input to the models. The most accurate and direct way to determine the permeability is through testing. By knowing the velocity, the pressure drop, and the viscosity of a fluid moving through a fabric, the permeability can be calculated. Because most RTM modelling has been done for closed-mould processes, the permeability in the plane of the fabric was typically of the greatest concern (Adams and Rebenfeld, 1991; Ferland *et al.*, 1996; Lai and Young, 1997).

For this reason, the majority of available permeability data is for flow in the plane. For the two-stage processes such as pressure bag moulding, the most important value is the permeability through the thickness. This is because the distribution channel covers the surface of the fabric, so all the in-plane flow occurs in the channel and the flow in the fabric is primarily though the thickness. For a process such as SCRIMP™ where there may be a large spacing between the flow channels, the in-plane permeability would be more important. The in-plane permeability can be either higher or lower than the through-thickness value, depending on the fabric type and compaction pressure. Parnas *et al.,* have found that in general the in-plane permeability in the direction of the fibres is 6–8 times larger than it is in the through-thickness direction (Parnas and Salem, 1993). However, if the flow in the plane is transverse to the fibres, the permeability could be expected to be close to the through-thickness value or possibly even less.

8.3.1 Saturated vs. unsaturated flow

Darcy's law was originally intended for modelling saturated flow of water through soil (Darcy, 1856). Because of this, it has some deficiencies when modelling unsaturated flow though a fabric. In order to use it to model this type of flow it must be modified slightly. In calculating the permeability, the velocity is determined by dividing the flow rate by the cross-sectional area. The area used is the total flow area of the fabric. This means that the velocity in Darcy's law is the superficial velocity, or the velocity averaged over the whole area. Due to the presence of the fabric, the actual flow area is less than the total area. This means that the actual velocity of the fluid through the preform is higher than the superficial velocity because the flow area is reduced. This reduction in flow area can be determined by knowing the fibre volume fraction of the fabric. Actually, the term commonly used is called the porosity (e) of the fabric which is one minus the fibre volume fraction. The modified equation becomes:

$$v_{\text{actual}} = -\frac{k}{\mu_e}\frac{\mathrm{d}p}{\mathrm{d}z}$$

Another additional term required to model unsaturated flow is the capillary pressure. This is a consequence of the wicking behaviour of the fabric caused by surface tension. This tends to pull the resin along, which results in a higher apparent pressure than the applied pressure. The $-\mathrm{d}P$ term will be replaced by ΔP, recognizing that the pressure drop is linear, and that the flow occurs from high to low pressure. Darcy's law is modified accordingly.

$$v_{\text{actual}} = -\frac{k}{\mu_e}\frac{(\Delta P_{\text{app}} + P_{\text{cap}})}{\mathrm{d}z}$$

where:

ΔP_{app} is the drop in fluid pressure
P_{cap} is the capillary pressure.

The capillary pressure is dependent on properties of the fabric and the resin. One equation for determining the surface tension as presented is:

$$P_{cap} = \frac{F}{D_f} \frac{1-e}{e} \sigma \cos \theta$$

where:

F is the form factor
σ is the surface tension
D_f is the diameter of a fibre
$\cos \theta$ is the wetting angle.

The fibre diameter, porosity and form factor are all properties of the fabric, while the surface tension is a property of the resin. The wetting angle is a property of the resin and fabric. Its value can vary depending on the measurement method. For the most accurate results in an infusion process, the dynamic contact angle is the most appropriate (Skartsis *et al.*, 1992). It is measured as the fluid is moving in relation to the solid interface. Both an advancing and a receding angle can be determined. However, the static contact angle gives a very good approximation for the resin systems used in RTM, and is easier to measure. Fortunately, the wetting angle depends only on the fabric material and not on the fabric architecture. Therefore, once the fabric properties are known for a given fabric, the capillary pressure can be calculated for any resin with that fabric if its surface tension and wetting angle are known. The form factor depends on the fabric architecture and whether the flow is along the fibres or transverse. Transverse flow typically has a form factor with a value from 1 to 2. The porosity is included because as the porosity decreases, the surface area to volume ratio increases, which increases the capillary pressure. Capillary pressure is not very temperature dependent, since both the contact angle and surface tension are very weak functions of temperature.

Although the capillary pressure is typically much smaller than the injection pressure, it can change the results of a test by a noticeable amount. Some researchers have claimed that the capillary effect was negligible in their permeability tests while others have claimed capillary pressures had a significant effect (Parnas and Phelan, 1991). The extent of this effect is going to vary depending on the fabric, the resin and the injection pressures. Previous researchers have conducted a study on the capillary pressures of a silicone oil and a corn syrup with a couple of fabrics. The largest capillary pressure they found was approximately 5 kPa for the silicone oil and less for the corn syrup, although they did not give a specific value.

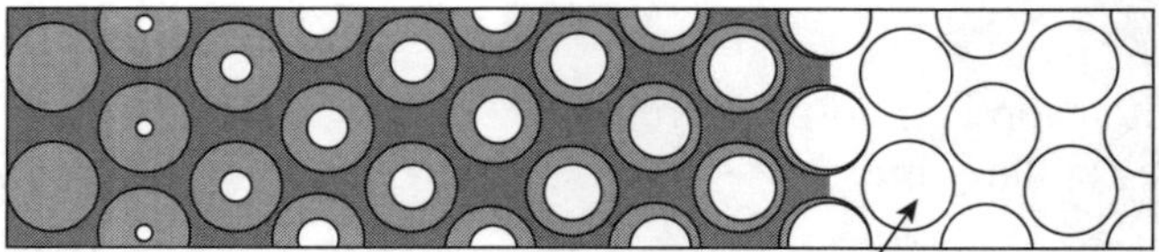

8.1 Illustration of dual-scale flow.

Another phenomenon of unsaturated flow arises from the fact that there is flow occurring between the fibre tows as well as within them. During a saturated permeability test, the flow in both these regions is factored into the total permeability. Because of the presence of macroscopic channels between tows and microscopic channels within them, a fabric is commonly referred to as a dual-scale porous medium. The consequence of this dual nature in unsaturated flow is that the flow in the macroscopic channels will advance much faster than the inside of the tows can be saturated. Cairns *et al.* (1999) found that the equivalent permeability of the channels between tows could be an order of magnitude larger than the permeability within the tow. This effect is shown in Fig. 8.1.

This can be a problem in modelling as well as for part quality and the effect on flow modelling is minor. This is mostly due to the fact that this occurs only at the flow front, and not in the saturated regions. It was determined, however, that this could have a large impact on part quality. This is why the use of a vacuum pump has become so critical in reducing porosity in RTM processes. The use of a vacuum reduces the amount of air that is trapped as the resin encircles a fibre tow.

8.3.2 Permeability as a function of weave type

There are hundreds of possible woven fabric combinations, which can be divided into biaxial and triaxial woven structures according to in-plane fibre orientation. The plane weave has the highest frequency of yarn interlacing, whereas the satin weave has the least, with the twill weave somewhere in between. Accordingly, woven fabrics have a higher level of structural integrity and greater ductility due to the crimp geometry produced by yarn interlacings. On the other hand, the satin weave has the highest level of fibre to fabric strength and modulus translation efficiency, due to the low level of yarn interlacing and yarn linearity. The low level of yarn integration in satin weave also allows freedom of yarn mobility, which contributes to higher fibre packing density and consequently a higher level of fibre volume fraction (Ko and Du, 1997). Open weave results in higher porosity and

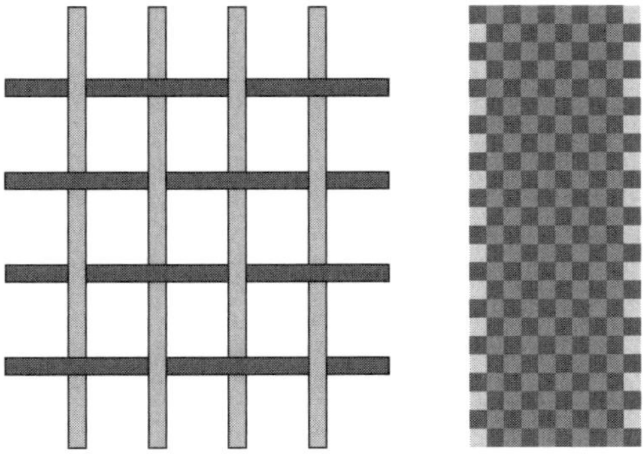

8.2 Difference between open weave and closed weave.

higher permeability. However, closed weave mats are difficult to penetrate, but offer relatively higher strength and module. For the 3-D multilayer woven fabric, the more porous or open preform structures have higher values of permeability and can be more easily filled. The difference between open weave and closed weave is shown in Fig. 8.2.

8.3.3 Direction of permeability measurement

The permeability tensor is represented by its components K_{xx}, K_{yy}, as the in-plane permeabilities, and K_t, as the transverse permeability (Advani and Calado, 1996). It is understood that effective (or directional) permeability in three dimensions follows the shape of an ellipsoid. Six independent measurements are required to determine the permeability tensor.

The first channel flow experiment to measure principal permeability was conducted in an arbitrary coordinate system x, y, z along the x-axis (Weitzenböck *et al.*, 1999b). For the permeability, there is some directional character in such anisotropic preforms, and three values for the permeability may be required: K_{xx}, K_{yy} and K_{zz}. In isotropic preforms, one value is sufficient: $K = K_{xx} = K_{yy} = K_{zz}$. The directional characteristics of K_{xx}, K_{yy} and K_{zz} are shown in Fig. 8.3.

8.3.4 Isotropic and anisotropic permeability

For isotropic materials a circular flow front can be observed in the radial flow experiment, while for anisotropic materials the flow front becomes elliptic.

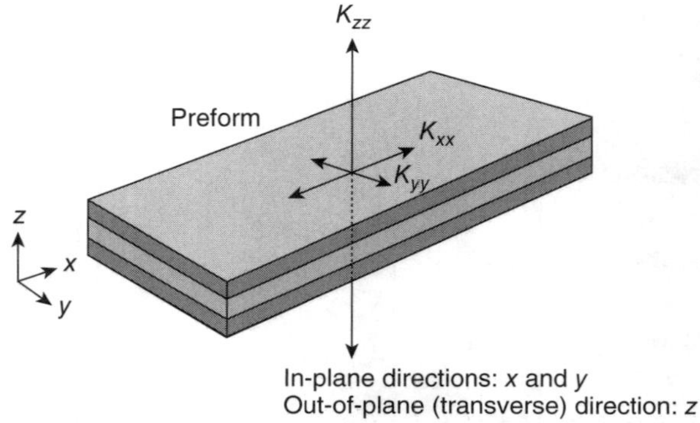

8.3 Directional characteristics of K_{xx}, K_{yy}, K_{zz}.

Isotropic permeability

To calculate isotropic permeability the pressure gradient within the mould has to be determined as a function of the flow front position. This is achieved by solving the Laplace equation in polar coordinates (Weitzenböck *et al.*, 1999a). If a preform is isotropic in the in-plane directions such that $K_{xx} = K_{yy}$, then the flow fronts progress as circles. Some fabrics, such as random fibre mats, four-harness woven fibreglass mat and symmetric bidirectional fabrics, produce isotropic preforms.

Anisotropic permeability

So far only flow in isotropic porous materials has been considered. However, for many fabrics commonly used in RTM an elliptical flow front is observed. As a consequence the second-order partial differential equation of the pressure distribution is no longer a Laplace equation, where K_{xx} and K_{yy} are two principal permeabilities (Weitzenböck *et al.*, 1999a). If the preform is anisotropic in the plane with K_{xx} different from K_{yy}, the flow fronts become elliptic and remain elliptic throughout filling. Furthermore, this situation can be encountered with unidirectional stitched mats.

8.3.5 Measurement of permeability

Generally speaking there are two different configurations for permeability measurement. These are the one-dimensional channel flow experiment and the two-dimensional radial flow experiment. For isotropic materials a circular flow front can be observed in the radial flow experiment, while for anisotropic materials the flow front becomes elliptic. The main attraction of the radial flow experiment is that both the principal permeability values and their orientation can be determined in a single experiment; in addition, no racetracking occurs during the experimental process, so it is widely used by researchers (Adams *et al.*, 1986; Hirt *et al.*, 1987; Chan and Hwang, 1991; Parnas and Salem, *et al.*, 1993). For constant inlet pressure three algorithms to calculate permeability data from flow front measurement have been developed (Adams *et al.*, 1988; Carter *et al.*, 1995; Rudd *et al.*, 1995). In all these approaches the orientation of the flow front ellipse needs to be determined visually, since the algorithms apply only to flow front measurements along the principal axes. For all methods permeability is assumed to be a symmetric tensor. As a consequence there is a set of two mutually orthogonal axes where all the off-diagonal tensor components become zero.

8.4 Monofilament permeability model

8.4.1 Characterization of pore microstructures

The permeability is a function of the preform architecture, pore structure and porosity, and different types of fibre preforms may have substantially different permeability at the same fibre volume fraction. The micrograph of fibre preform reveals considerably complicated microstructures of multilayer woven fabrics. However, our experimental data showed that the vari-ation in the alignment of the fibre layers and the pore size distribution was regular. Thus it is possible to find an analytical solution for permeability based on the concept of a unit cell.

Figure 8.4 displays the top view of the three types of multilayer woven fabrics fabricated by the monofilaments. Each fabric has five layers in total,

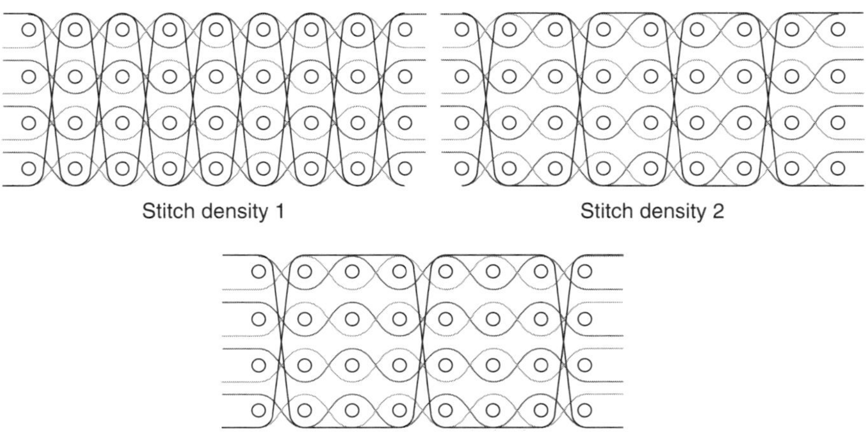

8.4 Schematic for three different stitch models of multilayer woven fabrics.

a layer being defined as containing both a warp and a weft yarn. All weaves are designed as plain weaves and weft yarns are used as stitches to interconnect adjacent layers. The differences between these designs are in the stitching density.

8.4.2 Permeability model of unit cell 1

From the cross-sectional view of three types of multilayer woven fabrics (Fig. 8.5), the structure of the preform can be idealized as a porous medium consisting of parallel fibres as shown in Fig. 8.6. Based on the assumption that the major contribution to the flow resistance comes from the narrow gap between the fibres, the permeability model therefore can be established as follows.

The fibres are arranged in a periodic pattern so that we only need to consider the flow in one 'representative unit cell'. The pattern that we are

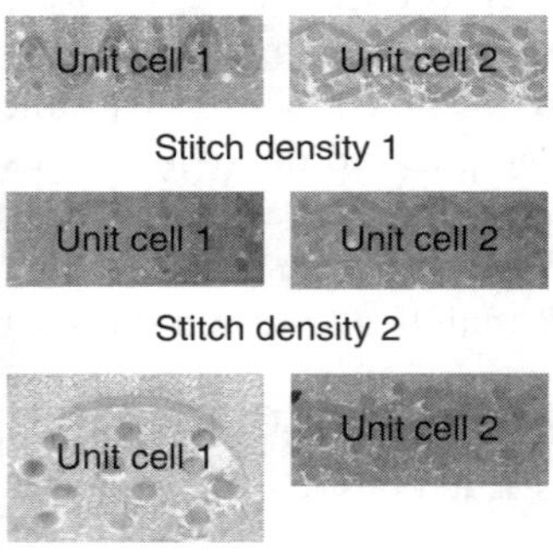

8.5 Cross-sectional view of three types of multilayer woven fabrics with different stitch density.

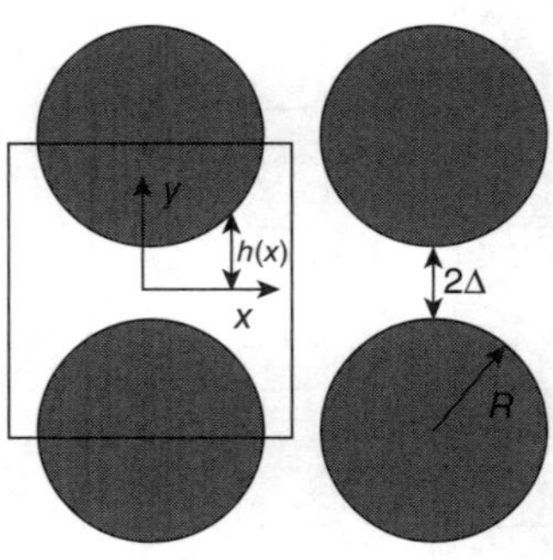

8.6 Definition sketch of the idealized monofilament reinforcement arrangement and the representative unit cell of quadratic fibre packing.

going to consider is the quadratic array. Let us assume that most of the resistance to flow which is perpendicular to the fibres comes from a small region close to the narrow gap formed between the fibres. If the fibres are very close to each other they form a channel with slowly varying area between them. Here we mean that the angle between the channel wall and the channel centre-line is small at all points along the channel (Liu Yi, 2006).

For the flow in the slow-varying channels, the cross-sectional area varying along the streamline direction is assumed to alter so slowly that inertia effects can be neglected (in the present case this is true for all geometries, since generally in RTM the Reynolds number of the resin is very small). If a constant pressure difference is maintained between two stations, the pressure gradient will vary slowly in the streamline direction:

$$\frac{\mathrm{d}p}{\mathrm{d}x} = -\frac{3}{2}\frac{\mu q}{h^3(x)} \tag{8.1}$$

where $h(x)$ is the half-height of the channel, p is the pressure, μ is the viscosity of the fluid q is the total flow rate. The total pressure drop between two stations a and b in the channel can be found from integration of Equation 8.1:

$$p_\mathrm{b} - p_\mathrm{a} = -\frac{3}{2}\mu q \int_\mathrm{a}^\mathrm{b} \frac{\mathrm{d}x}{h^3(x)} \tag{8.2}$$

From the definition sketch (Fig. 8.7) over the coordinate system and the channel half-height, the $h(x)$ needed in the integral can be calculated by:

$$h(x) = \Delta + R\left(1 - \sqrt{1 - \frac{x^2}{R^2}}\right) \tag{8.3}$$

which for $x << R$ can be written as

$$h(x) = \Delta + \frac{R}{2}\frac{x^2}{R^2} \tag{8.4}$$

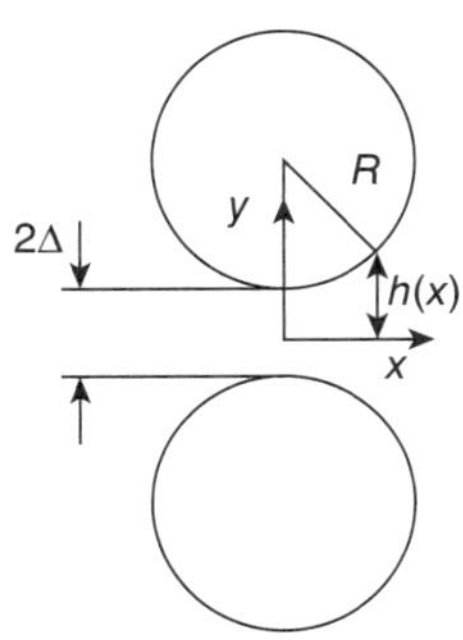

8.7 Definition sketch in the analysis of the flow between the fibres.

Substitution of $h(x)$ into Equation 8.2 gives

$$p_b - p_a = -\frac{3}{2}\mu q \int_a^b \frac{dx}{\left[\Delta + R\left(1 - \sqrt{1 - \frac{x^2}{R^2}}\right)\right]^3} \qquad 8.5$$

By integrating the above integral equation and making some simplification based on the boundary conditions, we can get the following equation:

$$p_b - p_a = -\frac{9\pi}{16}\frac{\mu q}{\Delta^3}\sqrt{2R\Delta} \qquad 8.6$$

which means that the pressure drop between two fibres in general can be written as:

$$\frac{\Delta p}{L} = -\frac{9\sqrt{2}\pi}{16L}\frac{\mu q}{R^2}\left(\frac{\Delta}{R}\right)^{-5/2} \qquad 8.7$$

Comparing Equation 8.7 with the Darcy law, we can get the equation to calculate the permeability of the porous medium:

$$K_1 = \frac{16}{9\pi\sqrt{2}}\left(\frac{\Delta}{R}\right)^{5/2} R^2 \qquad 8.8$$

Equation 8.8 can be rewritten in terms of fibre volume fraction if we notice that the fibre volume fraction can be calculated by:

$$V_f = \frac{\pi/4}{(1 + \Delta/R)^2} \qquad 8.9$$

from which we see that Δ/R is given by

$$\Delta/R = \sqrt{V_{f,max}/V_f} - 1$$

where $V_{f,max}$ is the maximum fibre volume fraction which is achieved when the adjacent fibres touch each other (Liu Yi, 2006). Substitution of Δ/R in Equation 8.8 finally yields:

$$K_1 = \frac{16}{9\pi\sqrt{2}}\left(\sqrt{\frac{V_{f,max}}{V_f}} - 1\right)^{5/2} R^2 \qquad 8.10$$

8.4.3 Evaluation of $V_{f,max}$

The maximum fibre volume fraction in the permeability expression of Equation 8.10 corresponds to the maximum space occupied by the fibres in the preforms. The architecture of the multilayer woven fabrics considered in the study consists of a layer to layer interlacement. For this architecture,

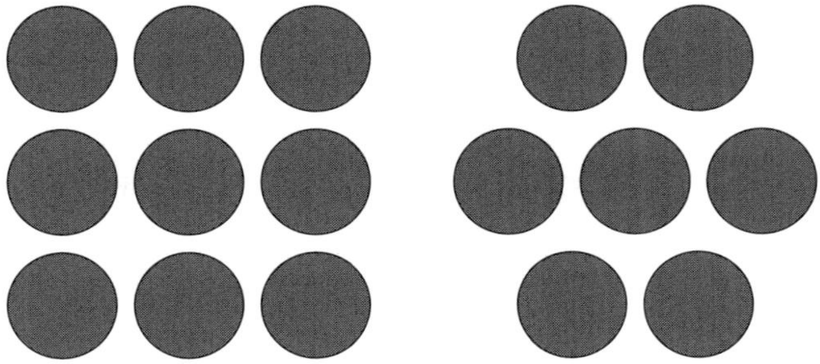

8.8 Definition sketch of quadratic and hexagonal arrays.

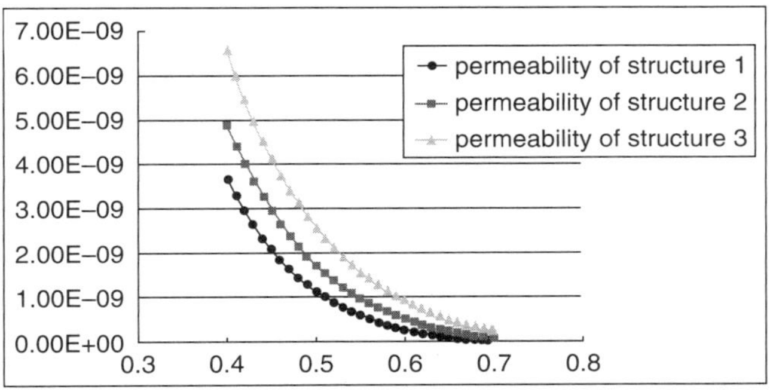

8.9 Permeability of three different structures of unit cells.

the maximum fibre volume fraction of these different stitch structures may be considered to vary from the quadratic array of Fig. 8.6 to the hexagonal array shown in Fig. 8.8.

For the quadratic array, the maximum fibre volume fraction is $\pi/4$; for the hexagonal array, the maximum fibre volume fraction is $\pi/2\sqrt{3}$. However, in this experiment, the maximum fibre volume fraction of the unit cell may be between these two values. A coefficient, C_f ($1 \le C_f \le 2/\sqrt{3}$), is thus introduced to define the maximum fibre volume fraction $V_{f,\max} = C_f \cdot \frac{\pi}{4}$. If the coefficient $C_f = 1$, this means that the fibres in the unit cell from a quadratic array, and if $C_f = 2/\sqrt{3}$, the fibres are arrayed in a hexagonal model. As shown in the experimental cross-images, for the structure of stitch density 1, $C_f = 1$; for the structure of stitch density 3, $C_f = 2/\sqrt{3}$; for the structure of stitch density 2, $C_f = 1.1$.

8.4.4 Predicted permeability of unit cell 1

So, from Equation 8.10, we can get the permeability of these three different structures of unit cell as shown in Fig. 8.9. This figure shows that

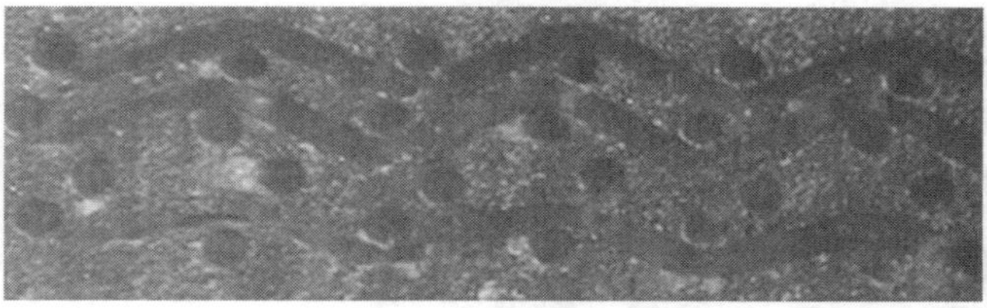

(a) Architecture of the structure

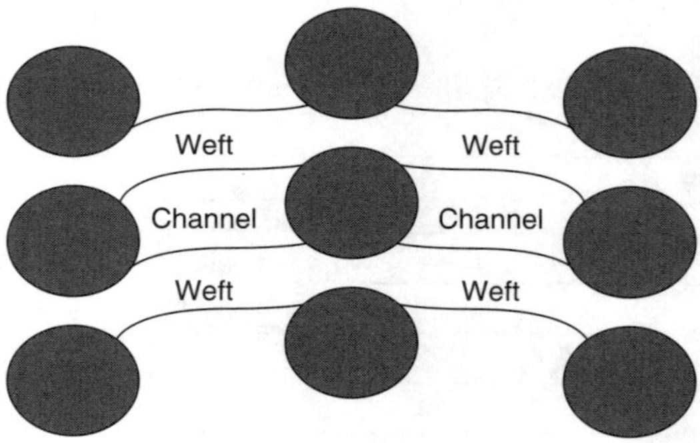

(b) An idealized structure

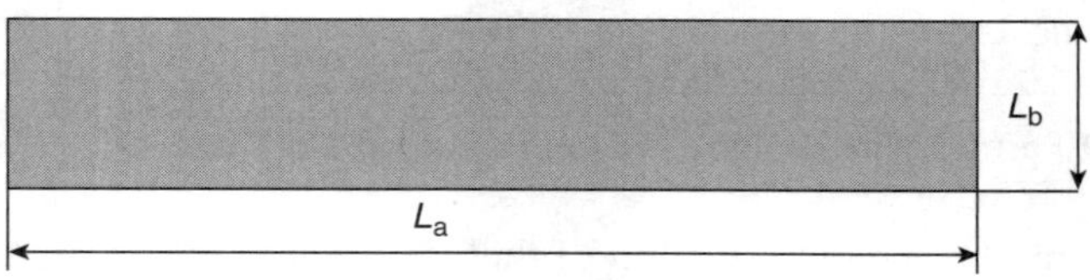

(c) An idealized unit cell

8.10 Simplified model of unit cell 2.

the permeability of stitch structure 3 is better than that of stitch structure 1. This is due to its easy-going pathway arrangement, and this is also testified by the subsequent experimental results as follows (Liu Yi, 2006).

8.4.5 Permeability model of unit cell 2

Unit cell 2 can be simplified as shown in Fig. 8.10. The permeability can be obtained from the expression for the equivalent permeability of a rectangular channel. In this study, the width of channel is assumed to be L_a; the height of the channel is L_b. The solution of the two-dimensional velocity field in an arbitrary duct geometry is governed by an equation of the Poisson form. Once the flow field is obtained, integration over the domain provides a relationship for the average flow rate with respect to the applied pressure gradient. For a rectangular duct with sides of length L_a and L_b,

$$u = -\frac{l_a^2}{12}\left(\frac{dp}{dx}\right)\left[1 - \frac{192 l_a}{\pi^2 l_b}\sum_{n=1,3,5...}^{\infty}\frac{\tanh(i\pi l_b/2l_a)}{i^5}\right] \quad 8.11$$

where u is fluid viscosity. This result can be compared to the Darcy law in one dimension:

$$u = -\frac{k_{equ}}{\mu}\frac{dp}{dx} \quad 8.12$$

Modelling the channel region as a porous medium of porosity 1.0, the average volume velocity may be equal to the average velocity provided by Equation 8.12. By comparison, we can obtain the permeability of this duct:

$$K_2 = -\frac{l_a^2}{12}\left[1 - \frac{192 l_a}{\pi^2 l_b}\sum_{n=1,3,5...}^{\infty}\frac{\tanh(i\pi l_b/2l_a)}{i^5}\right] \quad 8.13$$

8.4.6 Permeability of whole structure

As shown in Fig. 8.11, the permeability of layer n is represented by K_n, and the height is represented by M_n. The total flow rate through the multi-layer, q, and the total height M can be expressed as follows:

$$q = q_1 + q_2 + \ldots + q_n = \sum_{i=1}^{n} q_i \quad 8.14$$

$$M = M_1 + M_2 + \ldots + M_n = \sum_{i=1}^{n} M_i \quad 8.15$$

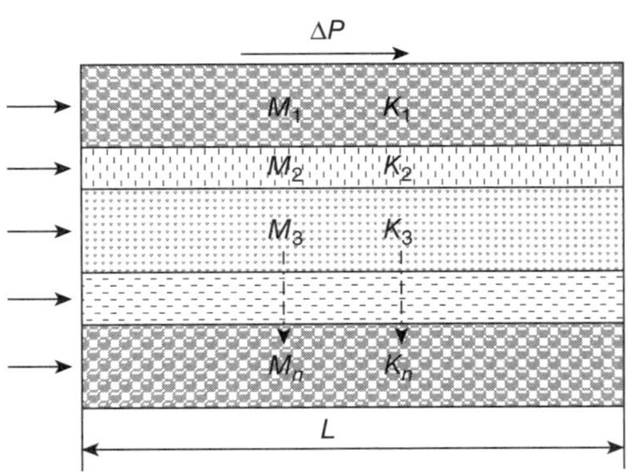

8.11 Schematic of multilayer laminar flow.

The pressure difference of each layer can be shown in the following equation:

$$\Delta P = \Delta P_1 = \Delta P_2 = \ldots = \Delta P_n \tag{8.16}$$

From the Darcy law,

$$q_i = K_i M_i \cdot \frac{\Delta P_i}{L} \tag{8.17}$$

we can get the total flow rate:

$$q = \sum_{i=1}^{n} q_i = \sum_{i=1}^{n} K_i M_i \cdot \frac{\Delta P_i}{L} = K_p \cdot M \cdot \frac{\Delta P}{L} \tag{8.18}$$

So the effective permeability of a multilayer flow channel K_p is given by

$$K_p = \sum_{i=1}^{n} K_i \cdot M_i \Big/ \sum_{i=1}^{n} M_i \tag{8.19}$$

Incorporating Equations 8.10 and 8.13 into Equation 8.19, we can get the final equation to calculate the effective permeability of multilayer woven fabrics (Liu Yi, 2006):

$$K_p = (K_1 M_1 + K_2 M_2)/(M_1 + M_2) \tag{8.20}$$

$$K_p = \frac{\dfrac{16}{9\pi\sqrt{2}}\left(\sqrt{\dfrac{V_{f,max}}{V_f}} - 1\right)^{5/2} R^2 \cdot M_1 - \dfrac{l_a^2}{12}\left(1 - \dfrac{192 l_a}{\pi^2 l_b} \sum_{n=1,3,5\ldots}^{\infty} \dfrac{\tanh(i\pi l_b/2l_a)}{i^5}\right) M_2}{(M_1 + M_2)} \tag{8.21}$$

8.5 Fractal permeability model

8.5.1 Fractal characterization of pore microstructures in fibre preforms

Figure 8.12 displays a cross-sectional view of multilayer woven preforms. There are two types of pores, a macropore (of the order of 10^{-3} m) between fibre tows, and the micropores (10^{-6}–10^{-5} m) inside the fibre tows. If several layers of fibre mats were stacked together, the macropores of each layer would form macro-channels with the sizes of 10^{-3} m, see Fig. 8.11, while micropores inside the fibre tows would form numerous tortuous micro-channels. Both the macro-channels and the micro-channels should follow the fractal scaling law given by Mandelbrot and Freeman (1982):

$$l_t(\lambda) = \lambda^{1-D_T} L_0^{D_T} \tag{8.22}$$

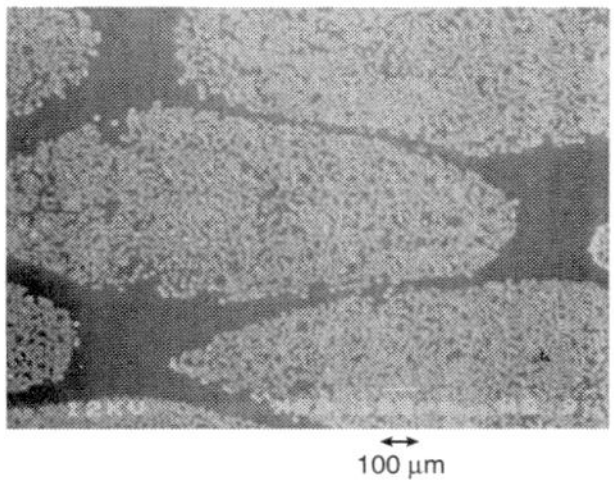

8.12 Micrograph of fibres and resin.

where D_T is the tortuosity fractal dimension, with $1 < D_T < 2$, representing the extent of convolutedness of capillary pathways for fluid flow through a medium, λ is the size of a pore channel in a fibre preform, and $L_t(\lambda)$ is its tortuous length along the flow direction.

For the macropores between fibre tows and the micropores inside fibre tows, the cumulative size distribution of pores in porous fibre preforms follows the fractal scaling law:

$$N(L \geq \lambda) = \left(\frac{\lambda_{\max}}{\lambda}\right)^{D_f} \quad 8.23$$

From Equation 8.23, it can be found that

$$-\mathrm{d}N = D_f \lambda_{\max}^{D_f} \lambda^{-(D_f+1)} \mathrm{d}\lambda \quad 8.24$$

Equation 8.24 describes the scaling relationship of the cumulative pore number (Liu Yi, 2006).

8.5.2 Fractal model for permeability

This model is related to the architectural parameters of preforms, the pore area fractal dimension D_f and the tortuosity fractal dimension D_T. The total volumetric flow rate, Q, through the unit cell is an accumulation of the flows through all the individual pore channels, including the macroscopic pore (gap or channel) between the fibre tows and the microscopic pores inside the fibre tows. The flow rate through a single pore channel is given by modifying the well-known lubrication flow theory (Wheatcraft and Taylor, 1988):

$$q(\lambda) = \frac{-1}{6\mu} A\lambda^2 \frac{\Delta p}{L(\lambda)} \quad 8.25$$

The total flow rate Q through the unit cell can be obtained by integrating the individual flow rate, $q(\lambda)$, over the entire range of pore sizes from the minimum pore size $\lambda = \lambda_{\min}$ to the maximum pore size $\lambda = \lambda_{\max}$ (Pitchumani and Ramakrishnan, 1999). According to Equations 8.22–8.24, we have

$$Q=-\int_{\lambda_{\min}}^{\lambda_{\max}} q(\lambda)\mathrm{d}N(\lambda) \tag{8.26}$$

$$Q=-\int_{\lambda_{\min}}^{\lambda_{\max}} \frac{-1}{6\mu} A\lambda^2 \frac{\Delta p}{L(\lambda)} D_f \lambda_{\max}^{D_f} \lambda^{-(D_f+1)}\mathrm{d}\lambda \tag{8.27}$$

$$Q=\frac{A\Delta p D_f}{6\mu L_0^{D_T}(D_T-D_f+1)} \lambda_{\max}^{1+D_T}\left(1-\left(\frac{\lambda_{\min}}{\lambda_{\max}}\right)^{D_T-D_f+1}\right) \tag{8.28}$$

Since $1 < D_T < 2$ and $1 < D_f < 2$, there exist $D_T - D_f + 1 > 0$ and $0<\left(\frac{\lambda_{\min}}{\lambda_{\max}}\right)^{D_T-D_f+1}<1$ in Equations 8.27 and 8.28. Due to $\left(\frac{\lambda_{\min}}{\lambda_{\max}}\right)\sim 10^{-2}$, Equation 8.28 can thus be reduced to

$$Q=\frac{A\Delta p D_f}{6\mu L_0^{D_T}(D_T-D_f+1)} \lambda_{\max}^{1+D_T} \tag{8.29}$$

Using the Darcy law, we obtain the permeability equation as follows:

$$K=\frac{\mu L_0 Q}{\Delta PA}=\frac{\mu L_0^{1-D_T} D_f \lambda_{\max}^{1+D_T}}{6\mu(D_T-D_f+1)} \tag{8.30}$$

Obviously, the permeability of fibre preforms is mainly determined by the macropore (gap or channel) between the fibre tows. The previous study indicated that the permeability contribution from micropores inside the fibre tows is negligible (Yu and Li, 2001). Thus, only the tortuosity of macropore channel pathways is included in the present consideration. It can be found that the macropore channels are approximately straight: see Fig. 8.11. Therefore, $D_T = 1$ is applied in this investigation, and Equation 8.30 can be thus reduced to

$$K=\frac{D_f \lambda_{\max}^2}{2-D_f} \tag{8.31}$$

in which D_f can be expressed as

$$D_f = 2-\frac{\ln\phi}{\ln(\lambda_{\min}/\lambda_{\max})} \tag{8.32}$$

So from this final equation, it can be predicted that with the increase of $\lambda_{\max}$, both D_f and K increase with it also (Liu Yi, 2006).

8.6 Conclusions

The effects of fabric microstructures and other properties of multilayer woven fabrics on permeability were examined in this chapter. A permeability model based on fractal theory has been presented to predict the permeability of preforms fabricated by porous yarns. Another model based on the unit cell of quadratic fibre packing has also been established to predict the permeability of 3-D multilayer woven fabrics fabricated with the monofilaments.

For the 3-D multilayer woven fabrics fabricated with the monofilaments, the permeability model was related to the architectures of fibre preforms and their fibre volume fraction changes. It is evident that the permeability of monofilament preforms is mainly determined by the arrangement of the channels between the fibre tows. At the same fibre volume fraction, fabrics with a loose stitch array have better permeability than those with a compact stitch array. This is due to the different flow resistance with regard to the different stitch array. For fabrics with loose stitch structure, the flow resistance from the stitch array is less than that of fabrics with tight stitch array, thus leading to better permeability of fabrics with loose stitch array.

For the 3-D multilayer woven fabrics fabricated by porous yarns, a fractal permeability model was developed based on the fractal characteristics of pores in the fibre preforms to describe the disordered pore structures of preforms. The permeability model was found to be related to the pore area fractal dimension, the tortuosity fractal dimension, and the architectural parameters of fibre preforms. For fabrics with a rather large maximum pore size, λ_{max}, the flow resistance is relatively lower, hence the resin can impregnate more easily within it, thus leading to better permeability and less void formation in fabrics.

8.7 References

Adams K L and Rebenfeld L (1991), Permeability characteristics of multilayer fibre reinforcements, Part II: Theoretical model, *Polymer Composites*, **12**, 3, 186–190.

Adams K L, Miller B and Rebenfeld L (1986), Forced in-plane flow of an epoxy resin in fibrous networks, *Polymer Engineering and Science*, **26**, 20, 1434–1441.

Adams K L, Russel W B and Rebenfeld L (1988), Radial penetration of a viscous liquid into a planar anisotropic porous medium, *International Journal of Multiphase Flow*, **14**, 2, 203–215.

Advani S G and Calado V M A (1996), Effective average permeability of multi-layer preforms in resin transfer molding, *Composites Science and Technology*, **56**, May, 519–531.

Ambrosi D and Preziosi L (1998), Modelling matrix injection through elastic porous preforms, *Composites, Part A*, **29**, 1–2, 5–18.

Batch G L and Cumiskey S (1990), Multilayer compaction and flow in composites processing, *45th Annual conference SPI*, 1-11/session 9-A.

Berdichevsky A L and Cai Z (1993), Preform permeability predictions by self-consistent method and finite element simulation, *Polymer Composites*, **14**, 2, 132–143.

Blake F C (1922), The resistance of packing to fluid flow, *Transactions of the American Institute of Chemical Engineers*, **14**, 415–421.

Bruschke M V and Advani S G (1993), Flow of generalized Newtonian fluids across a periodic array of cylinders, *Journal of Rheology*, **37**, 3, 479–498.

Cai Z and Berdichevsky A L (1993), An improved self-consistent method for estimating the permeability of a fiber assembly, *Polymer Composites*, **14**, 4, 314–323.

Cairns D S, Humbert D R and Mandell J F (1999), Modelling of resin transfer molding of composite materials with oriented unidirectional plies, *Composites, Part A: Applied Science and Manufacturing*, **30**, 375–382.

Carman P C (1937), Fluid flow through granular beds, *Transactions of the Institution of Chemical Engineers*, **15**, 150–166.

Carter E J, Fell A W and Summerscales J (1995), A simplified model to calculate the permeability tensor of an anisotropic fibre bed, *Composites Manufacturing*, **6**, 3/4, 228–235.

Chan A W and Hwang S T (1991), Anisotropic in-plane permeability of fabric media, *Polymer Engineering and Science*, **31**, 16, 1233–1239.

Chang C Y and Hourng L W (1998), Study on void formation in resin transfer molding, *Polymer Engineering and Science*, **38**, 5, 809–818.

Coulter J P and Guceri S I (1988), Resin transfer molding: process review, modeling and research opportunities, *Proceedings of Manufacturing International '88,* Atlanta, GA, 79–86.

Darcy H P G (1856), Les Fontaines Bibliques de la Ville de Dijon, Victor Dalmont, Paris.

Ferland P, Guittard D and Trochu F (1996), Concurrent methods for permeability measurement in resin transfer molding, *Polymer Composites*, **17**, 1, 149–158.

Gebart B R (1992), Permeability of unidirectional reinforcements for RTM, *Journal of Composite Materials*, **26**, 8, 1100–1133.

Guangbiao Xu and Fumei Wang (2005), Prediction of the permeability of woven fabrics, *Journal of Industrial Textiles*, **34**, 4, 243.

Hirt D E, Adams K L, Prud'Homme R K and Rebenfeld L (1987), In-plane radial fluid flow characterization of fibrous materials, *Journal of Thermal Insulation*, **10**, 153–172.

Ko F K and Du G-W (1997), Processing of textile preforms, in *Advanced Composites Manufacturing* (ed. T G Gutowski), John Wiley, New York, 157–206.

Kolodziej J A, Dziecilak R and Konczak Z (1998), Permeability tensor for heterogeneous porous medium of fibre type, *Transport in Porous Media*, **32**, 1–99.

Kozeny J (1927), Ueber Kapillare Leitung des Wassers im Boden, *Sitzungsberichte, Akademie der Wissenschaften in Wien*, **136**, 271–301.

Lai C L and Young W B (1997), Model resin permeation of fibre reinforcements after shear deformation, *Polymer Composites*, **18**, 5, 642–648.

Lee L J (1997), Liquid composite molding, in *Advanced Composites Manufacturing* (ed. T G Gutowski), Wiley-Interscience, New York, 393.

Lekakou C, Johari M A K and Bader M G (1996b), Compressibility and flow permeability of two-dimensional woven reinforcements in the processing of composites, *Polymer Composites*, **17**, 666–672.

Liu Yi (2006), Study of the permeability of multi-layer woven fabrics, Ph.D. thesis, Institute of Textiles and Clothing, Hong Kong Polytechnic University.

Loos A C, Weidermann M H and Kranbuchi D E (1991), Processing of advanced textile structural composites by RTM, *Proc. 5th Textile Structural Composites Symposium,* Drexel University, Philadephia, PA, 4–6 December.

Lundström T S (2000), The permeability of non-crimp stitched fabrics, *composites, Part A: Applied Science and Manufacturing*, **31**, 1345–1353.

Mandelbrot B B and Freeman W H (1982), *The Fractal Geometry of Nature*, W.H. Freeman, New York, 23–57.

Martin G Q and Son J S (1986), Fluid mechanics of mold filling for fibre reinforced plastics, *Proc. Second Conf. on Advanced Composites,* 18–20 November, Dearborn, MI, 149–157.

Mogavero J and Advani S G (1997), Experimental investigation of flow through multi-layer preforms, *Polymer Composites*, **18**, 5, October, 649.

Padaki N V, Alagirusamy R and Sugun B S (2006), Knitted preforms for composite applications, *Journal of Industrial Textiles*, **35**, 4, 295–321.

Parnas R S and Phelan Jr F R (1991), The effect of heterogeneous porous media on mold filling in resin transfer molding, *SAMPE Quarterly*, **22**, 53–60.

Parnas R S and Salem A J (1993), A comparison of the unidirectional and radial in-plane flow of fluids through woven composite reinforcements, *Polymer Composites*, **14**, 5, 383–394.

Pillai K M and Advani S G (1998), Numerical simulation of unsaturated flow in woven fibre preforms during the resin transfer molding process, *Polymer Composites*, **19**, 1, 71–80.

Pitchumani R and Ramakrishnan B (1999), A fractal geometry model for evaluating permeabilities of porous preforms used in liquid composite molding, *International Journal of Heat and Mass Transfer*, **42**, 2219–2232.

Rudd C D, Morris D J, Chick J P and Warrior N A (1995), Material characterization for SRIM, *4th Int. Conf. on Automated Composites (ICAC '95),* Nottingham, UK, **1**, 211–218.

Rudd C D, Long A C, McGeehin P and Smith P (1996), In-plane permeability determination for simulation of liquid composite molding of complex shapes, *Polymer Composites*, **17**, 1, 52–59.

Rudd C D, Long A C, Kendall K N and Mangin G E (1997), *Liquid Moulding Technologies*, Woodhead, Cambridge, UK.

Shih C H and Lee L J (1998), Effect of fibre architecture on permeability in liquid composite molding, *Polymer Composites*, **19**, 5, 629–639.

Simacek P and Advani S G (1996), Permeability model for a woven fabric, *Polymer Composites*, **17**, 6, 887–899.

Skartsis L, Kardos J L and Khomami B (1992), Resin flow through fibre beds during composite manufacturing processes, Part 1 and Part 2, *Polymer Engineering Science*, **32**, 4, 221–231.

Smith P, Rudd C D and Long A C (1997), The effect of shear deformation on the processing and mechanical properties of aligned reinforcements, *Composites Science and Technology*, **57**, 327–344.

Van der Westhuizen J and Du Plessis J P (1996), An attempt to quantify fibre bed permeability utilizing the phase average Navier–Stokes equation, *Composites, Part A*, **27A**, 263–269.

Wang C Y (1996), Stokes flow through an array of rectangular fibres, *International Journal of Multiphase Flow*, **22**, 1, 185–194.

Wang T J, Wu C H and Lee L J (1994), In-plane permeability measurement and analysis in liquid composite molding, *Polymer Composites*, **15**, 4, 278.

Weitzenböck J R, Shenoi R A and Wilson P A (1999a), Radial flow permeability measurement, Part A: Theory; Part B: Application, *Composites, Part A: Applied Science and Manufacturing*, **30**, 6, 781–796, 797–813.

Weitzenböck J R, Shenoi R A and Wilson P A (1999b), Measurement of principal permeability with the channel flow experiment, *Polymer Composites*, **20**, 2, 321–335.

Wheatcraft S W and Taylor S W (1988), An explanation of scale dependent dispersivity in heterogeneous aquifers using concepts of fractal geometry, *Water Resources Research*, **24**, 566–578.

Williams J G, Morris C E M and Ennis B C (1974), Liquid flow through aligned fibre beds, *Polymer Engineering and Science*, **14**, 6, 413–419.

Yu B M and Li J H (2001), Some fractal characters of porous media, *Fractals*, **9**, 3, 365–372.

9
Using multilayer woven fabrics in resin transfer moulding

Abstract: In this chapter a detailed theoretical analysis for in-plane impregnation in multilayer woven (MLW) fabrics is reported in order to understand the mechanism of void formation. Unlike the previous works, where the void is formed in the plane of one layer of woven fabric, the void formation in the cross-section of MLW fabrics is presented. Based on two simplified unit cells, which were identified from two typical multiple modes of MLW fabrics, a mathematical model is developed to analyze the formation and size of voids. The flow front and void formation processes are also numerically simulated using the control-volume method.

Key words: multilayer woven (MLW) fabrics, flow resistance of MLW fabrics, void formation, flow modelling of MLW fabrics, filling properties of MLW fabrics.

9.1 Introduction

Resin transfer moulding (RTM) is a high-performance net-shape manufacturing process for polymer and advanced high-performance textile composite parts, especially in the aircraft and automobile industries due to its relatively short cycle time, low labour requirements and equipment costs. It is particularly suitable for producing large and geometrically complex parts. However, at present, part fabrication by RTM is a very expensive proposition due to the typically long process development time, limiting the potential use of the process. RTM consists of the following steps: (a) preform fabrication, (b) preform placement inside the mould, (c) mould filling by resin injection, (d) curing, and finally (e) demoulding of the finished part. The inherent variations of the process make it difficult to predict what happens inside the mould during the filling and curing stages and therefore what the properties of the resulting part will be. Both filling and curing are critical steps, though successful mould filling without dry spots is the first step to having a defect-free part (Barooah *et al.*, 2001).

Ideally, the resin flow should progress so that the last part of the mould to be filled is near the vents, and the preform is wetted out completely without leaving any dry spots. The flow pattern in the mould depends on such parameters as preform porosity and permeability, and resin viscosity. These properties are required as input parameters for the physical models used to simulate mould filling. However, the permeability and porosity cannot be estimated accurately beforehand as compaction, preform preparation, etc. all affect their values. A common problem is the edge effect called racetracking which makes the resin flow faster along the edges of the preform. These random, unpredictable variations in the preform properties make it extremely difficult to predict resin flow progression accurately, which in turn makes it difficult to design an effective injection and venting scheme without trial and correction cycles.

Unfortunately, the fibre preform always has complicated and non-uniform microstructure, and hence its local permeability may differ by several orders of magnitude between inside and outside fibre tow. These non-uniform properties finally lead to the formation of air voids on the micro scale (Parnas and Phelan, 1991; Patel and Lee, 1995; Chen *et al.*, 1995).

Previous research (Yosida *et al.*, 1986; Harper *et al.*, 1987; Bowles and Frimpong, 1991; Feldgoise *et al.*, 1991; Ghiorse and Jurta, 1991) has revealed that the presence of voids is highly undesirable and has a deleterious effect on product mechanical properties including the interlaminar shear strength, compressive strength, impact resistance and fatigue life. The research has also revealed that the void formation and development are correlated to injection pressure, outlet pressure, resin properties (viscosity, surface tension), fabric characteristics (type and orientation of fibres, surface treatment), etc. Some studies (Hayward and Harris, 1990; Chen *et al.*, 1995) reported that having vacuum assistance will reduce the quantities of voids but cannot totally solve the problem. In recent years, the problem of void formation has received much attention and has been studied extensively. However, the variety of the architectures of fibre preforms is too large to allow the prediction of void formation based on one general model only. Furthermore, void formation in RTM seems to be inevitable and there is no proven way to eliminate voids completely (Chang and Hourng, 1998). Hence, understanding of the void formation mechanism is important and is necessary for the mould and fibre preform design.

As is well known, there are two scales of flow during RTM: one is the macro flow in the gaps between/around the tows, and the other is the micro flow within tows. Based on this concept, several investigators have developed models to predict void formation in RTM. Parnas and Phelan (1991) developed a model to predict the air entrapment for flow through unidirectional fibre mats with the global flow transverse to the fibres. Chen *et al.* (1995) also developed a model similar to that of Parnas and Phelan. The difference between these two models is that Chen *et al.* included the effects of capillary pressure on the impregnation of fibre tows. Also based on the concept of the two scales of flow, Chan and Morgan (1992) developed a model to analyze the resin flow and air entrapment during RTM of bidirectional non-woven fibre performs. Patel and co-workers carried out flow visualization experiments of void formation in unidirectional stitched, bidirectional stitched and four-harness woven fibreglass mat (Patel and Lee, 1995). They reported that fingering took place at the flow front because the permeabilities in the fibre tows and in the gaps between the fibre tows were different, and void formation was correlated to capillary number and the liquid-fibre–air contact angle. For axial flow, the microvoids were formed at $Ca^* > 10^{-3}$, and for transverse flow, the microvoids were formed at an even lower capillary number, ~10^{-4}. Once formed, the microvoids were difficult to purge and remained trapped even after bleeding the liquid at much higher flow rates than those at which they were formed (Rohatgi *et al.*, 1996).

Several researchers have carried out theoretical analyses to describe the mechanism of void formation during RTM. Binetruy and Hilaire (1998) developed an analytical model to describe the tow impregnation where the global flow is parallel to the fibre axis. Based on the study of the contribution of the axial and transverse flow mechanisms inside tows, this model shows that the main tow impregnation process is transverse to the fibre axis. A criterion has been established to indicate when the axial flow can be neglected to simplify the tow impregnation. Kang *et al.* (2000) proposed a mathematical model to describe the mechanisms of void formation when global flow is transverse to fibre tow. The model shows that for a given fibre preform, the effects of resin velocity and capillary pressure can be described by the capillary number. With proper calibration, it can predict the size and content of voids within fibre tows as well as between them.

In this chapter a detailed theoretical analysis for in-plane impregnation in multilayer woven fabrics (MWFs) is reported to understand the mechanism of void formation. Unlike the previous work (Patel and Lee, 1995), where the void is formed in the plane of one layer of the woven fabric, the void formation in the cross-section of MWFs has been presented. Based on two simplified unit cells, which were identified from two typical multiple modes of MWFs, a mathematical model is developed to analyze the formation and size of voids. The flow front and void formation processes are also numerically simulated using the control-volume method.

9.2 Flow resistance of multilayer woven fabrics

Flow resistance is an important property of textile materials. Earlier research showed that the efficiency of various processes in open-width wet textile finishing depends mainly on the amount of flow of the process fluid through the textile material. This flow depends not only on process parameters like fabric velocity and on machine parameters like roller diameter, but also on the flow resistance of the treated textile materials. It is clear that this flow resistance is a function of the geometry of the textile material, hence it important to establish a relation for the flow resistance of a woven textile fabric as a function of its geometry.

First, let us consider one layer of plain weave fabric. The typical architecture is displayed in Fig. 9.1(a). As described by Simacek and Advani (1996), there are two sub-domains within the perform. One consists of the fibre tows (warp and weft) which are woven together to create an interconnected network. Another is a network of empty pores and channels around yarns. Due to the non-uniform permeability in these two sub-domains, during RTM processes voids are formed not only in the plane of the fabric as described by Patel and Lee (1995), but also in the cross-section of multilayer woven fabric.

Three important geometrical parameters of this fabric are tow thickness h, tow width l_b and width of channel between adjacent tows l_c. Usually, the ratio l_b/h is about 5 or more for most reinforcement used in composites processing, and the cross-sections of fibre tows are not circular but elliptical as shown in Fig. 9.1(b). The present study focuses on resin impregnation and void formation in the cross-section of MWFs. For the purpose of simplicity, the configuration of the undulating yarns is approximated by linear segments and the cross-section of the fibre tow is considered to be rectangular; then an idealized cross-section of one-layer plain woven fabric for in-plane impregnation along the x-direction can be obtained as shown in Fig. 9.1(c).

When multiple layers of woven fabric are stacked together, due to the in-plane displacement, two typical modes can be observed as shown in

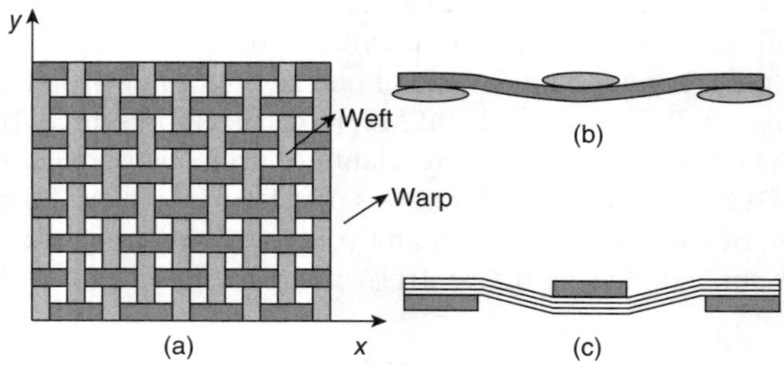

9.1 Architecture of the plain woven fabric.

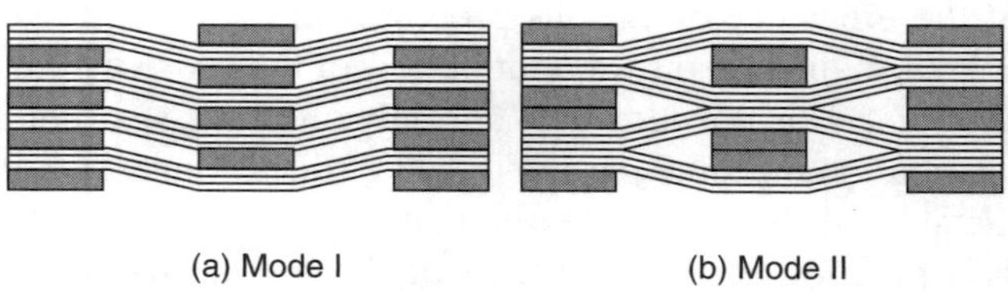

9.2 Two typical multiple modes of MWFs.

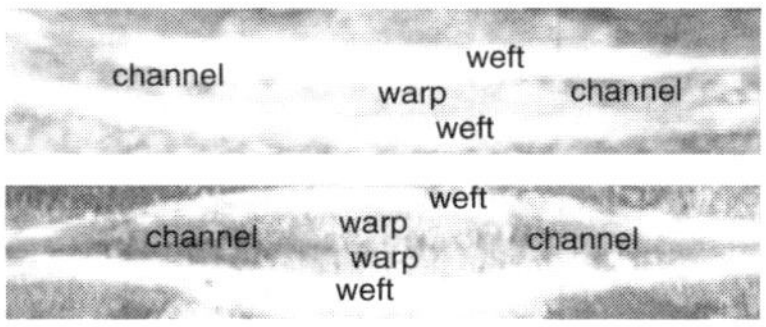

9.3 Typical cells of cross-section for MWFs.

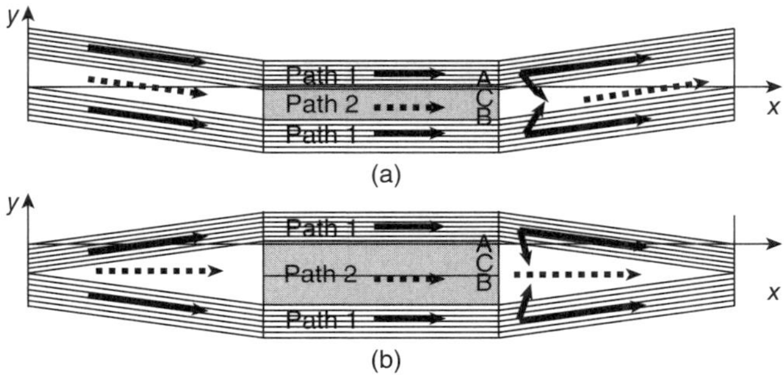

9.4 Unit cells for mode I and mode II.

Fig. 9.2. In mode I, there is no displacement between adjacent layers, while there is a displacement l_c in mode II. The above analysis is strongly supported by image analysis as shown in Fig. 9.3. Although the actual woven fabrics are 3-D, in this simulation we only consider the flow along the cross-section of woven fabrics. So this 2-D simulation model can simulate the flow along the cross-section of 3-D woven fabrics.

In both multiple modes, two unit cells for the void formation can be identified as shown in Fig. 9.4(a) and (b). It can be seen that the flow through the cross-section of the unit cell passes through not only the channel but also the weft and warp. Most resin would flow through the channels along the imposed pressure gradient, but it can also flow within warps and wefts. So we choose two paths as a representative cell, as shown in Fig. 9.4. One way is along the weft tow while the other is the flow in the channel and warp. As testified in the study of Patel and Lee (1995), the formation of voids is mainly due to the velocity difference of resin in different paths.

9.3 Flow modelling of multilayer woven fabrics

In the case of high injection pressure, the effect of capillaries is negligible; voids will form at the end of the warp tow as shown in Fig. 9.4. This will be discussed in more detail as follows. Since the thickness of fibre tows is far less than their width, the flow in path 1 can be simplified to flow along a straight fibre tow. Then the time t, required for the flow front to reach position A can be approximated by

$$t_1 = \frac{(l_c + l_b)^2}{2K_b^a(P_0 - P_f)} \qquad 9.1$$

where μ is the viscosity of the fluid, K_b^a is the axial permeability of the fibre tow, and P_0 and P_f are the pressures at the inlet and flow front, respectively. When the flow front in path 1 reaches position A, on the one hand resin will keep on flowing along the weft, and on the other hand resin also will flow transversely to position B due the high permeability of the channel between tows. The time T_1 for the flow front to reach position B can be obtained from the following analysis.

For the given coordinate system (see Fig. 9.4) the transverse velocity V_t at an arbitrary point C (y, x_c) between A and B takes the form

$$V_t = \frac{dy}{dt} = \frac{K_c(P_0 - P_f)}{y}\left(1 - \frac{x_c}{x_f}\right) \qquad 9.2$$

where K_c is the permeability of the channel, $x_c = l_c + l_b$, and x_f is the x-coordinate of the flow front along path 1 when transverse flow reaches point C, which is given by

$$x_f = \sqrt{\frac{2K_b^a(P_0 - P_f)}{\mu}t} \qquad 9.3$$

Inserting Equation 9.3 into Equation 9.2 and integrating it with the boundary conditions:

$y = 0$ at $t = t_1$

$y = h_0$ at $t = T_1$ ($h_0 = h/2$ for model I, h for model II)

we obtain

$$\frac{h_0^2}{2K_c(P_0 - P_f)} = T_1 - \frac{2(l_c + l_b)\sqrt{T_1}}{\sqrt{\dfrac{2K_b^a(P_0 - P_f)}{\mu}}} + \frac{(l_c + l_b)^2}{2K_b^a(P_0 - P_f)} \qquad 9.4$$

Then

$$T_1 = \left(\frac{l_c + l_b}{\sqrt{\dfrac{2K_b^a(P_0 - P_f)}{\mu}}} + \frac{h_0}{\sqrt{\dfrac{2K_c(P_0 - P_f)}{\mu}}}\right)^2 \qquad 9.5$$

For flow in path 2, the fluid will alternately flow within the channel and the fibre tows. We know that the transverse permeability in the fibre tow is much lower than that in the channel; at the same time the width l_c of the channel is always less than or equal to the width l_b of the tow. Then the time T_2 required for the flow front in path 2 to reach position B can be approximated by

$$T_2 = \frac{l_b(l_c + l_b)}{2K_b^t(P_0 - P_f)} \tag{9.6}$$

Dividing Equation 9.5 by Equation 9.6, this yields the time ratio

$$\frac{T_1}{T_2} = \left(\frac{l_c + l_b}{\sqrt{\dfrac{2K_b^a(P_0 - P_f)}{\mu}}} + \frac{h_0}{\sqrt{\dfrac{2K_c(P_0 - P_f)}{\mu}}} \right)^2 \times \frac{2K_b^t(P_0 - P_f)}{l_b(l_c + l_b)} \tag{9.7}$$

The above equation can be further simplified by the quantity analysis. Noticing that the ratio of $h_0/(l_c + l_b)$ is smaller (10^{-1} to 10^{-2}) and the ratio K_c/K_b^a is larger (10 to 10^2) for most preforms used in composite processing, the second term in Equation 9.5 is much smaller than the first term. Then Equation 9.7 can be simplified to

$$\frac{T_1}{T_2} = \frac{(l_c + l_b)^2}{2K_b^a(P_0 - P_f)} \cdot \frac{2K_b^t(P_0 - P_f)}{l_b(l_c + l_b)} = \frac{l_c + l_b}{l_b} \cdot \frac{K_b^t}{K_b^a} \tag{9.8}$$

Usually, the ratio K_b^a/K_b^t is about 20 for the same fibre tow. At the same time, $l_c \leq l_b$, so it seems that void formation in the cross-section of multilayer woven fabric is inevitable. The size of void mainly depends on the ratio K_b^a/K_b^t. In order to reduce the size and quantity of voids, it will be useful to use fibre tows with higher K_b^t for warp.

According to the simplified model, the flow through the cross-section of MWFs passes through five different basic sub-domains as shown in Fig. 9.5. In this thesis, the permeability tensors of these sub-domains are defined respectively as follows. For sub-domains a, b, c and d, the permeability for flow along and transverse to fibres is often evaluated from the Kozeny–Carman equation:

$$K = \frac{r_f^2}{4k} \cdot \frac{(1 - V_f)^3}{V_f^2} \tag{9.9}$$

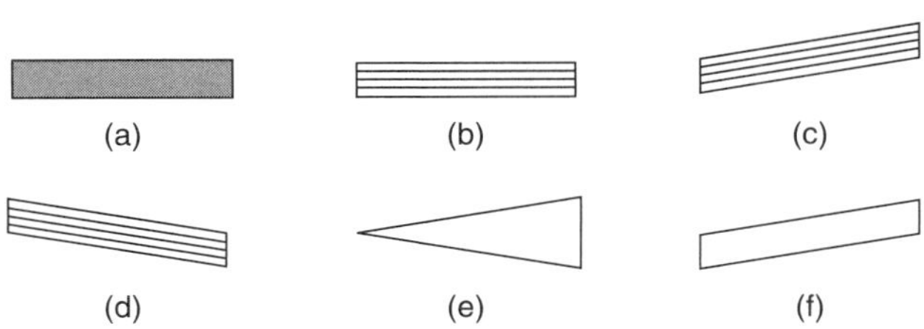

9.5 Sub-domains of MWFs.

where r_f is the fibre radius, V_f is the fibre volume fraction, and k is the Kozeny constant, which has been taken as $k_{//} = 0.5$ for flow along the fibres and $k_{\perp} = 10$ for flow transverse to the fibres (Williams *et al.*, 1974; Gutowski *et al.*, 1987b). Since this equation cannot predict the phenomenon that the transverse flow will stop at the maximum fibre volume fraction, a number of further modelling studies of permeability have been performed by Gutowski *et al.* (1987a), Gebart (1992) and Cai and Berdichevsky (1993). In this study, the fibre fractions employed were lower than $V_{f,max}$, hence permeability was evaluated according to Equation 9.9.

For sub-domains e and f, the permeability can be obtained from the expression for the equivalent permeability of a rectangular-shaped channel. In this study, the width of the channel is assumed to be l_b, and the height of the channel h_c is h for sub-domain f and 0 to h or $2h$ for sub-domain e. The velocity at the boundary of these channels is set to the Darcy velocity of fibre tows.

$$K_c = \frac{w_c^3}{h_c}\left[\frac{16}{\pi^5}\sum_{n=1,3,5...}^{\infty} -\frac{1}{n^5}\frac{\exp\left(n\pi\frac{h_c}{w_c}\right)}{\exp\left(n\pi\frac{h_c}{w_c}\right)+1} + \frac{1}{12}\frac{h_c}{w_c}\right] + \frac{K_b}{(1-V_f)} \quad 9.10$$

in which $w_c = l_b$ and

$$h_c = \begin{cases} h & \text{for sub-domain f} \\ 0 \text{ to h or 2h} & \text{for sub-domain e} \end{cases}$$

In Equation 9.10, K_b and V_f are permeability and volume fraction of fibre tows, respectively.

9.4 Modelling flow and void formation

The numerical simulations using the control-volume method of the flow fronts and void formation in both unit cells of MWFs are presented in this section. The preform in the mould is assumed to be rigid and is treated as porous medium, the flow in which is described by the following Darcy law:

$$\mathbf{U} = -\frac{1}{\mu}\mathbf{K}\cdot\nabla P \quad 9.11$$

in which μ is the viscosity of the resin. In a two-dimensional flow field,

$\mathbf{U} = (u_x, u_y)^T$ is the flow velocity vector

$$\nabla P = \left(\frac{P}{x}, \frac{P}{y}\right)^T \text{ is the pressure gradient vector}$$

$$\mathbf{K}=\begin{bmatrix} K_{xx} & K_{xy} \\ K_{yx} & K_{yy} \end{bmatrix} \text{ is the permeability tensor.}$$

Inserting Equation 9.11 into the continuity equation for an incompressible fluid gives the governing equation for pressure distribution in the flow field:

$$\nabla\cdot\left(\frac{1}{\mu}\begin{bmatrix} K_{xx} & K_{xy} \\ K_{yx} & K_{yy} \end{bmatrix}\begin{bmatrix} \dfrac{\partial p}{\partial x} \\ \dfrac{\partial p}{\partial y} \end{bmatrix}\right)=0 \qquad 9.12$$

The corresponding boundary conditions to be imposed can be stated as follows:

At injection gates: $P = P_0$ (for constant pressure injection)

At the mould wall: $\dfrac{P}{n_{\text{wall}}} = 0$

At the flow front: P = atmospheric pressure P_f (high injection pressure)

After the void is formed, the pressure in the void is assumed to obey the ideal gas law: $P_\rho = \rho RT$, where ρ is the density of air and R is the ideal gas constant.

It is noticed that Equation 9.12 is steady state. Although the mould filling process is time dependent, it can be treated as a series of steady-state problems by time increment. Within each time step, the pressure in the flow field can be determined by Equation 9.12. Generally, Equation 9.12 needs to be solved numerically. In the present simulation, the control-volume method is used.

The entire flow field is first discretized into a finite element mesh. In this simulation, a three-node triangular element is used. One node corresponds to one control volume. A control volume is composed of several sub-areas. The number of sub-areas is the same as the number of node-adjacent elements. The formation of the control volume is illustrated in Fig. 9.6 in which the region indicated by dashed lines is a control volume and the shaded region is a sub-area. Points i, j and k are nodes of a triangular element. Points a and b are the midpoints of element sides, and point o is the element centroid. Connecting points a, o and b, a sub-area is formed as the shaded region shown in Fig. 9.6. The procedure for forming other sub-areas is analogous.

Integrating Equation 9.12 over a control volume can lead to

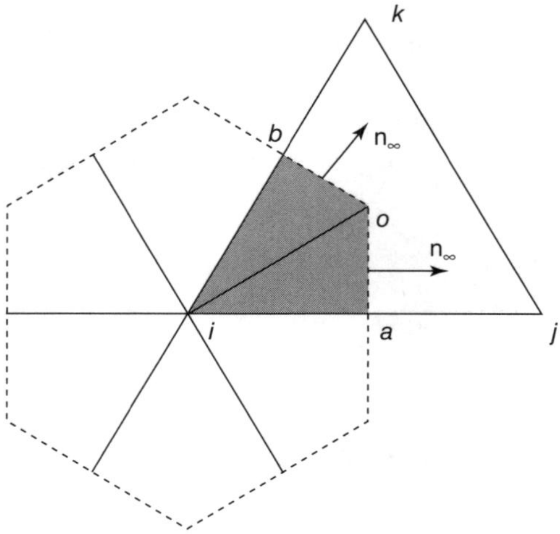

9.6 Control volume.

$$\iiint_V \nabla \cdot \left(\frac{1}{} \begin{bmatrix} K_{xx} & K_{xy} \\ K_{yx} & K_{yy} \end{bmatrix} \begin{bmatrix} \frac{p}{x} \\ \frac{p}{y} \end{bmatrix} \right) \mathrm{d}V = 0 \qquad 9.13$$

By using the divergence theorem, the above equation can be written as

$$\iint_S \left(\frac{1}{} \begin{bmatrix} K_{xx} & K_{xy} \\ K_{yx} & K_{yy} \end{bmatrix} \begin{bmatrix} \frac{p}{x} \\ \frac{p}{y} \end{bmatrix} \right) \cdot \mathbf{n}\, \mathrm{d}s = 0 \qquad 9.14$$

or

$$\int_C \frac{h_z}{} [n_x \quad n_y] \begin{bmatrix} K_{xx} & K_{xy} \\ K_{yx} & K_{yy} \end{bmatrix} \begin{bmatrix} \frac{p}{x} \\ \frac{p}{y} \end{bmatrix} \mathrm{d}L = 0 \qquad 9.15$$

in which h_z is the thickness of the preform, and the line C is along the boundary of the control volume. If a control volume consists of m sub-areas, the integration along C can be divided into m parts. For each part, by the assumption of linear distribution of pressure in the corresponding element, the pressure gradient can be written as (Yong *et al.*, 1991):

$$\begin{bmatrix} \frac{\partial p}{\partial x} \\ \frac{\partial p}{\partial y} \end{bmatrix} = \frac{1}{2A} \begin{bmatrix} \beta_1 & \beta_2 & \beta_3 \\ \gamma_1 & \gamma_2 & \gamma_3 \end{bmatrix} \begin{bmatrix} P_i \\ P_j \\ P_k \end{bmatrix} \qquad 9.16$$

where A is the area of the element, $\beta_i = y_j - y_k$, $\gamma_i = x_k - x_j$, and i, j and k permute in natural order.

Then the total integration of Equation 9.15 can be expressed as:

$$\sum_{i=1}^{m}\left\{\frac{h_z}{2A}\{l_{ao}[n_x \quad n_y]_{ao} + l_{bo}[n_x \quad n_y]_{bo}\}\cdot\begin{bmatrix} K_{xx} & K_{xy} \\ K_{yx} & K_{yy}\end{bmatrix}\begin{bmatrix}\beta_1 & \beta_2 & \beta_3 \\ \gamma_1 & \gamma_2 & \gamma_3\end{bmatrix}\begin{bmatrix}P_i \\ P_j \\ P_k\end{bmatrix}\right\} = 0 \qquad 9.17$$

where, l_{ao} and l_{bo} are length between points a and o, and b and o. Equation 9.17 is a linear algebraic equation, which can be written for all control volumes in the flow field. Together with the boundary condition, the pressure distribution in the flow field can then be solved.

Once the pressure field is determined, the velocity can be evaluated according to Equation 9.11; the flow front is then advanced using the FAN technique (Frederick and Phelan, 1997). In this method, the whole domain is divided into a fixed grid system and a scalar parameter f is introduced for each cell to represent the ratio of occupied volume to the total volume. As the flow front advances, all of the control volumes can be classified into three categories (see Fig. 9.7) as follows:

$f = 1$: filled region
$0 < f < 1$: flow front region
$f = 0$: empty region

For a selected volume f $(0 < f < 1)$ is calculated on the basis of the velocity field. The calculated volume of resin inflow is added to the original volume of resin in the flow front control volume. If the total resin volume in a control volume is equal to the volume of the control volume, that control volume is considered 'full' ($f = 1$). The time increment is selected in such a way that one control volume can be filled simultaneously. This restriction of the time increment ensures the stability of the quasi-steady-state approximation. The new flow front in each time step can be estimated according to the velocity vector in the flow front and the time increment after the pressure field is determined. After the value f is updated, another pressure

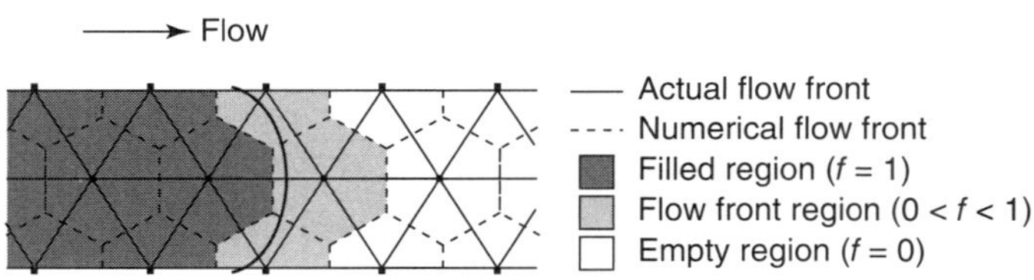

9.7 Illustration of the flow front advancing technique.

computation is performed for all the control volumes with $f = 1$. The procedure is repeated until the whole mould cavity is filled.

The time increment is determined in such a way that only one control volume is fully filled during the step, then

$$\Delta t = \min\left(\frac{V_i - V_i^{\mathrm{f}}(t)}{Q_i(t)}\right) \tag{9.18}$$

where V_i is the volume of control volume i, $V_i^{\mathrm{f}}(t)$ is the filled volume at time t, and $Q_i(t)$ is the flow rate into the control volume. Once the time increment is determined, the volume of fluid that flows into each control volume at the flow front can be calculated. Then the flow front is advanced.

$$f_i(t+\Delta t) = \frac{V_i^{\mathrm{f}}(t) + \Delta t \cdot Q_i(t)}{V_i} \tag{9.19}$$

A detailed flowchart of the numerical implementation is given in Fig. 9.8.

The numerical scheme outlined above has been implemented and tested by comparison with mould filling cases for which analytical solutions exist. The first case tested was boundary injection from one side of a rectangular mould: see Fig. 9.9. For constant injection pressure, the flow front position x_{f} at time t can be given by

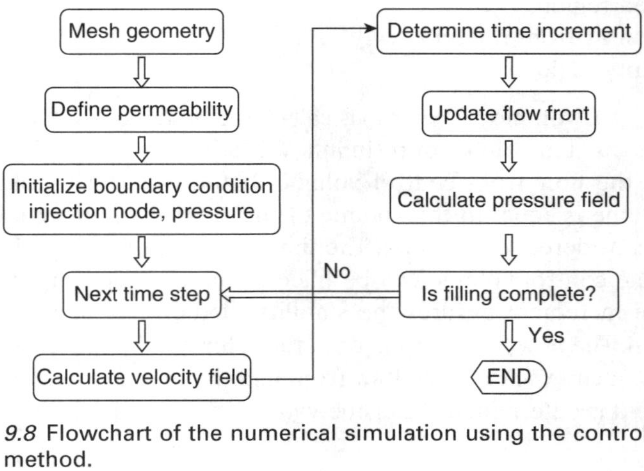

9.8 Flowchart of the numerical simulation using the control-volume method.

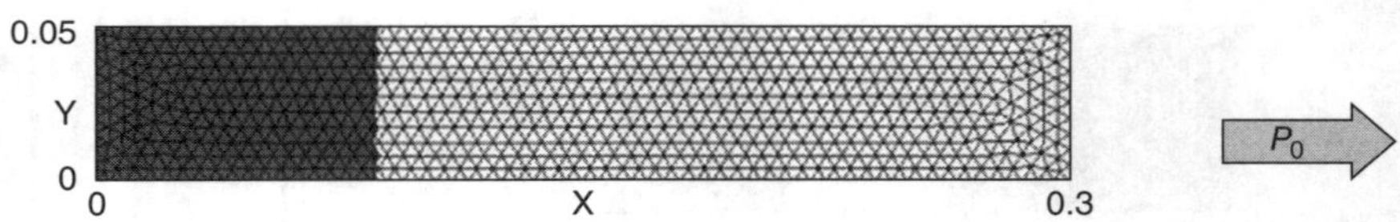

9.9 Boundary injection from one side of a rectangular mould.

$$x_f = \sqrt{\frac{2KP_0t}{}} \qquad 9.20$$

where P_0 is the injection pressure, K is the permeability of the preform, and μ is the viscosity of the fluid. Simulation was conducted for $P_0 = 200$ kPa, $K = 1.0 \times 10^{-10}$ m^2 and $\mu = 0.01$ Pa s. Predictions of flow front by simulation were in perfect agreement with the analytical results as shown in Fig. 9.10.

The second case tested was point injection from the centre of a square preform. The resin will flow in radial directions from the centre towards the preform edges. This results in an elliptical flow front, and the ratio a/b of its major axis to the minor axis is always equal to $\sqrt{K_1/K_2}$, where K_1 and K_2 are the principal permeabilities of the preform in two directions. The numerically simulated shapes of the flow front are shown in Fig. 9.11. For Fig. 9.11(a), there is $K_1 = K_2 = 1.0 \times 10^{-10}$ m^2; for Fig. 9.11(b), there is

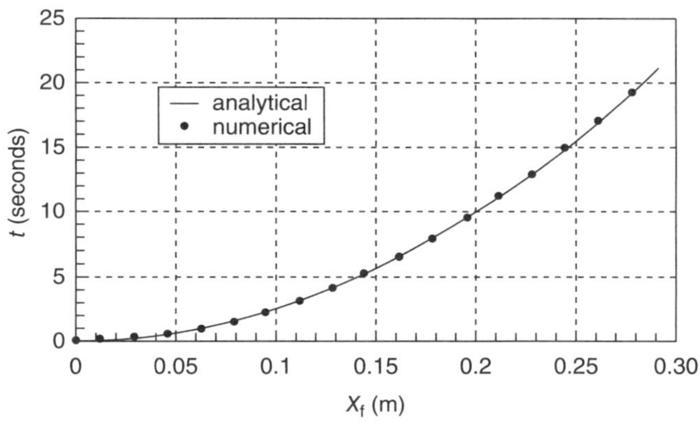

9.10 Comparison of analytical and numerical results of flow fronts.

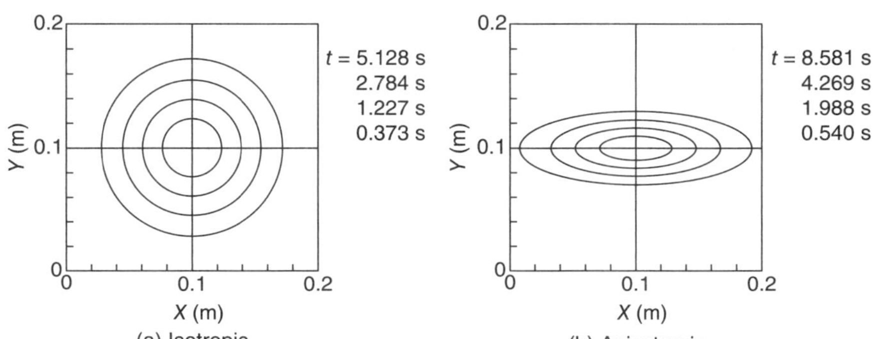

9.11 Flow fronts at different injection times of point injection in a square preform.

$K_1 = 1.0 \times 10^{-10}$ m^2, $K_2 = 1.0 \times 10^{-11}$ m^2. Excellent agreement can be observed between the numerical and analytical results for a/b.

It is noted that the flow fronts by simulation shown in Fig. 9.11 are not very smooth. The reason is that the flow front is identified by the filling factors on a fixed mesh, and it can be tacked to the resolution by using more elements in the flow domain.

9.4.1 Void formation in multilayer woven fabrics

In this section, the mould filling process and void formation in both unit cells of MWFs were simulated using the above numerical scheme. The fluid employed in the simulations was silicone oil of viscosity 0.1934 Pa s, which was taken from the experimental study of Patel and Lee (1995).

As described in section 9.2, since the thickness of the fibre tow is far less than the width, the unit cell of mode I (Fig. 9.4(a)) can be further simplified to Fig. 9.12, a cell with rectangular cross-section for both channel and warp/weft. This simplification was used by Yu and Lee to develop an in-plane permeability model of MWFs (Yu and Lee, 2000). Then a symmetry boundary condition can be imposed for the simulation of flow in both unit cells.

In order to define different permeabilities for different sub-domains, the flow field was first meshed by sub-domain to sub-domain using software ANSYS 5.7. In the simulation of the filling process, the boundary condition at the inlet can be either a constant volumetric flow rate or a constant injection pressure. Here, a constant injection pressure was taken. The pressures at inlet and flow front were set to be 5×10^5 Pa and 1×10^5 Pa, respectively. The fibre tows for weft and warp were the same with a fibre volume fraction of $V_{fb} = 0.45$ and a fibre radius of $r_f = 8 \times 10^{-6}$ m. Then the permeabilities of the fibre tow evaluated according to the Carman–Kozeny equation were $K_b^a = 2.63 \times 10^{-11}$ m^2 and $K_b^t = 1.32 \times 10^{-12}$ m^2.

Figure 9.13 shows the numerical simulation of flow fronts at various filling stages. The flow patterns in both unit cells are consistent with the analysis in section 9.2. In both unit cells, voids were formed at the ends of warp and the primary reason for void formation is the large difference in permeability between K_c, K_b^a and K_b^t. It can be observed that the void formed in the unit cell of mode I is smaller than that of mode II for the same parameters. This is understandable because there is only one warp between two adjacent

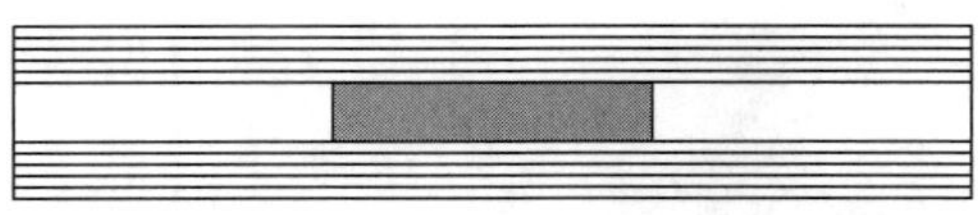

9.12 Simplified unit cells for mode I.

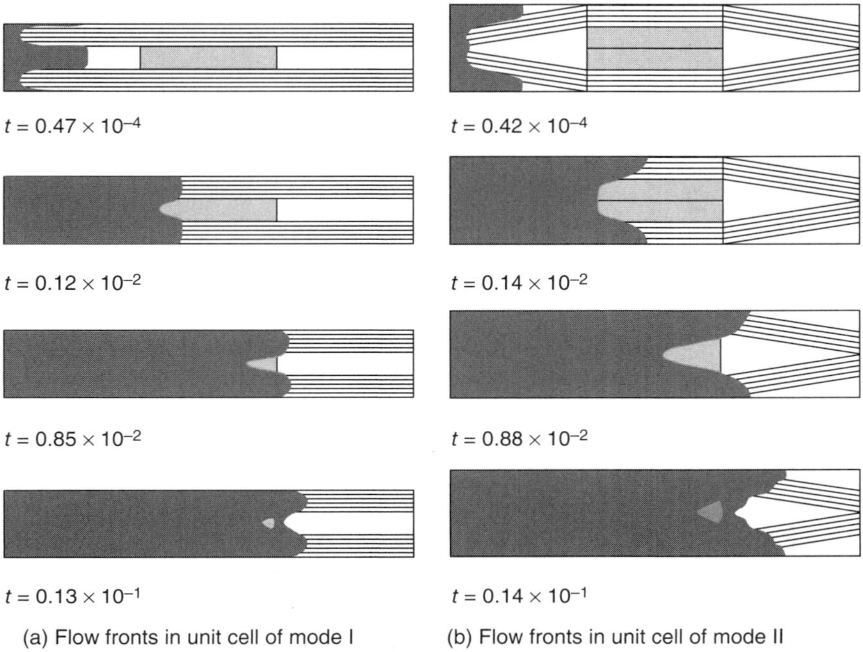

9.13 Flow fronts at different stages of mould filling.

wefts in mode I, but two for mode II. Thus, the impregnation in the warp of mode I is much accelerated by the relatively high flow velocity in adjacent wefts. Furthermore, the channel before warp in mode II is triangle-shaped, which also makes the flow in path 2 of mode II slower than that of mode I.

It is also noted that the simulated time for voids to be formed at the ends of warp, namely the time for the flow front in path 1 to reach position B, is less than that predicted according to Equation 9.5. This is because the impregnation in the weft between two adjacent warps is accelerated due the high permeability in the channel. In Equation 9.5, this effect is not considered.

According to the analytical result for the time ratio T_1/T_2, it can be observed that K_b^a/K_b^t seems to determine the size of the void. In this study, void formation processes on different values of K_b^a/K_b^t were simulated. Figure 9.14 presents the effect of changing K_b^a/K_b^t on the void size for both unit cells, which shows that larger K_b^a/K_b^t corresponds to larger void size. In addition, it seems that the size of void is proportional to the ratio of K_b^a/K_b^t as shown in Fig. 9.14. This agrees with our analytical result in section 9.2.

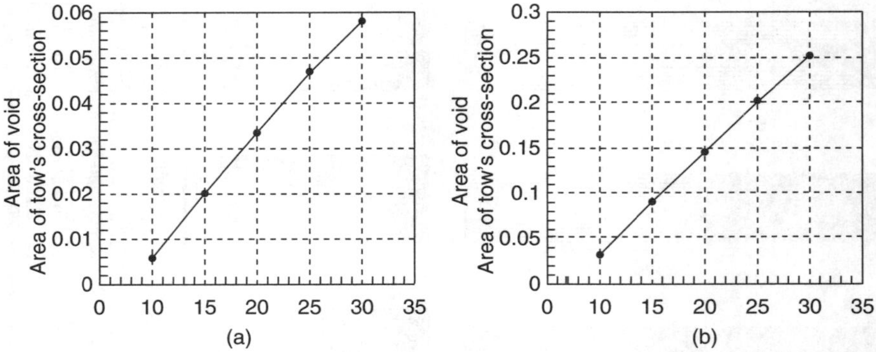

9.14 Size of void for different values of K_b^a/K_b^t. (a) unit cell of mode I, (b) unit cell of mode II.

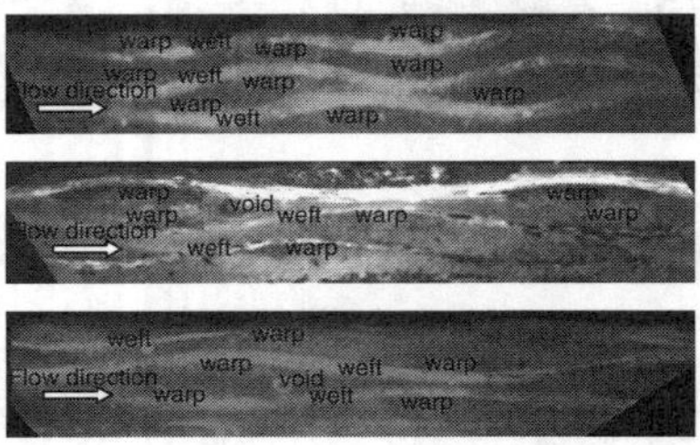

9.15 Experimental cross-sections of MWFs.

9.4.2 Experimental results

Experiments were performed to measure the formation of voids and their location. The reinforcement used in this experiment is plain woven fabrics. Three plies of glass fibre mat were stacked with low fibre volume fraction. The resin used was a two-part epoxy/amine resin, LY564/HY2954, from Ciba-Geigy (Hawthorne, NY). LY564 is the base resin of a bisphenol, an epoxy containing a reactive diluent. HY2954 is a hardener of 3,3′-dimethyl-4,4′-diaminodicyclohexyl methane. The mix ratio of LY564 to HY2954 (parts by weight) was 100:35. The resin had a viscosity of 0.617 Pa s at 25°C.

Resin was injected at constant pressure (0.5 MPa) into the mould. The longitudinal cross-section was examined using a microscope as shown in Fig. 9.15. From these images, two typical modes of the plain woven fabric can be clearly recognized and the formation of voids is mainly located at the end of warp.

9.5 Modelling the effect of stitch size, distribution and position

A key feature of 3-D multilayer woven fabrics is their interbundle stitches, but very limited information is available on the relationship between permeability and the effect of stitches. Shih and Lee (1998) proposed a parallel permeability model for flow through bidirectional stitched fabrics. The flow between the bundles is set by the Kozeny–Carman equation (Carman, 1937), and the effect of stitch was not specially mentioned in Shih and Lee's paper. Cairns *et al.* (1999) developed a model that incorporates Darcy's law within fibrous bundles and channels that exist between bundles. The paper reported that incorporating channel flow is an important feature for properly modelling the RTM process. With this model, pressure profiles, resin velocities and resin flow fronts are predicted accurately, and are available for manufacturing process development. The same materials were compared to fully stitched preforms. It was found that although the shapes for resin flow are similar between the experimental and analytical results, the stitch affects the permeability such that the unidirectional ply data do not accurately capture the times for resin flow.

Lundström (2000) also proposed a model for non-crimp stitched fabrics through theoretical analysis. The interbundle channel was assumed to be rectangularly shaped and the equivalent permeability is modelled by Equation 9.1. A comparison between the model and experimental data shows that this model works well in certain cases, while it fails in other cases to predict the permeability perfectly. Lundström suggested that one of the main reasons for the discrepancy was the assumption that the effect of stitch was negligible. Lundström also mentioned that study of both the dry fabric and the impregnated samples yielded stitches that were actually located in some of the channels. It is really important to make a quantitative study of the issue as to how stitches affect the effectiveness of the permeability. Such a study will be very helpful for designing stitched fabrics, exact pre-simulation of moulding processes, and the optimization of processing parameters.

The structure of 3-D multilayer woven fabrics is quite complicated and variable. In this section, we focus on its key feature, the interbundle stitch, and its effect on permeability. Several typical stitch structures were selected as the basic models, and the effect of stitches on permeability of the interbundle channel has been investigated through numerical simulation for various combinations of the stitch size, its off-centre position, slope, array, distribution density in the flow direction and the average Darcy velocity in the fibre bundle.

The results of numerical simulation are presented in this section. The effect of stitches on the equivalent permeability of the interbundle channel is presented by the value of K/K_0 where K and K_0 are the permeabilities of the channels with and without stitches, respectively. It should be particularly noted that K_0 is different for different boundary conditions and geometrical structures (Hu *et al.*, 2003).

9.5.1 The effect of stitch size on distribution density

In this section, we consider that there is only one stitch in one representative cell, and the stitch is located on the centreline of the channel and perpendicular to the flow direction. The corresponding computational cell is shown in Fig. 9.1(a). Based on this computational cell, the effect of stitch on the permeability of the interbundle channel has been investigated for various combinations of stitch size (R/a), distribution density in the flow direction (L/W), and the ratio of V_b to V_a, in which $V_a = Q/A$ is the average velocity in the channel corresponding to the given flow rate.

Figure 9.16 shows the variation of K/K_0 with R/a for different boundary conditions and distribution densities. As shown in these figures, it is evident that stitches may severely influence the permeability of the interbundle channel. For all simulated cases, the value of K/K_0 is less than 0.80 at R/a = 0.1, and is around 0.10 at R/a = 0.6. Obviously, the effect of stitches cannot be neglected when establishing an effective permeability model. We also note that there is similar variation of K/K_0 with R/a for different conditions, but the extent of the influence is different for different boundary conditions or different distribution densities.

In Fig. 9.17, the variation of K/K_0 with V_b/V_a for different values of R/a and the same L/W is presented. For all simulated cases, K/K_0 decreases with increasing V_b/V_a, especially for boundary condition BC3, where the effect of V_b/V_a on the value of K/K_0 is much higher than that for BC2. The value of K/K_0 is reduced about 50% when V_b/V_a is increased from 0.1 to 0.8 for BC3. Thus, the effect of stitches on the permeability of the interbundle channel is different with different boundary conditions. It is known that, for a given pressure drop Δp, the larger the average flow velocity V_b in the fibre bundle, the higher the permeability.

The stitch distribution density in the flow direction is expressed by the value of L/W. If W is kept constant, then larger L/W means sparser distribution. The effect of L/W on the value of K/K_0 is shown in Fig. 9.18. There is a fairly linear relationship between K/K_0 and R/a, but the slope of change is different for different cases. For the same boundary condition, the slope decreases with increasing R/a. This means that the effect of stitch distribution density decreases with increase of R/a.

All the simulations are performed with the condition that channel height H is equal to width W. Certainly, the variation of H may affect the value of K/K_0. When H tends to infinity, the flow in the channel can be simplified

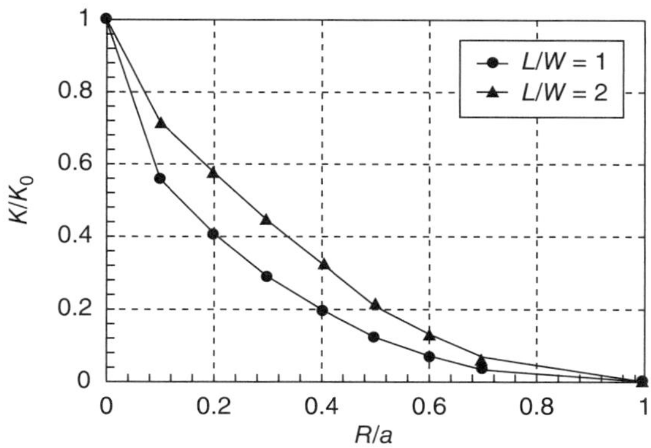

(a) BC1 ($V_b/V_a = 0.4$)

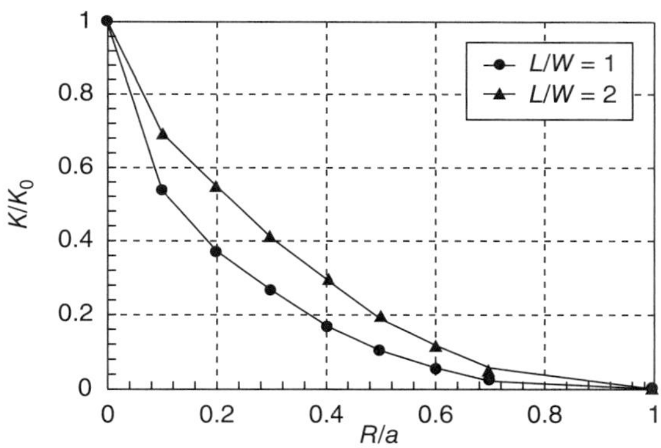

(b) BC2 ($V_b/V_a = 0.4$)

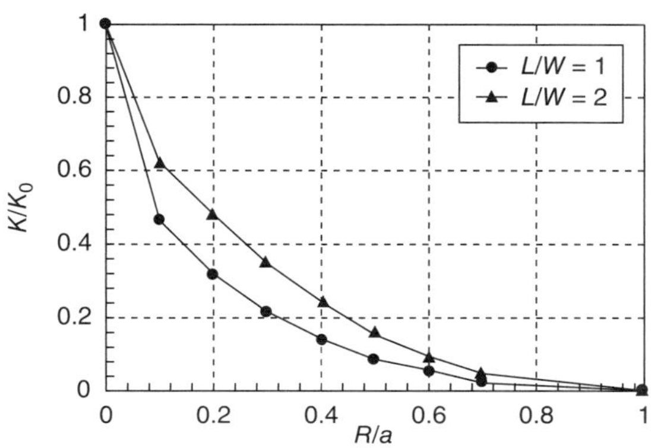

(c) BC3 ($V_b/V_a = 0.4$)

9.16 Change of K/K_0 with R/a on different boundary conditions.

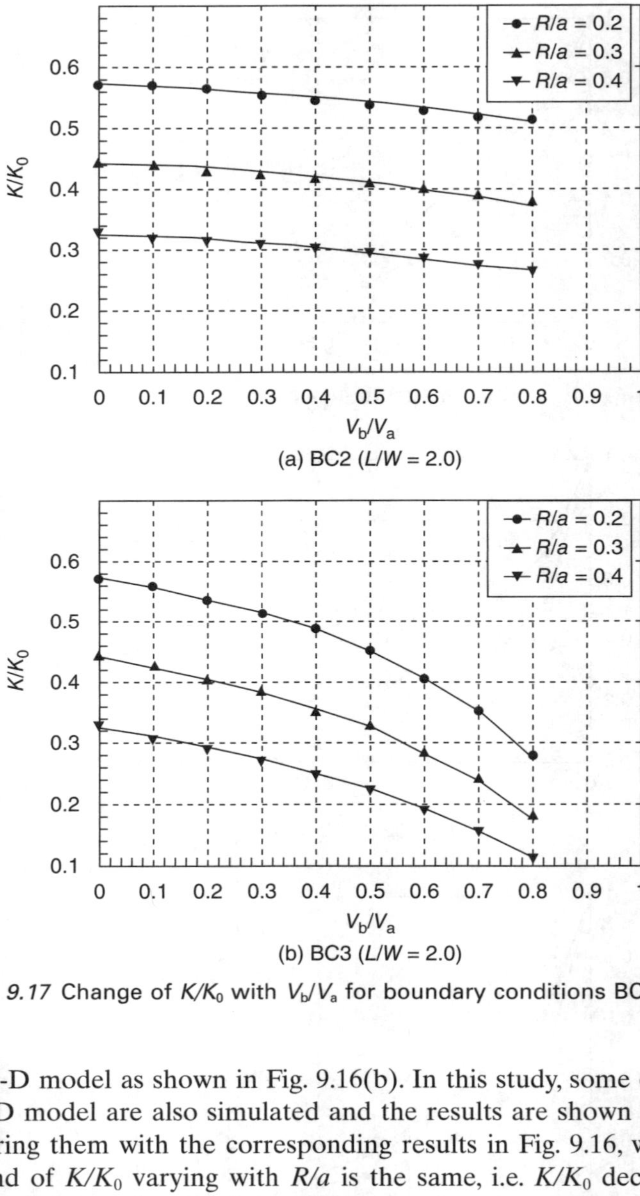

9.17 Change of K/K_0 with V_b/V_a for boundary conditions BC2 and BC3.

into a 2-D model as shown in Fig. 9.16(b). In this study, some cases based on a 2-D model are also simulated and the results are shown in Fig. 9.19. Comparing them with the corresponding results in Fig. 9.16, we find that the trend of K/K_0 varying with R/a is the same, i.e. K/K_0 decreases with increasing of R/a when other parameters are kept constant.

The impregnation process for different stitch radii was also simulated by finite element simulation. The schematic of the simulation unit cell is shown in Fig. 9.20 where the effect of stitch size is represented by the ratio between the stitch radius and the channel length (R/a). Based on the finite element method, the impregnation process of resin flow through the unit cell of the

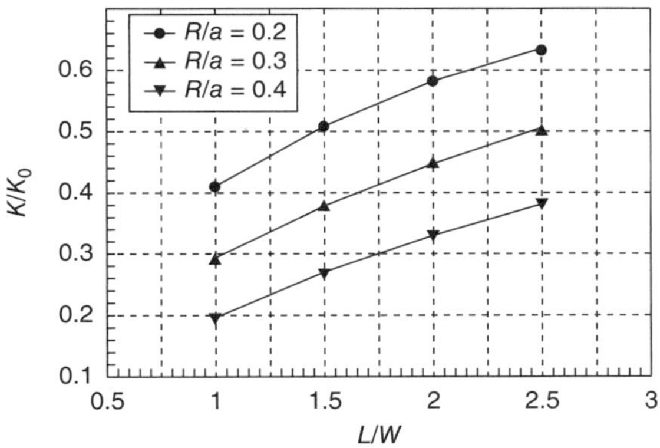

(a) BC1 ($V_b/V_a = 0.4$)

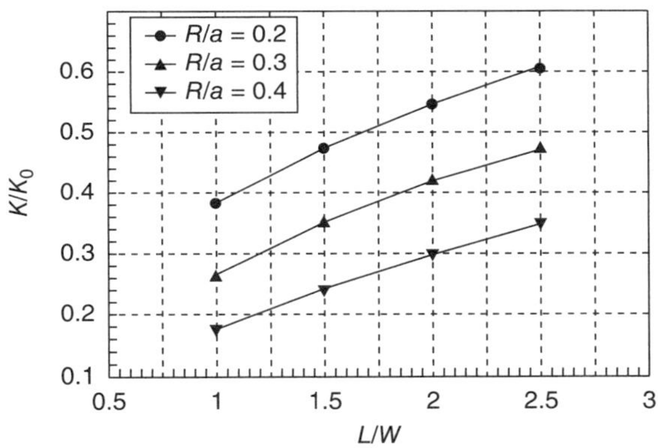

(b) BC2 ($V_b/V_a = 0.4$)

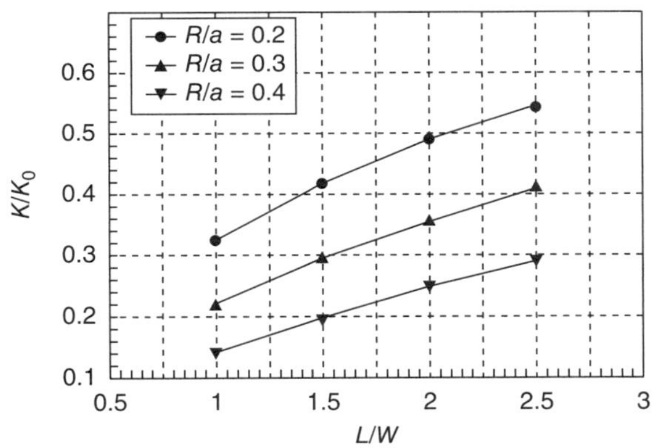

(c) BC3 ($V_b/V_a = 0.4$)

9.18 Change of K/K_0 with L/W on different boundary conditions.

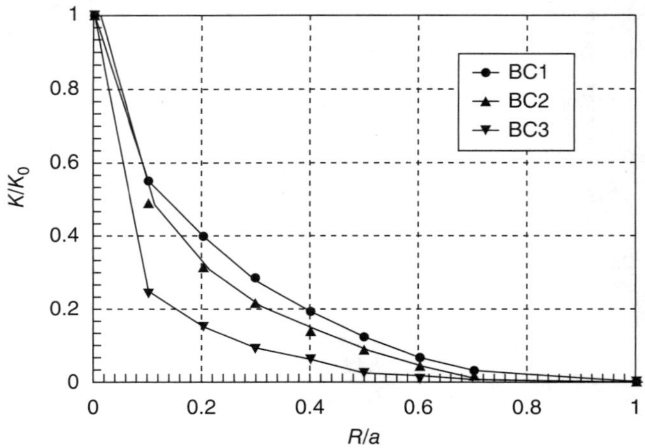

9.19 Results of 2-D simulation: change of K/K_0 with R/a ($L/W = 2.0$, $V_b/V_a = 0.4$).

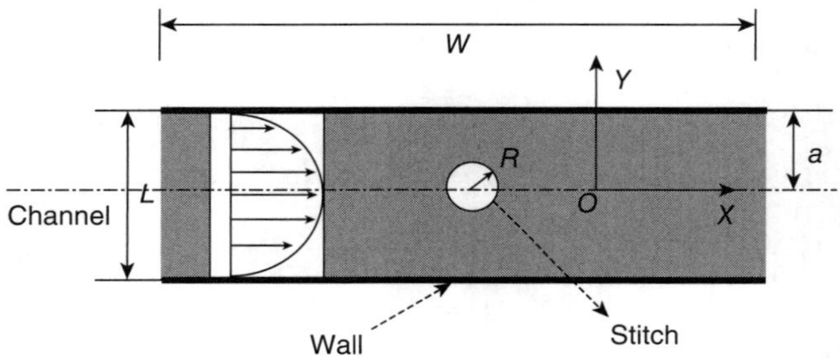

9.20 Unit cell of channel flow with different stitch sizes.

channel with stitch in the off-centre position is shown in the following simulation (Fig. 9.21).

9.5.2 The effect of stitch size on permeability

The effect of stitches on the equivalent permeability of the interbundle channel is represented by the value of K/K_0. The effect of stitch size on permeability is shown in Fig. 9.22, in which the x-coordinate is R/a, where R is the stitch radius and a is the channel width. As shown in Fig. 9.22, it is evident that stitch size may severely influence the permeability of the interbundle channel. The permeability decreases quickly with increase of stitch radius. These simulation results also correspond with the results of analysis by FLUENT as shown in Fig. 9.16.

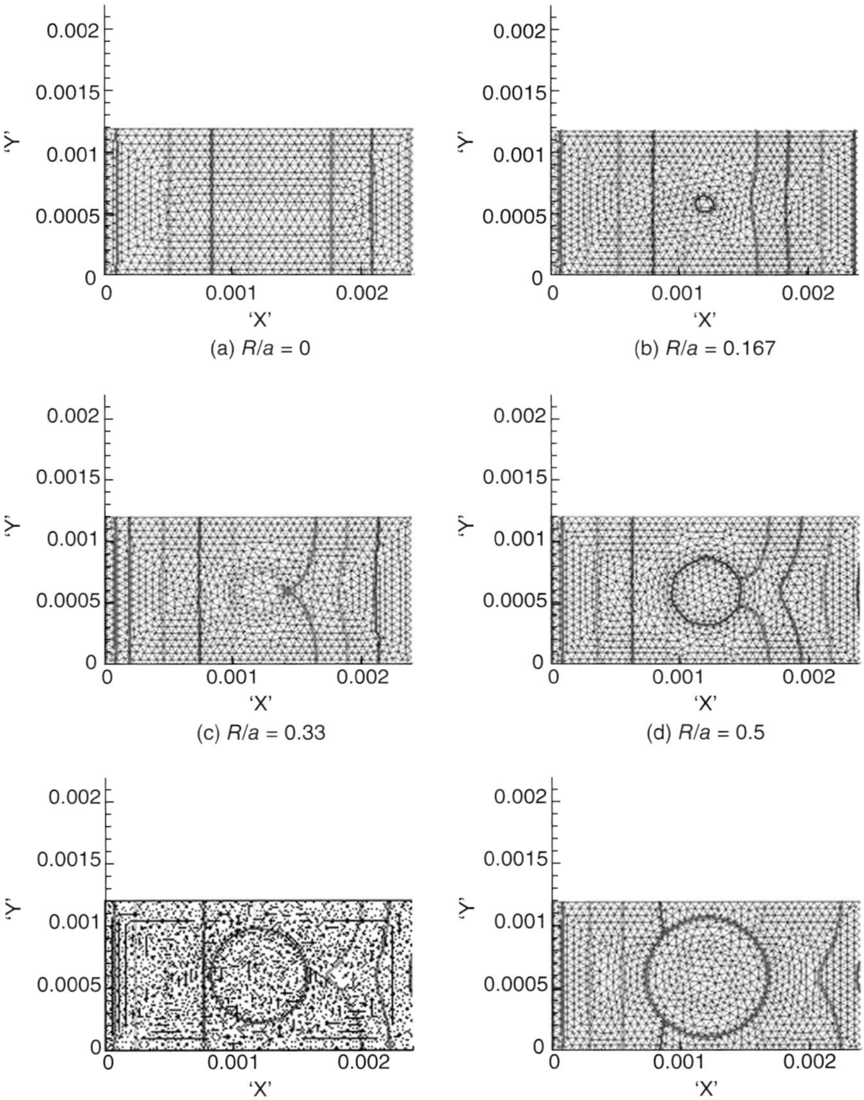

9.21 The simulation impregnation process of resin flow through the unit cell of the channel with different stitch sizes.

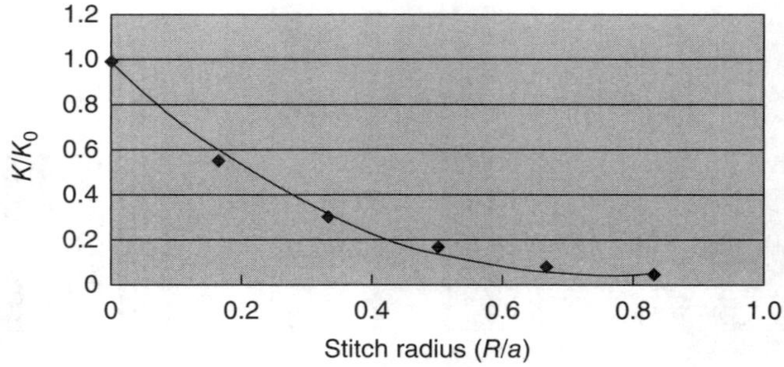

9.22 The effect of stitch size on permeability.

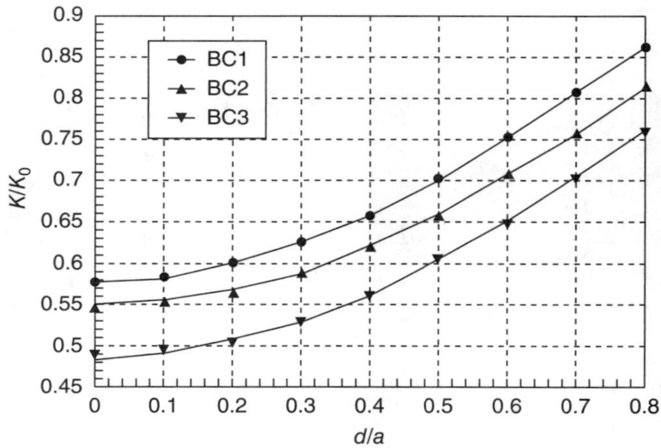

9.23 Change of K/K_0 with d/a.

9.5.3 The effect of stitch off-centre position in channel

Stitches in the interbundle channel do not always lie on the centreline. They are often off-centre. In this case, the upper half of the channel is selected as the computational cell. The simulation results are shown in Fig. 9.23, in which the x-coordinate is d/a where d is the distance between the stitch centre and the channel centreline. For the other parameters, $R/a = 0.2$, $L/W = 2.0$, and $V_b/V_a = 0.4$. Apparently, the effect of the stitch on the permeability of the channel is very different for different off-centre positions. When d/a increased from 0 to 0.8, i.e. the stitch location was moved from the centreline to close to the wall, the value of K/K_0 increased by about 50%, in other words, the effect decreased by 50%. So if we want to weaken the effect of the stitch on the permeability of the interbundle channels, and then

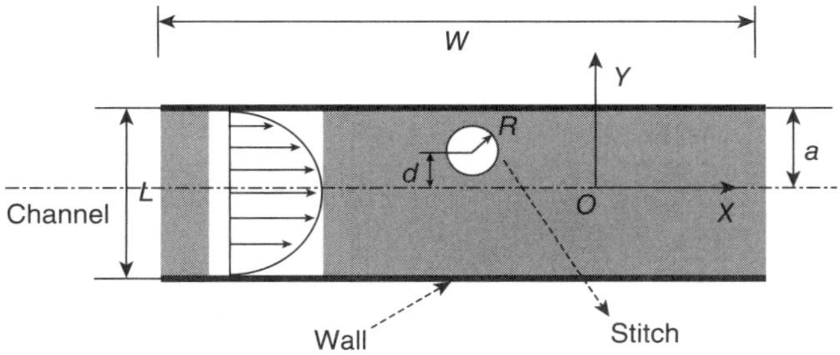

9.24 Unit cell of channel flow with stitch in the off-centre position.

to make the flow of resin in the channel smooth, it is better to keep the stitch off-centre in the interbundle channel.

The effect of the stitch off-centre position in the channel is also further testified by the simulated impregnation process analyzed by the finite element method as follows. A schematic of the simulation unit is shown in Fig. 9.24, where the stitch off-centre position in the channel is represented by the distance between the stitch centre and the channel centreline (d/a). For the simulated case in this section, L/W is 0.5. Based on the finite element method, the impregnation process of resin flow through the unit cell of the channel with stitch in the off-centre position is shown in Fig. 9.25(a)–(f).

Based on Darcy's law, the effective permeability of the unit cell can be calculated from the simulation impregnation process. The effect of stitches on the equivalent permeability of the interbundle channel is represented by the value of K/K_0, where K and K_0 are the permeabilities of the channel with and without stitches respectively. The effect of the stitch off-centre position on permeability is shown in Fig. 9.26, in which the x-coordinate is d/a, where d is the distance between the stitch centre and the channel centreline. Apparently, the effect of the stitch on the permeability of the channel is very different for different off-centre positions. When d/a increased from 0 to 0.7, i.e. the stitch location was moved from the centreline to close to the wall, the value of K/K_0 increased accordingly. These simulation results are consistent with the results of analysis by FLUENT.

In other cases, the stitch is always inclined towards the interbundle channel. So it is very necessary to make clear how the stitch slope will affect the equivalent permeability of the channel. The angle between the stitch orientation and the y-axis is defined as θ. In this section, the variation of K/K_0 with θ is investigated by keeping other parameters constant, such as R/a at 0.2, L/W at 2.0, and V_b/V_a at 0.4. As shown in Fig. 9.27, the permeability of the channel at $\theta = 0°$, $10°$, $20°$ and $30°$ is simulated for different boundary conditions. The results show that K/K_0 decreases with increasing θ, but the difference for a unit increment of θ is very small when $\theta \leq 10°$.

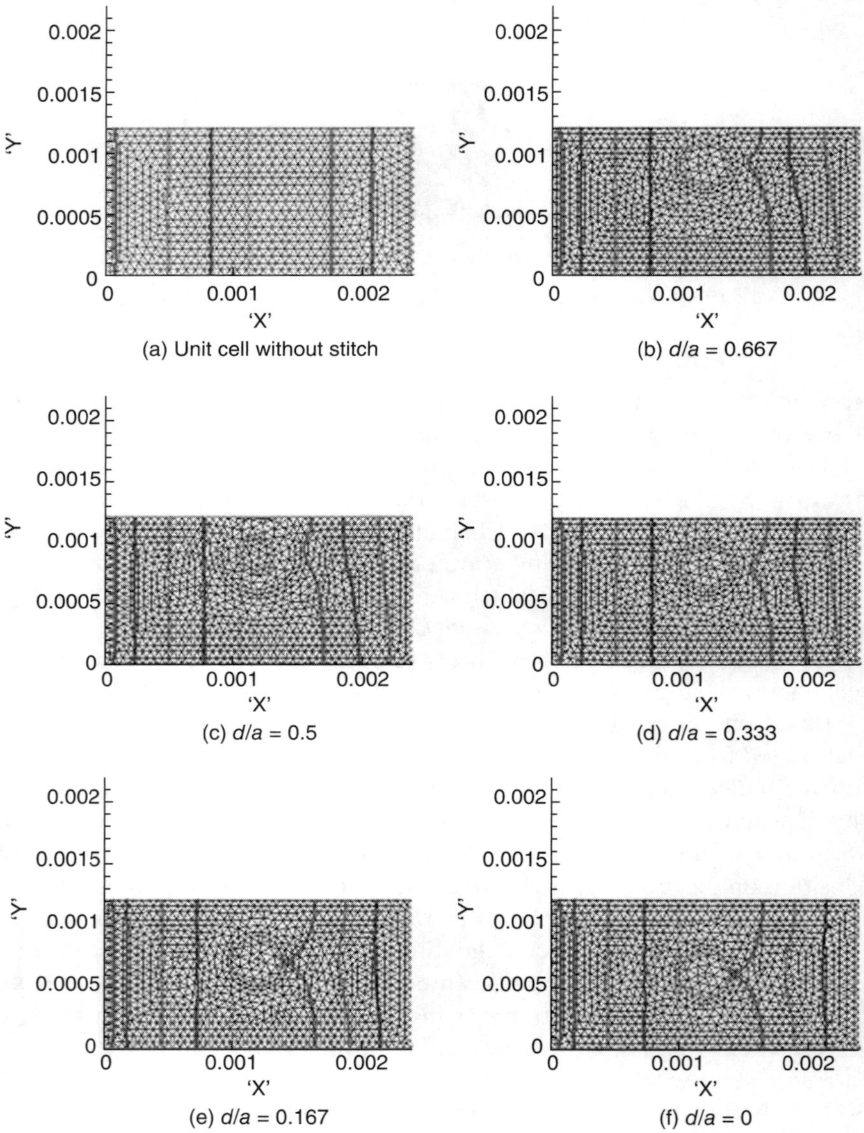

9.25 The simulation impregnation process of resin flow through the unit cell of the channel with stitch in the off-centre position.

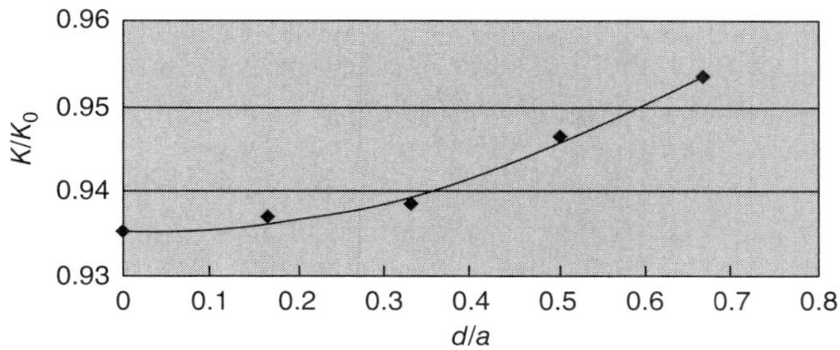

9.26 The effect of stitch off-centre position on permeability.

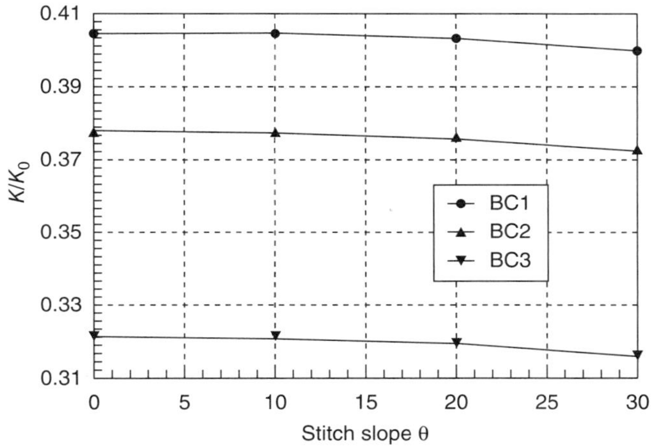

9.27 Change of K/K_0 with stitch slope θ.

This corresponds to the increment of surface area of the stitch in the channel with increasing θ. Therefore, it can be predicted that K/K_0 will decrease more quickly for larger θ. Comparing with other parameters, such as off-centre position and size, the effect of slope on the value of K/K_0 is relatively weaker.

9.5.4 The effect of stitch array

When there are two or more than two stitches in one representative cell, the effect of the stitches on the permeability of the channel must be different with different arrays. In this section, two situations are considered. Firstly, we consider that two stitches are located on the centreline, and arrayed up and down along the flow direction. We define this situation as array-1. Then two stitches are arrayed to be bilaterally symmetrical to the centreline of the channel. We define this situation as array-2. In both situations, the distance between the two stitches is marked as D. For all simulated cases in this section, R/a is 0.2, L/W is 2.0, and V_b/V_a is 0.4.

In the case of array-1, the variation of K/K_0 with D/L is shown in Fig. 9.28(a). In this case, when D/L is equal to 0.1, two stitches should remain in contact with each other. The simulation results show that K/K_0 at D/L = 0.1 is relatively higher, and decreases quickly at first and then slowly with increasing D/L. When D/L is equal to 0.5, the value of K/K_0 should be the

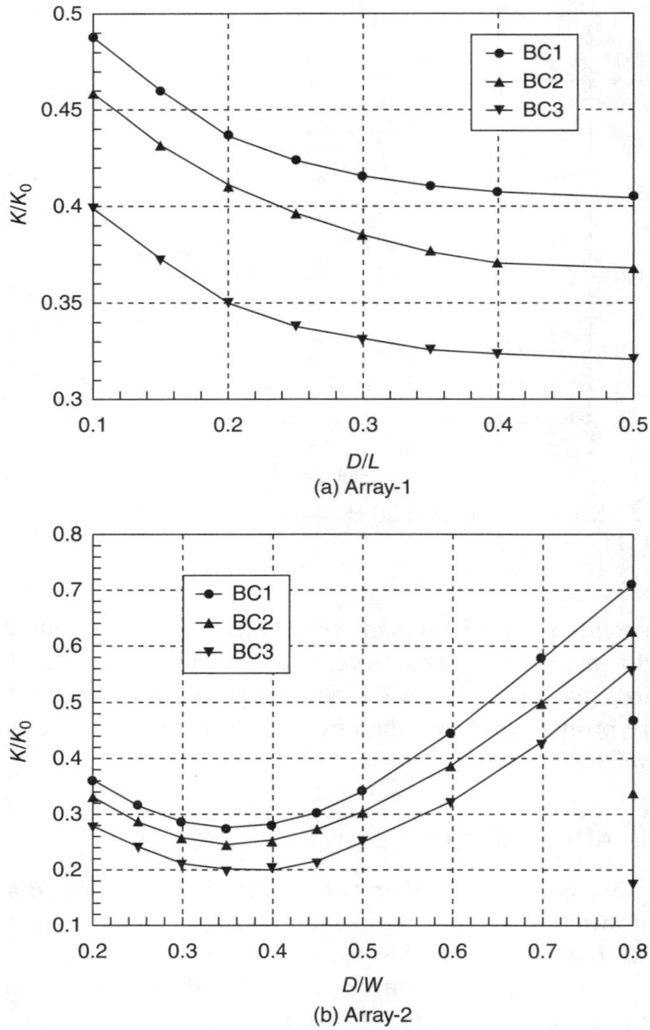

9.28 Change of K/K_0 with L/W in the cases of array-1 and array-2.

same as in the case where there is only one stitch in the representative cell for $L/W = 1.0$.

In the case of array-2, when D/W is equal to 0.2, two stitches also should remain in contact with each other, and when D/W is equal to 0.8, the stitch should touch the sidewall of the channel. The simulation results are shown in Fig. 9.28(b), which shows that the variation of K/K_0 with D/W is not monotonous. When two stitches move simultaneously from the centre to the sidewall of the channel, K/K_0 first decreases when D/W is equal to 0.35–0.40, then reaches its minimum, and then increases linearly. The difference between the minimum and maximum of K/K_0 is about three times. Therefore, a slight change in stitch array will induce a distinct variation of the permeability of the channel.

The impregnation process for the case of array-2 was also simulated by the finite element simulation. The schematic of the simulation unit is shown in Fig. 9.29, where the stitch off-centre position in the channel is represented by the ratio between the stitch centre distance and the channel length (d/b).

Based on the finite element method, the impregnation process of resin flow through the unit cell of the channel with stitch in the off-centre position (two stitches) is shown in Fig. 9.30(a)–(g).

The effect of stitches on the equivalent permeability of the interbundle channel is represented by the value of K/K_0. The effect of the stitch off-centre position on the permeability is shown in Fig. 9.31, in which the x-coordinate is d/b, where d is the half-length of the channel and b is the distance between the stitch centers. The stitch off-centre position can be expressed by d/b.

In this figure, when d/b is equal to 0.1667, two stitches should also keep touching each other, and when d/b is equal to 0.8333, the stitch should touch the sidewall of the channel. The simulation results are shown in Fig. 9.31, which shows that the change of K/K_0 with d/b is not monotonous. When two stitches approach closer, K/K_0 first decreases when d/b is equal to

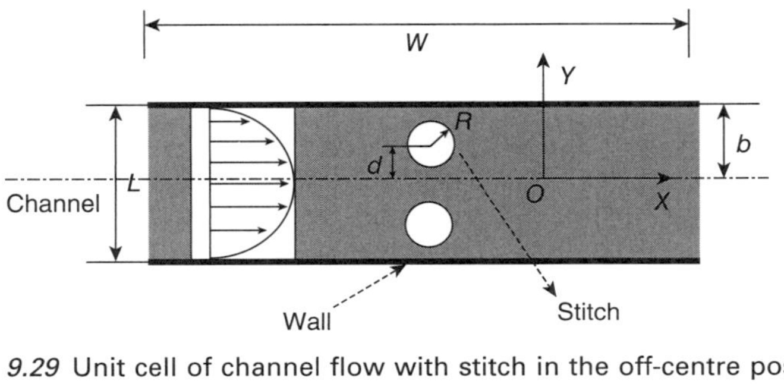

9.29 Unit cell of channel flow with stitch in the off-centre position (two stitches).

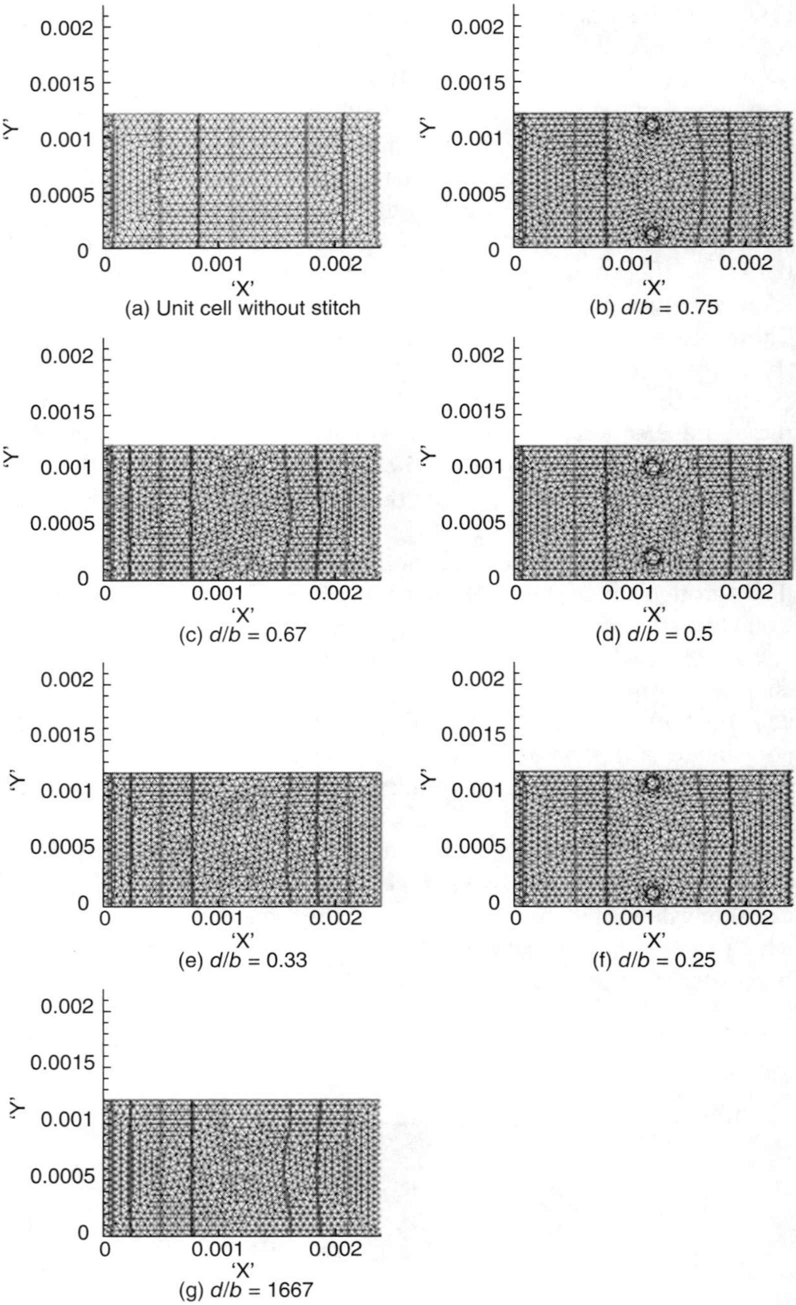

9.30 The simulation impregnation process of resin flow through the unit cell of the channel with stitch in the off-centre position (two stitches).

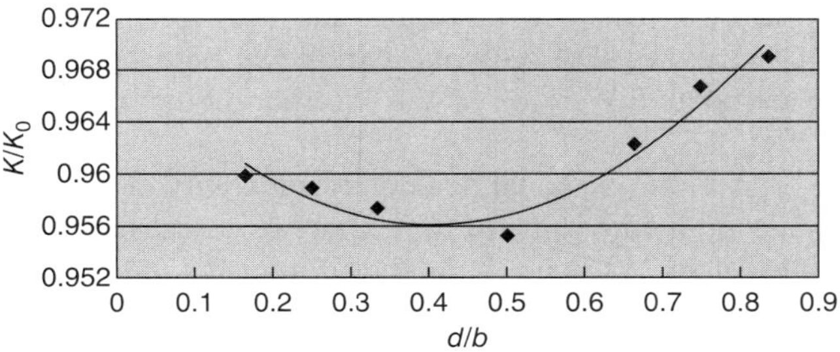

9.31 The effect of stitch off-centre position (two stitches) on permeability.

0.4–0.5, then reaches its minimum value, then with increase of d/b, K/K_0 increases linearly. These simulation results also correspond with the results of analysis by FLUENT.

9.6 Conclusions

In this chapter two simplified unit cells for in-plane impregnation in multilayer woven fabric have been suggested. In addition a mathematical model has been developed to analyze the void formation in both unit cells during RTM processes. The model recognizes non-uniform fibre permeability in different sub-domains of fibre preforms. Although a few simplifying assumptions have been introduced, the model still accounts for the key mechanisms of void formation in the cross-section of MWFs. The location of the void predicted by this model agrees quite well with the experiments. The model also shows that the ratio of weft axial permeability $K_{\rm b}^{\rm a}$ to warp transverse permeability $K_{\rm b}^{\rm t}$ largely determines the formation and size of voids.

The flows in both unit cells also were numerically simulated using the control-volume method. The presented numerical scheme was first tested by comparing it with mould-filling cases for which analytical solutions exist. The simulated results, including flow fronts and the effect of $K_{\rm b}^{\rm a}/K_{\rm b}^{\rm t}$, agree with the prediction according to the analytical model. The size of the void is proportional to the ratio $K_{\rm b}^{\rm a}/K_{\rm b}^{\rm t}$ (Hu *et al.*, 2004).

Numerical simulations of flow in the interbundle channel with stitch have been carried out, and the effect of the stitch on the equivalent permeability of the interbundle channel has been revealed. The results show that stitches may severely affect the permeability of channels, even if their size is relatively smaller. So this effect has to be considered when establishing an effective permeability model for multilayer fabrics with stitches.

The numerical simulation results also suggest that the effect of a stitch is quite different for different stitch size, off-centre position, slope, array, distribution density in the flow direction, and average Darcy velocity in fibre bundles. Therefore, with only one variable, usually the radius of the stitch, *R*, the permeability status of the interbundle channel is still very uncertain.

The effect of stitches on the equivalent permeability of the interbundle channel is presented by the value of K/K_0. The numerical simulation results suggest that the effect of stitches is quite different for different stitch size, off-centre position, slope, array, distribution density in the flow direction, and average Darcy velocity in fibre bundles. The value of K/K_0 will decrease with the increase of stitch size, slope and average Darcy velocity in fibre bundles, and will also decrease with the increase of off-centre distance. When there are two stitches in a representative cell, for an up and down array (array-1), the value of K/K_0 will decrease with increasing space between two stitches. For a left and right array (array-2), when two stitches move simultaneously from the centre to the sidewall of channel, K/K_0 first decreases to a minimum and then increases linearly.

9.7 References

Barooah P, Berker B and Sun J Q (2001), Integrated switching and feedback control for mold filling in resin transfer molding, *Journal of Manufacturing Science and Engineering*, **123**, May, 241.

Binetruy C and Hilaire B (1998), Tow impregnation model and void formation mechanisms during RTM, *Journal of Composite Materials*, **32**, 3, 223–245.

Bowles K J and Frimpong S (1991), Void effects on the interlaminar shear strength of unidirectional graphite-fiber-reinforced composites, *Journal of Composite Materials*, **26**, 1487–1509.

Cai Z and Berdichevsky A L (1993), An improved self-consistent method for estimating the permeability of a fiber assembly, *Polymer Composites*, **14**, 4, 314–323.

Cairns D S, Humbert D R and Mandell J F (1999), Modeling of resin transfer molding of composite materials with oriented unidirectional plies, *Composites, Part A: Applied Science and Manufacturing*, **30**, 375–382.

Carman P C (1937), Fluid flow through granular beds, *Transactions of the Institution of Chemical Engineers*, **15**, 150–166.

Chan A W and Morgan R J (1992), Sequential multiple port injection for resin transfer moulding of polymer composites, *SAMPE Quarterly*, **23**, 45–49.

Chang C Y and Hourng L W (1998), Study on void formation in resin transfer molding, *Polymer Engineering and Science*, **38**, 5, 809–818.

Chen Y T, Davis H T and Macosko C W (1995), Wetting of fiber mats for composites manufacturing: 1. Visualization experiments, *AIChE Journal*, **41**, 10, 2261.

Feldgoise S, Foley M F, Martin D and Bohan J (1991), The effect of microvoid content on composite shear strength, *Proc. 23rd International SAMPE Symposium*, 259–273.

Frederick R and Phelan J R (1997), Simulation of the injection process in resin transfer molding, *Polymer Composites*, **18**, 4, 460–476.

Gebart B R (1992), Permeability of unidirectional reinforcements for RTM, *Journal of Composite Materials,* **26**, 8, 1100–1133.
Ghiorse S R and Jurta R M (1991), Effects of low frequency vibration processing on carbon/epoxy laminates, *Composites*, **22**, 1, 3–8.
Gutowski T G, Morigaki T and Cai Z (1987a), The consolidation of laminate composites, *Journal of Composite Materials*, **21**, 172–188.
Gutowski T G, Cai Z, Bauer S, Boucher D, Kingery J and Wineman S (1987b), Consolidation experiments for laminate composites, *Journal of Composite Materials*, **21,** 650–669.
Harper B D, Staab G H and Chen R S (1987), A note on the effects of voids upon the hygral and mechanical properties of AS4/3502, *Journal of Composite Materials*, **21**, 280–289.
Hayward J S and Harris B (1990), Effect of process variables on the quality of RTM moulding, *SAMPE Journal*, **26**, 3, 39–46.
Hu J L, Liu Yi and Shao X M (2003), Effect of stitches on the permeability of interbundle channels in stitched fabrics, *Textile Research Journal*, **73**, 8, 691–699.
Hu J L, Liu Yi and Shao X M (2004), Study on void formation in multi-layer woven fabrics, *Composites: Part A, Applied Science and Manufacturing*, **35**, 595–603.
Kang M K, Lee W I and Hahn H T (2000), Formation of microvoids during resin-transfer molding process, *Composites Science and Technology,* **60**, 2427–2434.
Lundström T S (2000), The permeability of non-crimp stitched fabrics, *Composites, Part A: Applied Science and Manufacturing*, **31**, 1345–1353.
Parnas R S and Phelan Jr F R (1991), The effect of heterogeneous porous media on mold filling in resin transfer molding, *SAMPE Quarterly*, **22,** 53–60.
Patel N and Lee L J (1995), Effects of fiber mat architecture on void formation and removal in liquid composite molding, *Polymer Composites*, **16**, 5, 386–399.
Rohatgi V, Patel N and Lee L J (1996), Experimental investigation of flow-induced microvoids during impregnation of unidirectional stitched fiberglass mat, *Polymer Composites*, **17**, 161–170.
Shih C H and Lee L J (1998), Effect of fibre architecture on permeability in liquid composite molding, *Polymer Composites*, **19**, 5, 629–639.
Simacek P and Advani S G (1996), Permeability model for a woven fabric, *Polymer Composites*, **17**, 887–899.
Williams J G, Morris C E M and Ennis B C (1974), Liquid flow through aligned fiber beds, *Polymer Engineering and Science*, **14**, 6, 413–419.
Yong W B, Han K, Fong L H and Lee L J (1991), Flow simulation in molds with preplaced fiber mats, *Polymer Composites*, **12**, 6, 391–403.
Yosida H T, Ogasa T and Hayashi R (1986), Statistical approach to the relationship between ILSS and void content of CFRP, *Composites Science and Technology*, **25**, 3–18.
Yu B and Lee L J (2000), A simplified in-plane permeability model for textile fabrics, *Polymer Composites*, **21**, 5, 660–685.

Index